Collins

Student Book, **Foundation 2**

NEW GCSE MATHS
AQA Linear
Matches the 2010 GCSE Specification

Brian Speed • Keith Gordon • Kevin Evans • Trevor Senior

CONTENTS

INTRODUCTION

Welcome to Collins New GCSE Maths for AQA Linear Foundation Book 2.

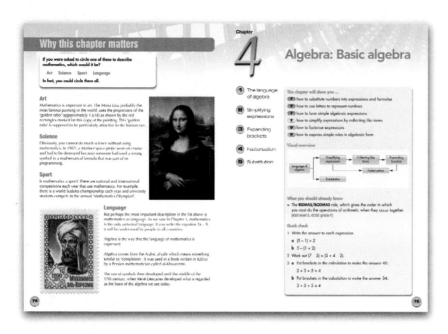

Why this chapter matters

Find out why each chapter is important through the history of maths, seeing how maths links to other subjects and cultures, and how maths is related to real life.

Chapter overviews

Look ahead to see what maths you will be doing and how you can build on what you already know.

Colour-coded grades

Know what target grade you are working at and track your progress with the colour-coded grade panels at the side of the page.

Use of calculators

Questions when you must or could use your calculator are marked with an icon. Explanations involving calculators are based on the *Casio fx-83ES.*

Grade booster

Review what you have learnt and how to get to the next grade with the Grade booster at the end of each chapter.

Worked examples

Understand the topic before you start the exercise by reading the examples in blue boxes. These take you through questions step by step.

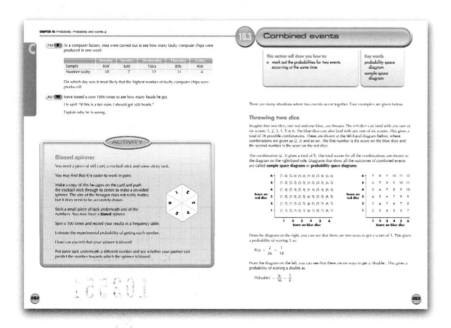

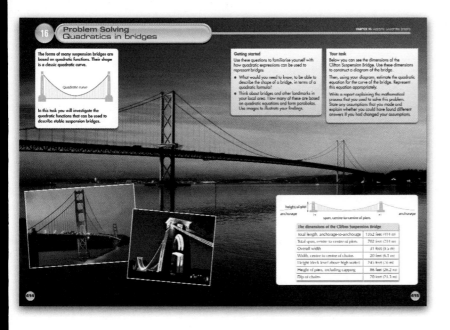

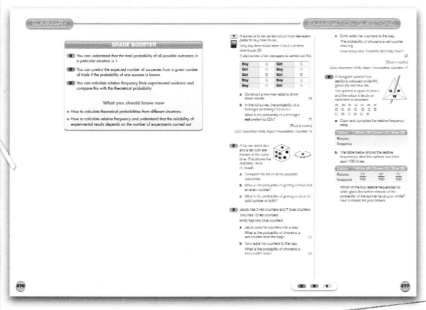

Functional maths

Practise functional maths skills to see how people use maths in everyday life. Look out for practice questions marked **FM**.

There are also extra functional maths and problem-solving activities at the end of most chapters to build and apply your skills.

New Assessment Objectives

Practise new parts of the curriculum (Assessment Objectives AO2 and AO3) with questions that assess your understanding marked **AU** and questions that test if you can solve problems marked **PS**. You will also practise some questions that involve several steps and where you have to choose which method to use; these also test AO2. There are also plenty of straightforward questions (AO1) that test if you can do the maths.

Exam practice

Prepare for your exams with past exam questions and detailed worked exam questions with examiner comments to help you score maximum marks.

Quality of Written Communication (QWC)

Practise using accurate mathematical vocabulary and writing logical answers to questions to ensure you get your QWC (Quality of Written Communication) marks in the exams. The Glossary and worked exam questions will help you with this.

Technology is increasingly important in our lives. It helps us do many things more efficiently than we could without it.

Modern **calculators** take away the need to perform long calculations by hand. They can help to improve accuracy – but a calculator is only as good as the person using it. If you press the buttons in the wrong order when doing a calculation then you will get the wrong answer. That is why learning to use a calculator effectively is important.

The earliest known calculating device was a **tally stick**, which was a stick with notches cut into it so that small numbers could be recorded.

In about 2000 BC the **abacus** was invented in Eygpt.

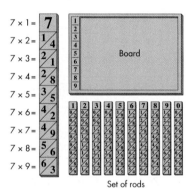

Set of rods

Abacuses are still used widely in China today and they were used widely for almost 3500 years, until John Napier devised a calculating aid called **Napier's bones**.

These led to the invention of the **slide rule** by William Oughtred in 1622. These stayed in use until the mid-1960s. Engineers working on the first ever moon landings used slide rules to do some of their calculations.

In the mid-sixteenth century the first **mechanical calculating machines** were produced. These were based on a series of cogs and gears and so were too expensive to be widely used.

The first **electronic computers** were produced in the mid-twentieth century. Once the transistor was perfected, the power increased and the cost and size decreased until the point where the average scientific calculator that students use in schools has more computing power than the first craft that went into space.

Number: Using calculators

1 Basic calculations and using brackets

2 Using a calculator to add and subtract fractions

3 Using a calculator to multiply and divide fractions

This chapter will show you ...

to **E** **C** how to use a calculator effectively

Visual overview

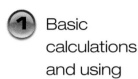

Basic calculations $(+, -, \times, \div)$ → Inputting fractions → Calculating with fractions

What you should already know

- How to add, subtract, multiply and divide with whole numbers, fractions and decimals **(KS3 level 5, GCSE grade E)**
- How to simplify fractions and decimals **(KS3 level 3, GCSE grade G)**
- How to convert improper fractions to mixed numbers or decimals and vice-versa **(KS3 level 4, GCSE grade F)**
- The rules of BIDMAS/BODMAS with decimals **(KS3 level 5, GCSE grade E)**

Quick check

1 Complete these calculations. Do not use a calculator.

  a $48 + 89$ **b** $102 - 37$ **c** 23×7 **d** $336 \div 8$

  e $3.6 + 2.9$ **f** $8.4 - 3.8$ **g** 3×4.5 **h** $7.8 \div 6$

2 a Convert these mixed numbers into improper fractions.

    i $2\frac{2}{5}$ **ii** $3\frac{1}{4}$ **iii** $1\frac{7}{9}$

  b Convert these improper fractions into mixed numbers.

    i $\frac{11}{6}$ **ii** $\frac{7}{3}$ **iii** $\frac{23}{7}$

3 Work these out without using a calculator.

  a $2 + 3 \times 4$ **b** $(2 + 3) \times 4$

  c $6 + 4 - 3^2$ **d** $6 + (4 - 3)^2$

4 Work these out without using a calculator.

  a $\frac{2}{3} + \frac{3}{4}$ **b** $\frac{1}{5} + \frac{2}{7}$

  c $\frac{4}{5} - \frac{1}{4}$ **d** $2\frac{1}{3} - 1\frac{2}{5}$

Basic calculations and using brackets

This section will show you how to:
- use some of the important keys, including the bracket keys, to do calculations on a calculator

Key words

brackets

equals

function key

key

shift key

Most of the calculations in this unit are carried out to find the final answer of an algebraic problem or a geometric problem. The examples are intended to demonstrate how to use some of the **function keys** on the calculator. Remember that some functions will need the **shift key** <kbd>SHIFT</kbd> to make them work. When you have **keyed** in the calculation, press the **equals** key <kbd>=</kbd> to give the answer.

Some calculators display answers to fraction calculations as fractions. There is always a key to change this to a decimal. In examinations, an answer given as a fraction or a decimal will always be acceptable unless the question asks you to round to a given accuracy.

Most scientific calculators can be set up to display the answers in the format you want.

EXAMPLE 1

These three angles are on a straight line.

To find the size of angle a, subtract the angles 68° and 49° from 180°.

You can do the calculation in two ways.

  $180 - 68 - 49$ or $180 - (68 + 49)$

Try keying each calculation into your calculator.

  $180 - 68 - 49$

  <kbd>1</kbd> <kbd>8</kbd> <kbd>0</kbd> <kbd>−</kbd> <kbd>6</kbd> <kbd>8</kbd> <kbd>−</kbd> <kbd>4</kbd> <kbd>9</kbd> <kbd>=</kbd>

The display will show 63.

  $180 - (68 + 49)$

  <kbd>1</kbd> <kbd>8</kbd> <kbd>0</kbd> <kbd>−</kbd> <kbd>(</kbd> <kbd>6</kbd> <kbd>8</kbd> <kbd>−</kbd> <kbd>4</kbd> <kbd>9</kbd> <kbd>)</kbd> <kbd>=</kbd>

Again, the display should show 63.

It is important that you can do this both ways.

You must use the correct calculation or use **brackets** to combine parts of the calculation.

A common error is to work out $180 - 68 + 49$, which will give the wrong answer.

> 68°
>
> a 49°

You will learn more about angles in Chapter 8.

RECALL   **FM** Functional Maths   **AU** (AO2) Assessing Understanding   **PS** (AO3) Problem Solving

EXAMPLE 2

Work out the area of this trapezium, where $a = 12.3$, $b = 16.8$ and $h = 2.4$.

To work out the area of the trapezium, you use the formula:

$$A = \frac{1}{2}(a + b)h$$

Remember, you should always substitute into a formula before working it out.

$$A = \frac{1}{2}(12.3 + 16.8)2.4$$

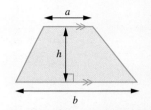

You will learn more about areas of shapes in Chapter 6.

Between the brackets and the numbers at each end there is an assumed multiplication sign, so the calculation is:

$$\frac{1}{2} \times (12.3 + 16.8) \times 2.4$$

Be careful starting. $\frac{1}{2}$ can be keyed in lots of different ways:

- As a division

0.5

- Using the fraction key and the arrows

 and the arrows

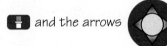

The display should show $\frac{1}{2}$.

Key in the full calculation, using the fraction key:

The display should show 34.92 or $\frac{873}{25}$.

Your calculator has a power key $x^\square$ and a cube key x^3.

EXAMPLE 3

Find the value of $4.5^3 - 2 \times 4.5$.

Try keying in:

| 4 | . | 5 | x^3 | − | 2 | × |

| 4 | . | 5 | = |

The display should show 82.125 or $\dfrac{657}{8}$.

> You often have to work out calculations like this in trial and improvement questions.
> You will learn more about trial and improvement in Chapter 2.

Most calculations involving circles will involve the number π (pronounced 'pi'), which has its own calculator button π.

> You will learn a lot more about π and circles in Chapter 9.

The decimal value of π goes on for ever. It has an approximate value of 3.14 but the value in a calculator is far more accurate and may be displayed as 3.1415926535 or π.

EXAMPLE 4

Work out: **a** $\pi \times 3.2^2$ **b** $2 \times \pi \times 4.9$

Give your answers to 1 decimal place.

a Try keying in:

| π | × | 3 | . | 2 | x^2 | = |

The display should show 32.16990877 or $\dfrac{256}{25}\pi$. (Convert this to a decimal.)

This is 32.2 to 1 decimal place.

b Try keying in:

| 2 | × | π | × | 4 | . | 9 | = |

The display should show 30.78760801 or $\dfrac{49}{5}\pi$. (Convert this to a decimal.)

This is 30.8 to 1 decimal place.

EXERCISE 1A

Use your calculator to work out the following.

Try to key in the calculation as one continuous set, without writing down any intermediate values.

1 Subtract these sets of numbers from 180.

    a 54, 81   **b** 21, 39, 68   **c** 51, 34, 29

2 Subtract these sets of numbers from 360.

    a 68, 92   **b** 90, 121, 34   **c** 32, 46, 46

3 **a** Subtract 68 from 180 and divide the answer by 2.

    b Subtract 46 from 360 and divide the answer by 2.

    c Subtract 52 from 180 twice.

    d Subtract 39 and 2 lots of 64 from 360.

4 Work these out.

    a $(10 - 2) \times 180 \div 10$

    b $180 - (360 \div 5)$

5 Work out:

    a $\frac{1}{2} \times (4.6 + 6.8) \times 2.2$

    b $\frac{1}{2} \times (2.3 + 9.9) \times 4.5$

6 Work out the following and give your answers to 1 decimal place.

    a $\pi \times 8.5$

    b $2 \times \pi \times 3.9$

    c $\pi \times 6.8^2$

    d $\pi \times 0.7^2$

FM 7 At Sovereign garage, Jon bought 21 litres of petrol for £21.52.

At the Bridge garage he paid £15.41 for 15 litres.

At which garage is the petrol cheaper?

AU **8** A teacher asked her class to work out $\dfrac{2.3 + 8.9}{3.8 - 1.7}$.

Abby keyed in:

(2 . 3 ÷ 8 . 9) ÷ 3 . 8 − 1 . 7 =

Bobby keyed in:

2 . 3 ÷ 8 . 9 ÷ 3 . 8 − 1 . 7 =

Col keyed in:

(2 . 3 ÷ 8 . 9) ÷ (3 . 8 − 1 . 7) =

Donna keyed in:

2 . 3 ÷ 8 . 9 ÷ (3 . 8 − 1 . 7) =

They each rounded their answers to 3 decimal places.

Work out the answer each of them found.

Who had the correct answer?

PS **9** Show that a speed of 31 metres per second is approximately 70 miles per hour.

You will need to know that 1 mile ≈ 1610 metres.

10 Work the value of each of these, if $a = 3.4$, $b = 5.6$ and $c = 8.8$.

  a abc

  b $2(ab + ac + bc)$

11 Work out the following giving your answer to 2 decimal places.

  a $\sqrt{(3.2^2 - 1.6^2)}$

  b $\sqrt{(4.8^2 + 3.6^2)}$

12 Work these out.

  a $7.8^3 + 3 \times 7.8$

  b $5.45^3 - 2 \times 5.45 - 40$

Using a calculator to add and subtract fractions

This section will show you how to:
- use a calculator to add and subtract fractions

Key words

fraction
improper fraction
key
mixed number
proper fraction
shift key

In this lesson, questions requiring calculation of **fractions** are set in a context linked to other topics, such as algebra or geometry.

You will recall from Chapter 2 in Book 1 that a fraction with the numerator bigger than the denominator is called an **improper fraction** or a *top-heavy fraction*.

You will also recall that a **mixed number** is made up of a whole number and a **proper fraction**.

For example:

$$\frac{14}{5} = 2\frac{4}{5} \text{ and } 3\frac{2}{7} = \frac{23}{7}$$

Using a calculator with improper fractions

Check that your calculator has a fraction key. Remember, for some functions, you may need to use the **shift key** .

To **key** in a fraction, press ⬛.

Input the fraction so that it looks like this **SHIFT**.

$\frac{9}{5}$ or $9\lrcorner 5$

Now press the equals key **=** so that the fraction displays in the answer part of the screen.

Pressing shift and the key **S⇔D** will convert the fraction to a mixed number.

$1\lrcorner4\lrcorner5$

This is the mixed number $1\frac{4}{5}$.

Pressing the equals sign again will convert the mixed number back to an improper fraction.

- Can you see a way of converting an improper fraction to a mixed number without using a calculator?

- Test your idea. Then use your calculator to check it.

Using a calculator to convert mixed numbers to improper fractions

To input a mixed number, press the shift key first and then the fraction key .

Pressing the equals sign will convert the mixed number to an improper fraction.

- Now key in at least 10 improper fractions and convert them to mixed numbers.

- Remember to press the equals sign to change the mixed numbers back to improper fractions.

- Now input at least 10 mixed numbers and convert them to improper fractions.

- Look at your results. Can you see a way of converting a mixed number to an improper fraction without using a calculator?

- Test your idea. Then use your calculator to check it.

EXAMPLE 5

A water tank is half full. One-third of the capacity of the full tank is poured out.

What fraction of the tank is now full of water?

The calculation is $\frac{1}{2} - \frac{1}{3}$.

Keying in the calculation gives:

The display should show $\frac{1}{6}$.

The tank is now one-sixth full of water.

EXAMPLE 6

Work out the perimeter of a rectangle $1\frac{1}{2}$ cm long and $3\frac{2}{3}$ cm wide.

To work out the perimeter of this rectangle, you can use the formula:

$$P = 2l + 2w$$

where $l = 1\frac{1}{2}$ cm and $w = 3\frac{2}{3}$ cm.

$$P = 2 \times 1\frac{1}{2} + 2 \times 3\frac{2}{3}$$

Keying in the calculation gives:

The display should show $10\frac{1}{3}$.

So the perimeter is $10\frac{1}{3}$ cm.

> You will learn more about perimeters of shapes in Chapter 6.

EXERCISE 1B

1 Use your calculator to work these out. Give your answers as fractions.

Try to key in the calculation as one continuous set, without writing down any intermediate values.

a $\dfrac{3}{4} + \dfrac{4}{5}$ **b** $\dfrac{5}{6} + \dfrac{7}{10}$ **c** $\dfrac{4}{5} + \dfrac{9}{20}$

d $\dfrac{3}{8} + \dfrac{9}{25}$ **e** $\dfrac{7}{20} + \dfrac{3}{16}$ **f** $\dfrac{5}{8} + \dfrac{9}{16} + \dfrac{3}{5}$

g $\dfrac{9}{20} - \dfrac{1}{12}$ **h** $\dfrac{3}{4} - \dfrac{7}{48}$ **i** $\dfrac{11}{32} - \dfrac{1}{6}$

j $\dfrac{4}{5} + \dfrac{9}{16} - \dfrac{2}{3}$ **k** $\dfrac{7}{16} + \dfrac{3}{8} - \dfrac{1}{20}$ **l** $\dfrac{3}{4} + \dfrac{2}{9} - \dfrac{3}{11}$

D

2 Use your calculator to work these out. Give your answers as mixed numbers.

Try to key in the calculation as one continuous set, without writing down any intermediate values.

a $4\frac{3}{4} + 1\frac{4}{5}$

b $3\frac{5}{6} + 4\frac{7}{10}$

c $7\frac{4}{5} + 8\frac{9}{20}$

d $9\frac{3}{8} + 2\frac{9}{25}$

e $6\frac{7}{20} + 1\frac{3}{16}$

f $2\frac{5}{8} + 3\frac{9}{16} + 5\frac{3}{5}$

g $6\frac{9}{20} - 3\frac{1}{12}$

h $4\frac{3}{4} - 2\frac{7}{48}$

i $8\frac{11}{32} - 5\frac{1}{6}$

j $12\frac{4}{5} + 3\frac{9}{16} - 8\frac{2}{3}$

k $9\frac{7}{16} + 5\frac{3}{8} - 7\frac{1}{20}$

l $10\frac{3}{4} + 6\frac{2}{9} - 12\frac{3}{11}$

3 A water tank is three-quarters full. Two-thirds of a full tank is poured out.

What fraction of the tank is now full of water?

4

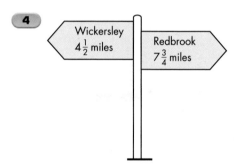

a What is the distance between Wickersley and Redbrook, using these roads?

b How much further is it to Redbrook than to Wickersley?

AU 5 Here is a calculation.

$$\frac{3}{25} + \frac{7}{10}$$

Imagine that you are trying to explain to someone how to use a calculator to do this.

Write down what you would say.

6 There are the same number of boys and girls in a school.

Because of snow $\frac{4}{5}$ of the boys are absent and $\frac{5}{12}$ of the girls are absent.

What fraction of the students are present?

PS 7 **a** Use your calculator to work out $\frac{18}{37} - \frac{23}{43}$.

b Explain how your answer tells you that $\frac{23}{43}$ is greater than $\frac{18}{37}$.

AU 8 Jon is working out $\frac{9}{32} + \frac{5}{7}$ without using a calculator.

He adds the numerators and the denominators to get an answer of $\frac{14}{39}$ which is not correct.

a Use a calculator to work out the correct answer.

b Work out $\frac{14}{39} - \frac{9}{32}$ on your calculator.

c Work out $\frac{14}{39} - \frac{5}{7}$ on your calculator.

d Explain why your answers to parts **b** and **c** show that $\frac{14}{39}$ is a fraction between $\frac{9}{32}$ and $\frac{5}{7}$.

AU 9 **a** Choose two other fractions to add together.

Write down the incorrect answer that Jon would get.

Repeat the steps of question **6** for these fractions.

b Is Jon's answer between your two fractions?

AU 10 To work out the perimeter of a rectangle the following formula is used.

$P = 2l + 2w$

Work out the perimeter when $l = 5\frac{1}{8}$ cm and $w = 4\frac{1}{3}$ cm.

PS 11 A shape is rotated 90° clockwise and then a further 60° clockwise.

What fraction of a turn is needed to return it to its original position?

Give both possible answers.

Using a calculator to multiply and divide fractions

This section will show you how to:
- use a calculator to multiply and divide fractions

Key words
fraction
key
shift key

In this lesson, questions requiring calculation of **fractions** will be set in a context linked to other topics such as algebra or geometry. Remember, for some functions, you may need to use the **shift key** ⊆ .

EXAMPLE 7

Work out the area of a rectangle of length $3\frac{1}{2}$ m and width $2\frac{2}{3}$ m.

You will learn more about perimeters of shapes in Chapter 6.

The formula for the area of a rectangle is:

    area = length × width

Keying in the calculation, where length = $3\frac{1}{2}$ and width = $2\frac{2}{3}$ gives:

SHIFT ▢ 3 ▶ 1 ▼ 2 ▶ ×

SHIFT ▢ 2 ▶ 2 ▼ 3 ▶ =

The display should show $9\frac{1}{3}$.

The area is $9\frac{1}{3}$ cm².

EXAMPLE 8

Work out the average speed of a bus that travels $20\frac{1}{4}$ miles in $\frac{3}{4}$ hour.

The formula for the average speed is:

$$\text{average speed} = \frac{\text{distance}}{\text{time}}$$

Use this formula to work the average speed of the bus, where distance is $20\frac{1}{4}$ and time is $\frac{3}{4}$.

Keying in the calculation gives:

The display should show **27**.

The average speed is 27 mph.

> You will learn more about distance, speed and time in Chapter 5.

EXERCISE 1C

1 Use your calculator to work these out. Give your answers as fractions.

Try to key in the calculation as one continuous set, without writing down any intermediate values.

a $\frac{3}{4} \times \frac{4}{5}$ **b** $\frac{5}{6} \times \frac{7}{10}$ **c** $\frac{4}{5} \times \frac{9}{20}$

d $\frac{3}{8} \times \frac{9}{25}$ **e** $\frac{7}{20} \times \frac{3}{16}$ **f** $\frac{5}{8} \times \frac{9}{16} \times \frac{3}{5}$

g $\frac{9}{20} \div \frac{1}{12}$ **h** $\frac{3}{4} \div \frac{7}{48}$ **i** $\frac{11}{32} \div \frac{1}{6}$

j $\frac{4}{5} \times \frac{9}{16} \div \frac{2}{3}$ **k** $\frac{7}{16} \times \frac{3}{8} \div \frac{1}{20}$ **l** $\frac{3}{4} \times \frac{2}{9} \div \frac{3}{11}$

2 The formula for the area of a rectangle is:

area = length × width

Use this formula to work the area of a rectangle of length $\frac{2}{3}$ m and width $\frac{1}{4}$ m.

3 Each of the steps on a ladder is $\frac{1}{5}$ m high. Ben needs to climb 3 m to fix the shed roof. How many steps will he go up the ladder?

AU 4 **a** Use your calculator to work out $\frac{3}{4} \times \frac{9}{16}$.

b Write down the answer to $\frac{9}{4} \times \frac{3}{16}$.

D

C

AU 5 **a** Use your calculator to work out $\frac{2}{3} \div \frac{5}{6}$.

   b Use your calculator to work out $\frac{2}{3} \times \frac{6}{5}$.

   c Use your calculator to work out $\frac{4}{7} \div \frac{3}{4}$.

   d Write down the answer to $\frac{4}{7} \times \frac{4}{3}$.

6 Use your calculator to work these out. Give your answers as mixed numbers.

Try to key in the calculation as one continuous set, without writing down any intermediate values.

a $4\frac{3}{4} \times 1\frac{4}{5}$

b $3\frac{5}{6} \times 4\frac{7}{10}$

c $7\frac{4}{5} \times 8\frac{9}{20}$

d $9\frac{3}{8} \times 2\frac{9}{25}$

e $6\frac{7}{20} \times 1\frac{3}{16}$

f $2\frac{5}{8} \times 3\frac{9}{16} \times 5\frac{3}{5}$

g $6\frac{9}{20} \div 3\frac{1}{12}$

h $4\frac{3}{4} \div 2\frac{7}{48}$

i $8\frac{11}{32} \div 5\frac{1}{6}$

j $12\frac{4}{5} \times 3\frac{9}{16} \div 8\frac{2}{3}$

k $9\frac{7}{16} \times 5\frac{3}{8} \div 7\frac{1}{20}$

l $10\frac{3}{4} \times 6\frac{2}{9} \div 12\frac{3}{11}$

7 The formula for the area of a rectangle is:

   area = length × width

Use this formula to work out the area of a rectangle of length $5\frac{2}{3}$ metres and width $3\frac{1}{4}$ metres.

8 The volume of a cuboid is $26\frac{3}{4}$ cm^3. It is cut into eight equal pieces.

Work out the volume of one of the pieces.

9 The formula for the distance travelled is:

   distance = average speed × time taken

Work out how far a car travelling at an average speed of $36\frac{1}{4}$ mph will travel in $2\frac{1}{2}$ hours.

10 Glasses are filled from litre bottles of water.

Each glass holds $\frac{1}{2}$ pint.

   1 litre = $1\frac{3}{4}$ pints.

How many litre bottles are needed to fill 10 glasses?

PS 11
FM The ribbon on a roll is $3\frac{1}{2}$ m long. Joe wants to cut pieces of ribbon that are each $\frac{1}{6}$ m long.

He needs 50 pieces.

How many rolls will he need?

GRADE BOOSTER

D You can use BIDMAS/BODMAS to carry out operations in the correct order

D You can use a calculator to add, subtract, multiply and divide fractions

C You can use a calculator to add, subtract, multiply and divide mixed numbers

What you should know now

● How to use a calculator effectively, including the brackets and fraction keys

1 Ahmed uses $\frac{2}{3}$ of a litre of milk each day.

He buys milk in 2-litre bottles.

What is the least number of bottles that he needs to buy for one week?

You **must** show your working.

AQA, Foundation, Module 3, June 2009, Question 18

 2 A train travels 350 miles in $4\frac{3}{4}$ hours. Work out the average speed of the train in miles per hour.

3 A painter has 40 litres of paint.

The paint is in 2.5-litre tins.

How many tins of paint does he have?

 4 The diagram shows a trapezium.

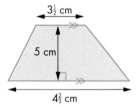

Work out the area of the trapezium.

Hint: Area $= \frac{1}{2}(a + b)h$

5 Matt counts 40 strides as he walks 30 m.

  a How long is each stride?

  b How many strides would he take if he walked 75 m?

  c He decides that to get enough exercise he will do 3000 strides.

    How far will he need to walk?

6 a A parallelogram has base $7\frac{1}{2}$ cm and perpendicular height $7\frac{1}{2}$ cm.

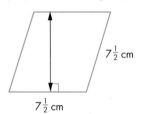

Work out the area.

Hint: Area of a parallelogram = base × perpendicular height

  b The perimeter of the parallelogram is $35\frac{1}{2}$ cm.

    How long is one of the sloping sides?

 7 Calculate $\dfrac{5.6 \times 7.8}{4.3 - 2.1}$

  a Write down your full calculator display. (1)

  b Write your answer to part **a** to 1 decimal place. (1)

(Total 2 marks)

AQA, Higher, Module 3, June 2008, Question 1

 8 Calculate $\sqrt{8.17^3 + 4.39^2}$

  a Give all the figures on your calculator display. (1)

  b Give your answer to an appropriate degree of accuracy. (1)

(Total 2 marks)

AQA, Specification A, Paper 2 November 2007, Question 1b

 9 Work out $\dfrac{21.6 \times 64}{35.1 + 9.57}$

  a Write down your full calculator display. (1)

  b Write your answer to 2 decimal places. (1)

(Total 2 marks)

AQA, Higher, Module 3, June 2009, Question 1

 10 Work out as a decimal $\dfrac{4.6^2}{8.6 - 2.7}$

  a Write down your full calculator display. (1)

  b Write your answer to three significant figures. (1)

(Total 2 marks)

AQA, Higher, Module 3, March 2008, Question 4

 11 a A cuboid has length $5\frac{1}{2}$ cm, width $3\frac{1}{2}$ cm and height 4 cm.

    Work out the volume.

    Hint: Volume of a cuboid = length × width × height

  b The volume of a cuboid is 50 cm^3

    One edge is $2\frac{1}{2}$ cm long.

    Work out a pair of possible lengths for the other edges.

    C **D**

Worked Examination Questions

AU **1** The perimeter of a rectangle is $32\frac{1}{2}$ cm.

Work out a pair of possible values for the length and the width of the rectangle.

Perimeter is 2 × length + 2 × width

Length + width = $32\frac{1}{2}$ ÷ 2

Length + width = $16\frac{1}{4}$ cm

Possible length and width are:

Length = 10 cm

Width = $6\frac{1}{4}$ cm

> You get 1 mark for method for writing down the correct formula and completing the first step of the calculation.

> You get a mark for an accurate calculation.

> Any two values with a sum of $16\frac{1}{4}$ would score the final mark.

Total: 3 marks

FM **2** A driver is travelling 200 miles.

He sets off at 10 am.

He stops for a 20-minute break.

His average speed when travelling is $42\frac{1}{2}$ mph.

He wants to arrive before 3 pm.

Is he successful?

Time travelling = distance ÷ average speed

= 200 ÷ $42\frac{1}{2}$

= $4\frac{12}{17}$ or 4.7058 ...

20 minutes = $\frac{1}{3}$ hour or 0.33

$4\frac{12}{17}$ + $\frac{1}{3}$ or 4.7058 ... + 0.33 ...

= $5\frac{2}{51}$ hours 5.039 hours

10 am to 3 pm is 5 hours so he arrives after 3 pm

> Substituting the correct figures into the formula gets 1 method mark.

> This calculation gets 1 mark for method.

> This answer gets 1 mark for accuracy.

> A statement giving the correct conclusion from correct working would get 1 mark for quality of written communication.

Total: 4 marks

You have been asked by your Business Studies teacher to set up a jewellery stall selling beaded jewellery at an upcoming Young Enterprise fair. There will be 50 stalls at this fair (many of which will be selling jewellery) and it is expected that there will be 500 attendees.

You will be competing against every other stall to sell your products to the attendees, either as one-off purchases or as bulk orders. In order to be successful in this you must carefully plan the design, cost and price of your jewellery, to ensure that people will buy your products and that you make a profit.

Getting started

Answer these questions to begin thinking about how beads can be used to make a piece of jewellery.

1 How many 6 mm beads are needed to make a bracelet?
2 How many 8 mm beads are needed to make an anklet?
3 How many 10 mm beads are needed to make a short necklace?
4 How many 10 mm beads are needed to make a long necklace?
5 You are asked to make a bracelet with beads of two different lengths. You decide to use 6 mm red beads and 8 mm blue beads. How many would you need if you used them alternately?

How to make beaded jewellery

Beads are sold in different sizes and wire is sold in different thicknesses, called the gauge. To make a piece of jewellery the beads are threaded onto the wire.

Step 1 Choose a gauge of wire and cut the length required.
Step 2 Put a fastening on one end.
Step 3 Thread on beads of different sizes in a pattern.
Step 4 Put a fastener on the other end.

Your jewellery is now complete.

Cost of materials

Here are the costs of the raw materials that you will need to make your jewellery.

6 mm beads	10p each
8 mm beads	12p each
10 mm beads	15p each
24-gauge wire	10p per centimetre
20-gauge wire	8p per centimetre

Fasteners for both ends: 30p per item of jewellery

Advice

For bracelets and anklets use 20-gauge wire.

For necklaces use 24-gauge wire.

Beads are available in three lengths: 6 mm, 8 mm and 10 mm.

Beads are available in three colours, green, blue and red.

Standard lengths for bracelets and necklaces

Bracelet	17 cm
Anklet	23 cm
Short necklace	39 cm
Long necklace	46 cm

Your task

With a partner, draw up a business plan for your jewellery stall, to ensure you produce high quality beaded jewellery that will turn a good profit. In your plan, you should include:

- an outline of who you expect to buy your jewellery (your 'target market')
- a design for at least one set of jewellery that will appeal to your target market
- a list of all the materials you will need
- the cost of your designs
- a fair price at which to sell your jewellery
- a discounting plan for bulk orders, or if you must reduce your prices on the day
- an expected profit.

Use all the information given on these pages to create your business plan.

Be sure to justify your plan, using appropriate mathematics and describing the calculations that you have done.

Present your business plan as a report to the Young Enterprise committee.

Have you ever wondered why we write numbers in two different ways, either as fractions or as decimals?

The answer is that sometimes it is easier to use one and sometimes it is easier to use the other.

For example, we use decimals when we write an amount of money but we use fractions when we talk about parts of an hour.

ROYAL GALA APPLE	
SEA BASS	1.39
RHUBARB	9.77
BABY POTATOES	2.39
TROPICAL FRUIT + SOYA	1.49
PRAWNS	3.69
CELERY HEARTS	6.99
CHICKEN	8.95
BLKTAIL FREE RANGE EGGS	10.99
GREEN PEPPERS	4.35
MILK SKIMMED	3.99
WILD ROCKET	1.82
ORANGES	1.99
RASPBERRIES	1.37
TOMATO BABY PLUM	2.99
MOD THAI CHILLES	1.79
CORIANDER POT	0.99
RED ONIONS	1.99
BEEF FILET	1.29
SMK MACKEREL	17.99
PROBIOTIC YOGHURT LF	1.99
SALAD CRESS	0.43
SALAD ONIONS	0.29
GFF GUACAMOLE	0.75
NOODLES	2.09
AVOCADO PEAR	2.99
SALMON FILLETS	1.29
ORANGES	4.39

Some fractions like $\frac{1}{2}$ or $\frac{3}{4}$ or $\frac{5}{8}$ are easy to write as decimals. Those examples are 0.5, 0.75 and 0.625.

However, some fractions are not easy to write as decimals.

The simplest example is $\frac{1}{3}$.

$\frac{1}{3}$ is approximately 0.33 or 0.333 but neither of these is exactly $\frac{1}{3}$.

We can easily show that:

$\frac{1}{3} \times 3 = 1$ but $0.33 \times 3 = 0.99$ and $0.333 \times 3 = 0.999$ and both of those are slightly less than 1.

The only way we can write $\frac{1}{3}$ as a decimal is as 0.333333.... and you have to imagine that going on for ever. It is called a recurring decimal.

It is easier to write one third as $\frac{1}{3}$ rather than as 0.333333…

Lots of other fractions are like this; $\frac{4}{7}$ and $\frac{2}{9}$ are examples.

$\frac{4}{7} = 0.571428571428571428571428 \ldots$

Can you see the repeating pattern? How will it go on?

Some fractions have very long repeating patterns.

$\frac{1}{17} = 0.05882352941176470588235294117647058823529411764705882 \ldots$

$\frac{1}{29} = 0.03448275862068965517241379310344827586206896551724137931 \ldots$

How long is the repeating pattern in each case?

Think about the square root of 2 which we write as √2.

Remember this means the number which, when multiplied by itself, gives an answer of 2.

Your calculator will tell you that √2 = 1.41421356 …

Can we write √2 as a fraction? The answer is no, not exactly. Decimal notation is more useful for numbers like that.

17 twelfths can be written as $\frac{17}{12}$ or $1\frac{5}{12}$.

Use a calculator that works with fractions to show that $1\frac{5}{12} \times 1\frac{5}{12} = 2\frac{1}{144}$

Do you agree that $1\frac{5}{12}$ is very close to √2?

Fractions and decimals are just one property of number. You have already seen many other properties in Book 1 such as multiples of whole numbers, prime numbers and square roots, and will see many more in this chapter.

Number: Number properties 2

1 Multiplying and dividing by powers of 10

2 Prime factors, LCM and HCF

3 Rules for multiplying and dividing powers

This chapter will show you ...

D how to use powers of 10

C how to break a number down into its prime factors

C how to work out the least common multiple of two numbers

C how to work out the highest common factor of two numbers

C how to use rules for multiplying and dividing powers

Visual overview

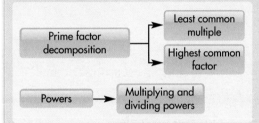

What you should already know

● Multiplication tables up to 10 × 10 **(KS3 level 4, GCSE grade G)**

Quick check

Write down the answers to the following.

1 a 2 × 3		**b** 4 × 3		**c** 5 × 3	
d 6 × 3		**e** 7 × 3		**f** 8 × 3	
2 a 2 × 4		**b** 4 × 4		**c** 5 × 4	
d 6 × 4		**e** 7 × 4		**f** 8 × 4	
3 a 2 × 5		**b** 9 × 5		**c** 5 × 5	
d 6 × 5		**e** 7 × 5		**f** 8 × 5	
4 a 2 × 6		**b** 9 × 6		**c** 8 × 8	
d 6 × 6		**e** 7 × 9		**f** 8 × 6	
5 a 2 × 7		**b** 9 × 7		**c** 8 × 9	
d 6 × 7		**e** 7 × 7		**f** 8 × 7	

Multiplying and dividing by powers of 10

This section will show you how to:
- multiply and divide by powers of 10

Key words

power

product

When you write a million in figures, how many zeros does it have? What is a million as a power of 10? This table shows some of the pattern of the powers of 10.

Number	0.001	0.01	0.1	1	10	100	1000	10 000	100 000
Powers	10^{-3}	10^{-2}	10^{-1}	10^{0}	10^{1}	10^{2}	10^{3}	10^{4}	10^{5}

What pattern is there in the top row?

What pattern is there in the powers in the bottom row?

Multiplication

In the first part of this section, you will look at multiplications. Multiplication calculations are also called **products**. For example, the product of 5 and 7 is the same as $5 \times 7 = 35$.

The easiest number to multiply by is zero, because any number multiplied by zero gives zero.

The next easiest number to multiply by is one, because any number multiplied by one stays the same.

After that it is a matter of opinion, but it is generally accepted that multiplying by 10 is simple. Try these on your calculator.

    a 7×10 **b** 7.34×10 **c** 43×10

    d 0.678×10 **e** 0.007×10 **f** 34.5×10

Can you see the rule for multiplying by 10? You may have learned that when you multiply a number by 10, you add a zero to the number. This is **only** true when you start with a whole number. It is **not true** for a decimal. The rule is:

- Every time you multiply a number by 10, move the digits in the number one place to the left.

Check to make sure that this happened in examples **a** to **f** above.

It is almost as easy to multiply by 100. Try these on your calculator.

    a 7×100 **b** 7.34×100 **c** 43×100

    d 0.678×100 **e** 0.007×100 **f** 34.5×100

This time you should find that the digits move two places to the left.

You can write 100, 1000, 10 000 as powers of 10. For example,

$$100 = 10 \times 10 = 10^2$$

$$1000 = 10 \times 10 \times 10 = 10^3$$

$$10\,000 = 10 \times 10 \times 10 \times 10 = 10^4$$

You should know the connection between the number of zeros and the power of 10. Try these on your calculator. Look for the connection between the calculation and the answer.

a 12.3×10 **b** 3.45×1000 **c** 3.45×10^3

d $0.075 \times 10\,000$ **e** 2.045×10^2 **f** 6.78×1000

g 25.67×10^4 **h** 34.21×100 **i** $0.032\,4 \times 10^4$

Division

Can you find a similar connection for division by multiples of 10? Try these on your calculator. Look for the connection between the calculation and the answer.

a $12.3 \div 10$ **b** $3.45 \div 1000$ **c** $3.45 \div 10^3$

d $0.075 \div 100$ **e** $2.045 \div 10^2$ **f** $6.78 \div 1000$

g $25.67 \div 10^4$ **h** $34.21 \div 100$ **i** $0.032\,4 \div 10^4$

You can use this principle to multiply multiples of 10, 100 and so on. You use this method in estimation. You should have the skill to do this mentally so that you can check that your answers to calculations are about right. (Approximation of calculations is covered on pages 352–356.)

Use a calculator to work out these multiplications.

a $200 \times 300 =$ **b** $100 \times 40 =$ **c** $2000 \times 3000 =$

d $200 \times 50 =$ **e** $200 \times 5000 =$ **f** $300 \times 40 =$

Can you see a way of doing them without using a calculator or pencil and paper? Dividing is almost as simple. Use a calculator to do these divisions.

a $400 \div 20 =$ **b** $200 \div 50 =$ **c** $1000 \div 200 =$

d $300 \div 30 =$ **e** $250 \div 50 =$ **f** $30\,000 \div 600 =$

Once again, there is an easy way of doing these 'in your head'. Look at these examples.

$300 \times 4000 = 1\,200\,000$ $5000 \div 200 = 25$ $200 \times 50 = 10\,000$

$60 \times 5000 = 300\,000$ $400 \div 20 = 20$ $30\,000 \div 600 = 500$

In 200×3000, for example, you multiply the non-zero digits ($2 \times 3 = 6$) and then write the total number of zeros in both numbers at the end, to give $600\,000$.

$$200 \times 3000 = 2 \times 100 \times 3 \times 1000 = 6 \times 100\,000 = 600\,000$$

For division, you divide the non-zero digits and then cancel the zeros. For example:

$$400\,000 \div 80 = \frac{400\,000}{80} = \frac{^5400\,000}{_1 80} = 5000$$

EXERCISE 2A

1 Write down the value of each product.

   a 3.1 × 10   **b** 3.1 × 100   **c** 3.1 × 1000   **d** 3.1 × 10 000

2 Write down the value of each product.

   a 6.5 × 10   **b** 6.5 × 10^2 **c** 6.5 × 10^3 **d** 6.5 × 10^4

3 In questions 1 and 2 there is a connection between the multipliers. What is the connection? (It isn't that the first number is the same.)

4 This list of answers came from a set of questions very similar to those in questions 1 and 2. Write down what the questions must have been, using numbers written out in full and powers of 10. (There is a slight catch!)

   a 73   **b** 730   **c** 7300   **d** 730 000

5 Write down the value of each of the following.

   a 3.1 ÷ 10   **b** 3.1 ÷ 100   **c** 3.1 ÷ 1000   **d** 3.1 ÷ 10 000

6 Write down the value of each of the following.

   a 6.5 ÷ 10   **b** 6.5 ÷ 10^2 **c** 6.5 ÷ 10^3 **d** 6.5 ÷ 10^4

7 In questions 5 and 6 there is a connection between the divisors. What is it?

8 This list of answers came from a set of questions very similar to those in questions 5 and 6. Write down what the questions must have been, using numbers written out in full and powers of 10.

   a 0.73   **b** 0.073   **c** 0.0073   **d** 0.000 073

9 Without using a calculator, write down the answers to these.

   a 2.5 × 100   **b** 3.45 × 10   **c** 4.67 × 1000

   d 34.6 × 10   **e** 20.789 × 10   **f** 56.78 × 1000

   g 2.46 × 10^2 **h** 0.076 × 10   **i** 0.076 × 10^3

   j 0.897 × 10^5 **k** 0.865 × 1000   **l** 100.5 × 10^2

   m 0.999 × 10^6 **n** 234.56 × 10^2 **o** 98.7654 × 10^3

   p 43.23 × 10^6 **q** 78.679 × 10^2 **r** 203.67 × 10^1

   s 76.43 × 10   **t** 34.578 × 10^5 **u** 0.003 4578 × 10^5

10 Without using a calculator, write down the answers to these.

a $2.5 \div 100$

b $3.45 \div 10$

c $4.67 \div 1000$

d $34.6 \div 10$

e $20.789 \div 100$

f $56.78 \div 1000$

g $2.46 \div 10^2$

h $0.076 \div 10$

i $0.076 \div 10^3$

j $0.897 \div 10^5$

k $0.865 \div 1000$

l $100.5 \div 10^2$

m $0.999 \div 10^6$

n $234.56 \div 10^2$

o $98.7654 \div 10^3$

p $43.23 \div 10^6$

q $78.679 \div 10^2$

r $203.67 \div 10^1$

s $76.43 \div 10$

11 Without using a calculator, write down the answers to these.

a 200×300

b 30×4000

c 50×200

d 60×700

e 70×300

f 10×30

g 3×50

h 200×7

i 200×500

j 100×2000

k 20×1400

l 30×30

m $(20)^2$

n $(20)^3$

o $(400)^2$

p 30×150

q 40×200

r 50×5000

12 Without using a calculator, write down the answers to these.

a $2000 \div 400$

b $3000 \div 60$

c $5000 \div 200$

d $6000 \div 200$

e $2100 \div 300$

f $9000 \div 30$

g $300 \div 50$

h $2100 \div 70$

i $2000 \div 500$

j $10\,000 \div 2000$

k $2800 \div 1400$

l $3000 \div 30$

m $2000 \div 50$

n $80\,000 \div 400$

o $400 \div 20$

p $3000 \div 150$

q $400 \div 200$

r $5000 \div 5000$

s $4000 \div 250$

t $300 \div 2$

u $6000 \div 500$

v $30\,000 \div 2000$

w $2000 \times 40 \div 2000$

x $200 \times 20 \div 800$

y $200 \times 6000 \div 30\,000$

z $20 \times 80 \times 600 \div 3000$

AU 13 You are given that $16 \times 34 = 544$.

a Write down the value of 160×340.

b What is $544\,000 \div 34$?

PS 14 Write the following calculations in order, starting with the one that gives the smallest answer.

5000×4000 600×8000 $200\,000 \times 700$ $30 \times 90\,000$

FM 15 In 2009 there were £20 notes to the value of £28 000 million in circulation. How many £20 notes is this?

Prime factors, LCM and HCF

This section will show you how to:
- identify prime factors
- identify the least common multiple (LCM) of two numbers
- identify the highest common factor (HCF) of two numbers

Key words

factor tree

highest common factor

index notation

least common multiple

prime factor

product

product of prime factors

Start with a number, such as 110, and find two numbers that, when multiplied together, give that number, for example, 2 × 55. Are they both prime? No, 55 isn't. So take 55 and repeat the operation, to get 5 × 11. Are these both prime? Yes. So,

110 = 2 × 5 × 11

The **prime factors** of 110 are 2, 5 and 11.

This method is not very logical and you need to know your multiplication tables well to use it. There are, however, two methods that you can use to make sure you do not miss any of the prime factors.

EXAMPLE 1

Find the prime factors of 24.

Divide 24 by any prime number that goes into it. (2 is an obvious choice.)

Now divide the answer (12) by a prime number. As 12 is even, again 2 is the obvious choice.

Repeat this process until you finally have a prime number as the answer.

So, written as a **product of its prime factors**, 24 = 2 × 2 × 2 × 3.

A quicker and neater way to write this answer is to use **index notation**, expressing the answer using powers. (Powers are dealt with on pages 284–286.)

In index notation, as a product of its prime factors, 24 = 2^3 × 3.

```
2 | 24
2 | 12
3 |  6
      2
```

EXAMPLE 2

Find the prime factors of 96.

As a product of prime factors, 96 is $2 \times 2 \times 2 \times 2 \times 2 \times 3 = 2^5 \times 3$.

2	96
2	48
2	24
2	12
2	6
	3

The method shown below is called a **factor tree**.

You start by splitting the number into a **product** of two factors. Then you split these factors, and carry on splitting, until you reach prime numbers.

EXAMPLE 3

Find the prime factors of 76.

Stop splitting the factors here because 2, 2 and 19 are all prime numbers.

So, as a product of prime factors, 76 is $2 \times 2 \times 19 = 2^2 \times 19$.

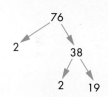

EXAMPLE 4

Find the prime factors of 420.

You can work it upside down, to make an upright tree.

So, as a product of prime factors:

$420 = 2 \times 5 \times 2 \times 3 \times 7 = 2^2 \times 3 \times 5 \times 7$

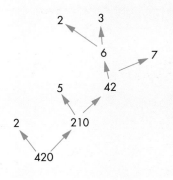

EXAMPLE 5

50 written as the product of prime factors is 2×5^2.

Write the following numbers as products of prime factors in index notation.

    a 150   **b** 500   **c** 200

    a This is 3×50. So, $150 = 2 \times 3 \times 5^2$

    b This is 10×50. So, $500 = 2^2 \times 5^3$.

    c This is 4×50. So, $200 = 2^3 \times 5^2$.

EXERCISE 2B

1 Copy and complete these factor trees.

a

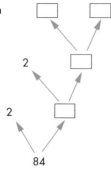

$$84 = 2 \times 2 \ldots \ldots$$

b

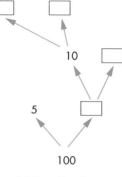

$$100 = 5 \times 2 \ldots \ldots$$

c

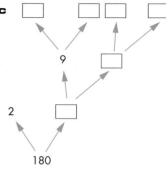

$$180 = 2 \ldots \ldots \ldots \ldots$$

d

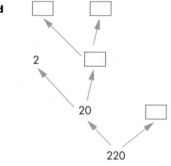

$$220 = 2 \ldots \ldots \ldots$$

e

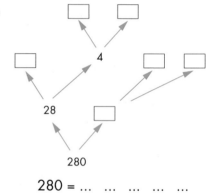

$$280 = \ldots \ldots \ldots \ldots \ldots$$

f

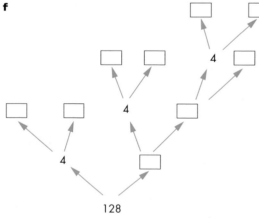

$$128 = \ldots \ldots \ldots \ldots \ldots \ldots \ldots$$

g

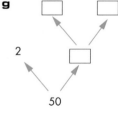

$$50 = \ldots \ldots \ldots$$

h

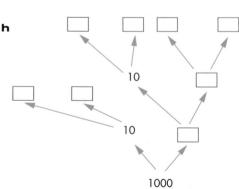

$$1000 = \ldots \ldots \ldots \ldots \ldots \ldots$$

i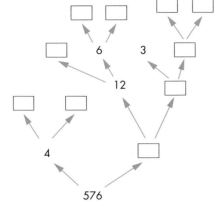

$576 = \ldots \quad \ldots \quad \ldots \quad \ldots \quad \ldots \quad \ldots \quad \ldots \quad \ldots$

j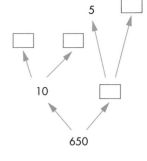

$650 = \ldots \quad \ldots \quad \ldots \quad \ldots$

2 Use index notation to rewrite your answers to question 1, parts **a** to **j**.
For example:

$$100 = 2 \times 2 \times 5 \times 5 = 2^2 \times 5^2$$

and $\quad 540 = 2 \times 2 \times 3 \times 3 \times 3 \times 5 = 2^2 \times 3^3 \times 5$

3 Write the numbers from 1 to 50 as products of their prime factors. Use index notation. For example:

$1 = 1 \qquad 2 = 2 \qquad 3 = 3$

$4 = 2^2 \qquad 5 = 5 \qquad 6 = 2 \times 3 \qquad \ldots$

> **HINTS AND TIPS**
>
> Use your previous answers to help you.
> For example, $9 = 3^2$ so as $18 = 2 \times 9$, $18 = 2 \times 3^2$.

4 a What is special about the numbers 2, 4, 8, 16, 32, …?

b What are the next two terms in this series?

c What are the next three terms in the series 3, 9, 27, …?

d Continue the series 4, 16, 64, …, for three more terms.

e Rewrite all the series in parts **a**, **c** and **d** in index notation. For example, the first series is:

$$2^1, 2^2, 2^3, 2^4, 2^5, 2^6, 2^7, \ldots$$

AU 5 a Express 60 as a product of prime factors.

b Write your answer to part **a** in index form.

c Use your answer to part **b** to write 120, 240 and 480 as a product of prime factors in index form.

C

PS **6** $1001 = 7 \times 11 \times 13$

$1001^2 = 1\,002\,001$

$1001^3 = 1\,003\,003\,001$

a Write $1\,002\,001$ as a product of prime factors, in index form.

b Write $1\,003\,003\,001$ as a product of prime factors, in index form.

c Write $100\,110$ as a product of prime factors, in index form.

FM **7** Harriet wants to share £40 between three of her grandchildren. Explain why it is not possible for them to get equal shares.

Least common multiple

The **least common multiple** (or *lowest common multiple*), or LCM of two numbers is the smallest number that appears in the multiplication tables of both numbers.

For example, the LCM of 3 and 5 is 15, the LCM of 2 and 7 is 14 and the LCM of 6 and 9 is 18.

There are two ways of working out the LCM.

EXAMPLE 6

Find the LCM of 18 and 24.

Write out the 18 times table: 18, 36, 54, ⑦2, 90, 108, …

Write out the 24 times table: 24, 48, ⑦2, 96, 120, …

Numbers that appear in both tables are *common multiples*. You can see that 72 is the smallest (least) number that appears in both tables, so it is the least common multiple.

EXAMPLE 7

Find the LCM of 42 and 63.

Write 42 in prime factor form: $42 = 2 \times 3 \times 7$

Write 63 in prime factor form: $63 = 3^2 \times 7$

Write down the smallest number, in prime factor form (that includes all the prime factors of both 42 and 63).

$2 \times 3^2 \times 7$ (This includes $2 \times 3 \times 7$ and $3^2 \times 7$.)

Then work it out:

$2 \times 3^2 \times 7 = 2 \times 9 \times 7 = 18 \times 7 = 126$

Highest common factor

The **highest common factor** or HCF of two numbers is the biggest number that divides exactly into both of them.

For example, the HCF of 24 and 18 is 6, the HCF of 45 and 36 is 9 and the HCF of 15 and 22 is 1.

EXAMPLE 8

Find the HCF of 28 and 16.

Write out the factors of 28: {1, 2, ④, 7, 14, 28}

Write out the factors of 16: {1, 2, ④, 8, 16}

Numbers that appear in both sets of factors are *common factors*. You can see that 4 is the biggest (highest) number that appears in both lists, so it is the highest common factor.

EXAMPLE 9

a Find the HCF of 48 and 120.

b Write 48 and 120 as products of prime factors.

c What connection is there between the answers to **a** and **b**?

a Write out the factors of 48 and 120.

48 = {1, 2, 3, 4, 6, 8, 12, 16, 24, 48}

120 = {1, 2, 3, 4, 5, 6, 8, 10, 12, 15, 20, 24, 30, 40, 60, 120}

So, the HCF is 24.

b $48 = 2 \times 2 \times 2 \times 2 \times 3 = 2^4 \times 3$

$120 = 2 \times 2 \times 2 \times 3 \times 5 = 2^3 \times 3 \times 5$

c 48 and 120 are connected by the number 24. 24 is $2 \times 2 \times 2 \times 3 = 2^3 \times 3$ as the product of prime factors, which is in both the product of prime factors of 48 and 120.

EXERCISE 2C

1 Find the LCM of each pair of numbers.

   a 4 and 5   **b** 7 and 8   **c** 2 and 3   **d** 4 and 7

   e 2 and 5   **f** 3 and 5   **g** 3 and 8   **h** 5 and 6

2 What connection is there between the LCMs and the pairs of numbers in question 1?

3 Find the LCM of each pair of numbers.

   a 4 and 8   **b** 6 and 9   **c** 4 and 6   **d** 10 and 15

AU 4 Does the connection you found in question 2 still work for the numbers in question 3? If not, can you explain why not?

5 Find the LCM of each pair of numbers.

   a 24 and 56   **b** 21 and 35   **c** 12 and 28   **d** 28 and 42

   e 12 and 32   **f** 18 and 27   **g** 15 and 25   **h** 16 and 36

FM 6 Cheese slices are in packs of eight.
Bread rolls are in packs of six.

What is the smallest number of each pack that needs to be bought to have the same number of cheese slices and bread rolls?

7 Find the HCF of each pair of numbers.

   a 24 and 56   **b** 21 and 35   **c** 12 and 28   **d** 28 and 42

   e 12 and 32   **f** 18 and 27   **g** 15 and 25   **h** 16 and 36

   i 42 and 27   **j** 48 and 64   **k** 25 and 35   **l** 36 and 54

PS 8 In prime factor form $1250 = 2 \times 5^4$ and $525 = 3 \times 5^2 \times 7$.

   a Which of these are common multiples of 1250 and 525?

      i $2 \times 3 \times 5^3 \times 7$ **ii** $2^3 \times 3 \times 5^4 \times 7^2$ **iii** $2 \times 3 \times 5^4 \times 7$ **iv** $2 \times 3 \times 5 \times 7$

   b Which of these are common factors of 1250 and 525?

      i 2×3 **ii** 2×5 **iii** 5^2 **iv** $2 \times 3 \times 5 \times 7$

PS 9 The HCF of two numbers is 6.

The LCM of the same two numbers is 72.

What are the numbers?

Rules for multiplying and dividing powers

This section will show you how to:

- use rules for multiplying and dividing powers

Key word

power

Look what happens when you multiply numbers that are written as **powers** of the same number or variable (letter).

For example:

$$3^3 \times 3^5 = (3 \times 3 \times 3) \times (3 \times 3 \times 3 \times 3 \times 3)$$
$$= 3^8$$

$$a^2 \times a^3 = (a \times a) \times (a \times a \times a)$$
$$= a^5$$

Can you see the rule? You can find these products just by *adding* the powers.
For example,

$$2^3 \times 2^4 \times 2^5 = 2^{12}$$
$$a^3 \times a^4 = a^{3+4} = a^7$$

Now look what happens when you divide numbers that are written as powers of the same number or letter (variable).

For example,

$$7^6 \div 7 = (7 \times 7 \times 7 \times 7 \times 7 \times 7) \div (7)$$
$$= 7 \times 7 \times 7 \times 7 \times 7$$
$$= 7^5$$

$$a^5 \div a^2 = (a \times a \times a \times a \times a) \div (a \times a)$$
$$= a \times a \times a$$
$$= a^3$$

Can you see the rule? You can do these divisions just by *subtracting* the powers.
For example,

$$a^4 \div a^3 = a^{4-3}$$
$$= a^1$$
$$= a$$

$$b^7 \div b^4 = b^3$$

EXERCISE 2D

1 Write these as single powers of 5.

  a $5^2 \times 5^2$ **b** $5^4 \times 5^6$ **c** $5^2 \times 5^3$ **d** 5×5^2 **e** $5^6 \times 5^9$

  f 5×5^8 **g** $5^2 \times 5^4$ **h** $5^6 \times 5^3$ **i** $5^2 \times 5^6$

2 Simplify these (write them as single powers of x).

  a $x^2 \times x^6$ **b** $x^5 \times x^4$ **c** $x^6 \times x^2$ **d** $x^3 \times x^2$ **e** $x^6 \times x^6$

  f $x^5 \times x^8$ **g** $x^7 \times x^4$ **h** $x^2 \times x^8$ **i** $x^{12} \times x^4$

3 Write these as single powers of 6.

  a $6^5 \div 6^2$ **b** $6^7 \div 6^2$ **c** $6^3 \div 6^2$ **d** $6^4 \div 6^4$ **e** $6^5 \div 6^4$

  f $6^5 \div 6^2$ **g** $6^4 \div 6^2$ **h** $6^4 \div 6^3$ **i** $6^5 \div 6^3$

4 Simplify these (write them as single powers of x).

  a $x^7 \div x^3$ **b** $x^8 \div x^3$ **c** $x^4 \div x$ **d** $x^6 \div x^3$ **e** $x^{10} \div x^4$

  f $x^6 \div x$ **g** $x^8 \div x^6$ **h** $x^8 \div x^2$ **i** $x^{12} \div x^3$

AU 5 **a** Write down the value of $216 \div 216$.

  b Write $6^3 \div 6^3$ as a single power of 6.

  c $6^3 = 216$

  Use parts **a** and **b** to write down the value of 6^0.

AU 6 **a** Write down the value of $625 \div 625$.

  b Write $5^4 \div 5^4$ as a single power of 5.

  c $5^4 = 625$

  Use parts **a** and **b** to write down the value of 5^0.

AU 7 What do you notice about your answers to questions **5** and **6**?

AU 8 $4^a \times 4^b = 4^7$

  Write down one pair of possible values for a and b.

PS 9 **a** A common error is to write $x^a + x^a = x^{2a}$.

  Find two whole numbers for a and x for which this is true.

  b Another common error is to write $x^a \times x^b = x^{ab}$.

  Find a whole number for x for which this is true.

> **HINTS AND TIPS**
>
> There is only one possible answer, and both a and x are less than 5.

> **HINTS AND TIPS**
>
> There is only one possible answer

1 **a** Write down the value of: **i** 2^3 **ii** 3^3.

b Complete the number pattern below.

1^2 $= 1 = 1^3$

$(1 + 2)^2$ $= 9 = 1^3 + 2^3$

$(1 + 2 + 3)^2$ $= 36 = 1^3 + 2^3 + 3^3$

$(1 + 2 + ... + ...)^2 = ... = 1^3 + 2^3 + ...$
$+ ......$

2 Simplify:

a $c \times c \times c \times c$

b $d^3 \times d^2$

c $\dfrac{e}{e^8}$

3 **a** Write 28 as the product of its prime factors.

b Find the lowest common multiple (LCM) of 28 and 42.

4 Tom, Sam and Matt are counting drum beats.

Tom hits a snare drum every 2 beats.

Sam hits a kettle drum every 5 beats.

Matt hits a bass drum every 8 beats.

Tom, Sam and Matt start by hitting their drums at the same time. How many beats is it before Tom, Sam and Matt next hit their drums at the same time?

5 **a** Work out the value of $5^7 \div 5^4$.

b a and b are prime numbers.
$ab^3 = 54$
Find the values of a and b.

c Find the highest common factor (HCF) of 54 and 135.

6 Use these number cards to make each of the following.

Each card can only be used once throughout all parts of the question.

a A two-digit multiple of 7

b The HCF of 20 and 35

c A two-digit prime number

7 **a** Write 64 as a power of 2.

b Explain how you can tell that 640 000 has only two prime factors.

8 Small pies are sold in packs of 4

Bread sticks are sold in packs of 10

What is the least number of each pack that needs to be bought to have the same number of pies and bread sticks?

Why this chapter matters

You will have seen percentages in many places and situations.

Why use percentages?

Because:

- basic percentages are quite easy to understand
- they are a good way of comparing fractions
- percentages are used a lot in everyday life.

Who uses them?

Shops often have sales where they offer a certain percentage off their standard prices. By understanding percentages you will be able to quickly work out where the real bargains are.

Each month (or year) banks pay their customers interest on their savings. The interest rate the banks offer is expressed as a percentage. Also, banks charge interest on loans. Again, the rate of interest is also expressed as a percentage. With loans, banks add a percentage of the total money owed onto the debt each month so unless you pay off the loan, the money you owe will continue to get bigger.

Some salespeople, such as this car salesman, earn commission on every sale they make. Their commission is paid as a percentage of the retail price of every item they sell. Commission acts as an incentive to sell lots of items – all those percentages can add up to a lot of money!

The government will often use percentages to demonstrate changes in social and economic circumstances. For example, they may tell the public that 7% of the population of working age is unemployed or that value added tax (VAT) will be set at 20%.

You will have often received marks on a test in the form of a percentage.

Can you think of other examples? You will find several everyday uses in this chapter.

3

Number: Percentages and ratio

1 Equivalent percentages, fractions and decimals

2 Calculating a percentage of a quantity

3 Increasing or decreasing quantities by a percentage

4 Expressing one quantity as a percentage of another quantity

5 Ratio

This chapter will show you ...

- **G** what is meant by percentage
- **E** how to convert between decimals and fractions
- **E** what a ratio is
- **E** how to do calculations involving percentages
- **D** how to use your calculator to work out percentages by using a multiplier
- to **D** **C** how to divide an amount according to a given ratio
- **C** how to work out percentage increases and decreases

Visual overview

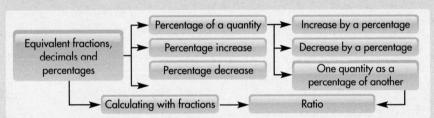

What you should already know

- Multiplication tables up to 10×10 **(KS3 level 4, GCSE grade G)**
- How to simplify fractions **(KS3 level 5, GCSE grade G)**
- How to calculate with fractions **(KS3 level 5, GCSE grade F)**
- How to multiply decimals by 100 (move the digits two places to the left) **(KS3 level 5, GCSE grade G)**
- How to divide decimals by 100 (move the digits two places to the right) **(KS3 level 5, GCSE grade G)**

Quick check

1 Simplify these fractions.

   a $\frac{12}{32}$ **b** $\frac{20}{45}$ **c** $\frac{28}{48}$ **d** $\frac{36}{60}$

2 Work out these amounts.

   a $\frac{2}{3}$ of 27   **b** $\frac{5}{8}$ of 32   **c** $\frac{1}{4} \times 76$ **d** $\frac{3}{5} \times 45$

3 Work out these amounts.

   a 12×100 **b** $34 \div 100$ **c** 0.23×100 **d** $4.7 \div 100$

This section will show you how to:
- convert percentages to fractions and decimals and vice versa

Key words

decimal

decimal equivalents

fraction

percentage

100% means the *whole* of something. So if you want to, you can express *part* of the whole as a **percentage**.

Per cent means 'out of 100'.

So, any percentage can be converted to a **fraction** with denominator 100.

For example:

$32\% = \dfrac{32}{100}$ which can be simplified by cancelling to $\dfrac{8}{25}$

Also, any percentage can be converted to a **decimal** by dividing the percentage number by 100. This means moving the digits two places to the right.

For example:

$65\% = 65 \div 100 = 0.65$

Any decimal can be converted to a percentage by multiplying by 100%.

For example:

$0.43 = 0.43 \times 100\% = 43\%$

Any fraction can be converted to a percentage by making the denominator into 100 and taking the numerator as the percentage.

For example:

$\dfrac{2}{5} = \dfrac{40}{100} = 40\%$

Fractions can also be converted to percentages by dividing the numerator by the denominator and multiplying by 100%.

For example:

$\dfrac{2}{5} = 2 \div 5 \times 100\% = 40\%$

Knowing the percentage and **decimal equivalents** of the common fractions is extremely useful. So, do try to learn them.

$$\frac{1}{2} = 0.5 = 50\% \qquad \frac{1}{4} = 0.25 = 25\% \qquad \frac{3}{4} = 0.75 = 75\% \qquad \frac{1}{8} = 0.125 = 12.5\%$$

$$\frac{1}{10} = 0.1 = 10\% \qquad \frac{1}{5} = 0.2 = 20\% \qquad \frac{1}{3} = 0.33 = 33\frac{1}{3}\% \qquad \frac{2}{3} = 0.67 = 67\%$$

The following table shows how to convert from one to the other.

Convert from percentage to:	
Decimal	Fraction
Divide the percentage by 100, for example $52\% = 52 \div 100$ $= 0.52$	Make the percentage into a fraction with a denominator of 100 and simplify by cancelling down if possible, for example $52\% = \frac{52}{100} = \frac{13}{25}$

Convert from decimal to:	
Percentage	Fraction
Multiply the decimal by 100%, for example $0.65 = 0.65 \times 100\%$ $= 65\%$	If the decimal has 1 decimal place put it over the denominator 10, if it has 2 decimal places put it over the denominator 100, etc. Then simplify by cancelling down if possible, for example $0.65 = \frac{65}{100} = \frac{13}{20}$

Convert from fraction to:	
Percentage	Decimal
If the denominator is a factor of 100 multiply numerator and denominator to make the denominator 100, then the numerator is the percentage, for example $\frac{3}{20} = \frac{15}{100} = 15\%$ or convert to a decimal and change the decimal to a percentage, for example $\frac{7}{8} = 7 \div 8 = 0.875 = 87.5\%$	Divide the numerator by the denominator, for example $\frac{9}{40} = 9 \div 40 = 0.225$

EXAMPLE 1

Convert the following to decimals: **a** 78% **b** 35% **c** $\frac{3}{25}$ **d** $\frac{7}{40}$.

a $78\% = 78 \div 100 = 0.78$ **b** $35\% = 35 \div 100 = 0.35$

c $\frac{3}{25} = 3 \div 25 = 0.12$ **d** $\frac{7}{40} = 7 \div 40 = 0.175$

EXAMPLE 2

Convert the following to percentages: **a** 0.85 **b** 0.125 **c** $\frac{7}{20}$ **d** $\frac{3}{8}$.

a $0.85 = 0.85 \times 100\% = 85\%$ **b** $0.125 = 0.125 \times 100\% = 12.5\%$

c $\frac{7}{20} = \frac{35}{100} = 35\%$ **d** $\frac{3}{8} = 3 \div 8 \times 100\% = 0.375 \times 100\% = 37.5\%$

EXAMPLE 3

Convert the following to fractions: **a** 0.45 **b** 0.4 **c** 32% **d** 15%.

a $0.45 = \frac{45}{100} = \frac{9}{20}$ **b** $0.4 = \frac{4}{10} = \frac{2}{5}$

c $32\% = \frac{32}{100} = \frac{8}{25}$ **d** $15\% = \frac{15}{100} = \frac{3}{20}$

EXERCISE 3A

1 Write each percentage as a fraction in its simplest form.

   a 8%   **b** 50%   **c** 25%

   d 35%   **e** 90%   **f** 75%

2 Write each percentage as a decimal.

   a 27%   **b** 85%   **c** 13%

   d 6%   **e** 80%   **f** 32%

3 Write each decimal as a fraction in its simplest form.

   a 0.12   **b** 0.4   **c** 0.45

   d 0.68   **e** 0.25   **f** 0.625

4 Write each decimal as a percentage.

   a 0.29   **b** 0.55   **c** 0.03

   d 0.16   **e** 0.6   **f** 1.25

5 Write each fraction as a percentage.

   a $\frac{7}{25}$ **b** $\frac{3}{10}$ **c** $\frac{19}{20}$

   d $\frac{17}{50}$ **e** $\frac{11}{40}$ **f** $\frac{7}{8}$

6 Write each fraction as a decimal.

   a $\frac{9}{15}$ **b** $\frac{3}{40}$ **c** $\frac{19}{25}$

   d $\frac{5}{16}$ **e** $\frac{1}{20}$ **f** $\frac{1}{8}$

7 Of the 300 members of a social club 50% are men. How many members are women?

8 Gillian came home and told her dad that she got 100% of her spellings correct. She told her mum that there were 25 spellings to learn. How many spellings did Gillian get wrong?

9 Every year a school library likes to replace 1% of its books. One year the library had 2000 books. How many did it replace?

10 a If 23% of pupils go home for lunch, what percentage do not go home for lunch?

b If 61% of the population takes part in the National Lottery, what percentage do not take part?

c If 37% of members of a gym are males, what percentage of the members are females?

11 I calculated that 28% of my time is spent sleeping and 45% is spent working. How much time is left to spend doing something else?

12 In one country, 24.7% of the population is below the age of 16 and 13.8% of the population is aged over 65. How much of the population is aged from 16 to 65 inclusive?

13 Approximately what percentage of each bottle is filled with water?

 **a**    **b**    **c**

PS 14 Helen made a cake for James. The amount of cake left each day is shown in the diagram.

What percentage has been eaten each day?

Monday   Tuesday   Wednesday   Thursday   Friday

15 Convert each fraction into a percentage.

a $\frac{1}{5}$ **b** $\frac{1}{4}$ **c** $\frac{3}{4}$ **d** $\frac{9}{20}$ **e** $\frac{7}{50}$

f $\frac{1}{2}$ **g** $\frac{3}{5}$ **h** $\frac{7}{40}$ **i** $\frac{11}{20}$ **j** $\frac{13}{10}$

16 Convert each fraction into a percentage. Give your answers to one decimal place.

a $\frac{1}{3}$ **b** $\frac{1}{6}$ **c** $\frac{2}{3}$ **d** $\frac{5}{6}$ **e** $\frac{2}{7}$

f $\frac{47}{60}$ **g** $\frac{31}{45}$ **h** $\frac{8}{9}$ **i** $\frac{73}{90}$ **j** $\frac{23}{110}$

17 Change each of these decimals into a percentage.

a 0.07 **b** 0.8 **c** 0.66 **d** 0.25 **e** 0.545

f 0.82 **g** 0.3 **h** 0.891 **i** 1.2 **j** 2.78

18 Chris scored 24 marks out of a possible 40 in a maths test.

a Write this score as a fraction.

b Write this score as a decimal.

c Write this score as a percentage.

 19 a Convert each of the following test scores into a percentage. Give each answer to the nearest whole number.

Subject	Result	Percentage
Mathematics	38 out of 60	
English	29 out of 35	
Science	27 out of 70	
History	56 out of 90	
Technology	58 out of 75	

b If all the tests are of the same standard, which was the best result?

 20 There were two students missing from my class of 30. What percentage of my class were away?

 21 In one season, Paulo Di Canio had 110 shots at goal. He scored with 28 of these shots. What percentage of his shots resulted in goals?

22 Copy and complete the table.

Percentage	Decimal	Fraction
34%		
	0.85	
		$\frac{3}{40}$
45%		
	0.3	
		$\frac{2}{3}$
84%		
	0.45	
		$\frac{3}{8}$

FM 23 The manager of a garage wants to order 27 000 litres of fuel. A fuel tanker holds 30 000 litres when full. How much fuel should the manager order as a fraction or percentage of a full tanker load?

Calculating a percentage of a quantity

This section will show you how to:
- calculate a percentage of a quantity

Key words
multiplier
quantity

To calculate a percentage of a **quantity**, you multiply the quantity by the percentage. The percentage may be expressed as either a fraction or a decimal. When finding percentages without a calculator, base the calculation on 10% (or 1%) as these are easy to calculate.

EXAMPLE 4

Calculate: **a** 10% of 54 kg   **b** 15% of 54 kg.

a 10% is $\frac{1}{10}$ so $\frac{1}{10}$ of 54 kg = 54 kg ÷ 10 = 5.4 kg

b 15% is 10% + 5% = 5.4 kg + 2.7 kg = 8.1 kg

EXAMPLE 5

Calculate 12% of £80.

10% of £80 is £8 and 1% of £80 is £0.80

12% = 10% + 1% + 1% = £8 + £0.80 + £0.80 = £9.60

Using a percentage multiplier

You have already seen that percentages and decimals are equivalent so it is easier, particularly when using a calculator, to express a percentage as a decimal and use this to do the calculation.

For example, 13% is a **multiplier** of 0.13, 20% a multiplier of 0.2 (or 0.20) and so on.

EXAMPLE 6

Calculate 45% of 160 cm.

45% = 0.45, so 45% of 160 = 0.45 × 160 = 72 cm

Find 52% of £460.

52% = 0.52

So, 0.52 × 460 = 239.2

This gives £239.20.

Remember to always write a money answer with 2 decimal places.

EXERCISE 3B

1 What multiplier is equivalent to a percentage of:

    a 88%   **b** 30%   **c** 25%   **d** 8%   **e** 115%?

2 What percentage is equivalent to a multiplier of:

    a 0.78   **b** 0.4   **c** 0.75   **d** 0.05   **e** 1.1?

3 Calculate the following.

    a 15% of £300   **b** 6% of £105   **c** 23% of 560 kg

    d 45% of 2.5 kg   **e** 12% of 9 hours   **f** 21% of 180 cm

    g 4% of £3   **h** 35% of 8.4 m   **i** 95% of £8

    j 11% of 308 minutes   **k** 20% of 680 kg   **l** 45% of £360

FM 4 The manager of a school canteen estimates that 40% of students will buy a school lunch. There are 1200 students in the school. She knows that her estimates are usually accurate to within 2%. Including this 2% figure, what is the greatest number of lunches she will have left over?

5 An estate agent charges 2% commission on every house he sells. How much commission will he earn on a house that he sells for £120 500?

6 A department store had 250 employees. During one week of a flu epidemic, 14% of the store's employees were absent.

    a What percentage of the employees went into work?

    b How many of the employees went into work?

7 It is thought that about 20% of fans at a rugby match are women. For a match at Twickenham there were 42 600 fans. How many of these do you think would be women?

8 At St Pancras railway station, in one week 350 trains arrived. Of these trains, 5% arrived early and 13% arrived late. How many arrived on time?

AU 9 For the FA Cup Final that was held at Wembley, each year the 90 000 tickets were split up as follows.

Each of the teams playing received 30% of the tickets.

The referees' association received 1% of the tickets.

The other 90 teams received 10% of the tickets among them.

The FA associates received 20% of the tickets among them.

The rest were for the special celebrities.

How many tickets went to each set of people?

FM 10 A school estimates that for a school play 60% of the students will attend. There are 1500 students in the school. The caretaker is told to put out one seat for each person expected to attend plus an extra 10% of that amount in case more attend. How many seats does he need to put out?

> **HINTS AND TIPS**
>
> It is not 70% of the number of students in the school.

11 A school had 850 pupils and the attendance record in the week before Christmas was:

Monday 96%   Tuesday 98%   Wednesday 100%   Thursday 94%   Friday 88%

How many pupils were present each day?

12 Calculate the following.

a 12.5% of £26

b 6.5% of 34 kg

c 26.8% of £2100

d 7.75% of £84

e 16.2% of 265 m

f 0.8% of £3000

13 Air consists of 80% nitrogen and 20% oxygen (by volume). A man's lungs have a capacity of 600 cm³. How much of each gas will he have in his lungs when he has just taken a deep breath?

14 A factory estimates that 1.5% of all the garments it produces will have a fault in them. One week the factory produces 850 garments. How many are likely to have a fault?

15 An insurance firm sells house insurance and the annual premiums are usually set at 0.3% of the value of the house. What will be the annual premium for a house valued at £90 000?

PS 16 Average prices in a shop went up by 3% last year and 3% this year. Did the actual average price of items this year rise by more, the same amount, or less than last year?

Explain how you decided.

3.3 Increasing or decreasing quantities by a percentage

This section will show you how to:	Key word
• increase and decrease quantities by a percentage	multiplier

Increasing by a percentage

There are two methods for increasing a quantity by a percentage.

Method 1

Work out the increase and add it on to the original amount.

EXAMPLE 7

Increase £6 by 5%.

Work out 5% of £6: (5 ÷ 100) × 6 = £0.30

Add the £0.30 to the original amount: £6 + £0.30 = £6.30

Method 2

Use a **multiplier**. An increase of 6% is equivalent to the original 100% *plus* the extra 6%. This is a total of 106% and is equivalent to the multiplier 1.06

EXAMPLE 8

Increase £6.80 by 5%.

A 5% increase is a multiplier of 1.05

So £6.80 increased by 5% is 6.80 × 1.05 = £7.14

EXERCISE 3C

1 What multiplier is used to increase a quantity by:

    a 10%   **b** 3%   **c** 20%   **d** 7%   **e** 12%?

2 Increase each of the following by the given percentage. (Use any method you like.)

    a £60 by 4%   **b** 12 kg by 8%   **c** 450 g by 5%   **d** 545 m by 10%

    e £34 by 12%   **f** £75 by 20%   **g** 340 kg by 15%   **h** 670 cm by 23%

    i 130 g by 95%   **j** £82 by 75%   **k** 640 m by 15%   **l** £28 by 8%

3 Kevin, who was on a salary of £27 500, was given a pay rise of 7%. What is his new salary?

4 In 2005 the population of Melchester was 1 565 000. By 2010 it had increased by 8%. What was the population of Melchester in 2010?

AU 5 A small firm made the same pay increase of 5% for all its employees.

    a Calculate the new pay of each employee listed below. Each of their salaries before the increase is given.

        Bob, caretaker, £16 500   Jean, supervisor, £19 500
        Anne, tea lady, £17 300   Brian, manager, £25 300

    b Explain why the actual pay increases are different for each employee?

FM 6 A bank pays 7% interest on the money that each saver keeps in the bank for one year. Alison keeps £385 in this bank for one year. At the end of the year she wants to buy some furniture advertised for £550. She is offered a 10% discount on the furniture. If she uses all her savings, including interest, from the bank, how much more money will she need to buy the furniture?

7 In 1980 the number of cars on the roads of Sheffield was about 102 000. Since then it has increased by 90%. Approximately how many cars are there on the roads of Sheffield now?

8 An advertisement for a breakfast cereal states that a special offer packet contains 15% more cereal for the same price as a normal 500 g packet. How much breakfast cereal is there in a special offer packet?

9 A headteacher was proud to point out that, since he had arrived at the school, the number of students had increased by 35%. How many students are now in the school, if there were 680 when the headteacher started at the school?

10 At a school disco there are always about 20% more girls than boys. If there were 50 boys at a recent disco, how many girls were there?

FM 11 The Government adds a tax called VAT to the price of most goods in shops. At the moment, it is 17.5% on all electrical equipment.

Calculate the price of the following electrical equipment after VAT of 17.5% has been added.

Equipment	Pre-VAT price
TV set	£245
Microwave oven	£72
CD player	£115
Personal stereo	£29.50

PS 12 A television costs £400 before the VAT is added.

If the rate of VAT goes up from 15% to 20%, by how much will the cost of the television increase?

Decreasing by a percentage

There are two methods for decreasing a quantity by a percentage.

Method 1

Work out the decrease and subtract it from the original amount.

EXAMPLE 9

Decrease £8 by 4%.

Work out 4% of £8: (4 ÷ 100) × 8 = £0.32

Subtract the £0.32 from the original amount: £8 − £0.32 = £7.68

Method 2

Use a multiplier. A 7% decrease is 7% less than the original 100% so it represents
100% – 7% = 93% of the original. This is a multiplier of 0.93

EXAMPLE 10

Decrease £8.60 by 5%.

A decrease of 5% is a multiplier of 0.95

So £8.60 decreased by 5% is 8.60 × 0.95 = £8.17

EXERCISE 3D

1 What multiplier is used to decrease a quantity by:

    a 8%   **b** 15%   **c** 25%   **d** 9%   **e** 12%?

2 Decrease each of the following by the given percentage. (Use any method you like.)

    a £10 by 6%   **b** 25 kg by 8%   **c** 236 g by 10%   **d** 350 m by 3%

    e £5 by 2%   **f** 45 m by 12%   **g** 860 m by 15%   **h** 96 g by 13%

    i 480 cm by 25%   **j** 180 minutes by 35%   **k** 86 kg by 5%   **l** £65 by 42%

3 A car valued at £6500 last year is now worth 15% less. What is its value now?

4 A new P-plan diet guarantees that you will lose 12% of your weight in the first month.
How much should the following people weigh after one month on the diet?

    a Gillian, who started at 60 kg   **b** Peter, who started at 75 kg

    c Margaret, who started at 52 kg

FM 5 A motor insurance firm offers no-claims discounts off the full premium, as follows.

    1 year no claim   15% discount off the full premium

    2 years no claim   25% discount off the full premium

    3 years no claim   45% discount off the full premium

    4 years no claim   60% discount off the full premium

Mr Speed and his family are all offered motor insurance from this firm:

    Mr Speed has four years' no-claim discount and the full premium would be £440.

    Mrs Speed has one year's no-claim discount and the full premium would be £350.

    James has three years' no-claim discount and the full premium would be £620.

    John has two years' no-claim discount and the full premium would be £750.

Calculate the actual amount each member of the family has to pay for the motor insurance.

6 A large factory employed 640 people. It had to streamline its workforce and lose 30% of the workers. How big is the workforce now?

7 On the last day of the Christmas term, a school expects to have an absence rate of 6%. If the school population is 750 pupils, how many pupils will the school expect to see on the last day of the Christmas term?

8 A particular charity called *Young Ones* said that since the start of the National Lottery they have had a decrease of 45% in the amount of money raised by scratch cards. If before the Lottery the charity had an annual income of £34 500 from their scratch cards, how much do they collect now?

9 Most speedometers in cars have an error of about 5% from the true reading. When my speedometer says I am driving at 70 mph:

a what is the lowest speed I could be doing

b what is the highest speed I could be doing?

10 You are a member of a club that allows you to claim a 12% discount off any marked price in shops. What will you pay in total for the following goods?

Sweatshirt £19

Track suit £26

FM 11 **AU** **a** I read an advertisement in my local newspaper last week that stated: "By lagging your roof and hot water system you will use 18% less fuel." Since I was using an average of 640 units of gas a year, I thought I would lag my roof and my hot water system. How much gas would I expect to use now?

b I actually used 18% more gas than I expected to use.

Did I use less gas than last year, more gas than last year, or the same amount of gas as last year?

Show how you work out your answer.

PS 12 Shops add VAT to the basic price of goods to find the selling price that customers will be asked to pay. In a sale, a shop reduces the selling price by a certain percentage to set the sale price. Calculate the sale price of each of these items.

Item	Basic price	VAT rate	Sale discount	Sale price
TV	£220	17.5%	14%	
DVD player	£180	17.5%	20%	

AU 13 **PS** A shop advertises garden ornaments at £50 but with 10% off in a sale. It then advertises an extra 10% off the sale price.

Show that this is not a decrease in price of 20%.

Expressing one quantity as a percentage of another quantity

This section will show you how to:
- express one quantity as a percentage of another
- work out percentage change

Key words

percentage change
percentage decrease
percentage increase
percentage loss
percentage profit

You can express one quantity as a percentage of another by setting up the first quantity as a fraction of the second, making sure that the *units of each are the same.* Then you convert the fraction into a percentage by multiplying by 100%.

EXAMPLE 11

Express £6 as a percentage of £40.

Set up the fraction and multiply by 100%.

$(6 \div 40) \times 100\% = 15\%$

EXAMPLE 12

Express 75 cm as a percentage of 2.5 m.

First, change 2.5 m to 250 cm to work in a common unit.

So the problem now becomes: Express 75 cm as a percentage of 250 cm.

Set up the fraction and multiply by 100%.

$(75 \div 250) \times 100\% = 30\%$

Percentage change

$$\text{Percentage change} = \frac{\text{change}}{\text{original amount}} \times 100\%$$

You can use this method to calculate **percentage profit (percentage increase)** or **percentage loss (percentage decrease)** in a financial transaction.

EXAMPLE 13

Jabeer buys a car for £1500 and sells it for £1800. What is Jabeer's percentage profit?

Jabeer's profit is £300.

$$\text{Percentage profit} = \frac{\text{profit}}{\text{original amount}} \times 100\%$$

$$= \frac{300}{1500} \times 100 = 20\%$$

Using a multiplier

Find the multiplier by dividing the change by the original quantity, then change the resulting decimal to a percentage.

EXAMPLE 14

Express 5 as a percentage of 40.

Set up the fraction or decimal: $5 \div 40 = 0.125$

Convert the decimal to a percentage: $0.125 = 12.5\%$

EXERCISE 3E

1 Express each of the following as a percentage. Give suitably rounded figures where necessary.

   a £5 of £20   **b** £4 of £6.60   **c** 241 kg of 520 kg

   d 3 hours of 1 day   **e** 25 minutes of 1 hour   **f** 12 m of 20 m

   g 125 g of 600 g   **h** 12 minutes of 2 hours   **i** 1 week of a year

   j 1 month of 1 year   **k** 25 cm of 55 cm   **l** 105 g of 1 kg

2 Liam went to school with his pocket money of £2.50. He spent 80p at the tuck shop. What percentage of his pocket money had he spent?

3 In Greece, there are 3 654 000 acres of agricultural land. Olives are grown on 237 000 acres of this land. What percentage of the agricultural land is used for olives?

4 During the wet year of 1981, it rained in Manchester on 123 days of the year. What percentage of days were wet?

5 Find the percentage profit on the following. Give your answers to one decimal place.

Item	Retail price (selling price)	Wholesale price (price the shop paid)
a CD player	£89.50	£60
b TV set	£345.50	£210
c Computer	£829.50	£750

6 Before Anton started to diet, he weighed 95 kg. He now weighs 78 kg. What percentage of his original weight has he lost?

7 In 2009 the Melchester County Council raised £14 870 000 in council tax. In 2010 it raised £15 597 000 in council tax. What was the percentage increase?

8 When Blackburn Rovers won the championship in 1995, they lost only four of their 42 league games. What percentage of games did they *not* lose?

AU 9 In the year 1900 Britain's imports were as follows.

British Commonwealth	£109 530 000
USA	£138 790 000
France	£53 620 000
Other countries	£221 140 000

a What percentage of the total imports came from each source? Give your answers to 1 decimal place.

b Add up your answers to part **a**. What do you notice? Explain your answer.

AU 10 Calum and Stacey take the same tests. Both tests are out of the same mark.

Here are their results.

	Test A	Test B
Calum	12	17
Stacey	14	20

Whose result has the greater percentage increase from test A to test B?

Show your working to explain your answer.

FM 11 A shopkeeper wants to make 40% profit on his total sales of three items of clothing.

Item	Wholesale price	Retail Price
Shirt	£15	£19.99
Trousers	£18	£24.99
Pair of socks	£1.50	£2

If he sells 5 shirts, 1 pair of trousers and 10 pairs of socks, will he achieve his target?

EXERCISE 3F

This exercise includes a mixture of percentage questions, which you should answer without using a calculator.

1 Copy and complete this table.

	Fraction	Decimal	Percentage
a	$\frac{3}{5}$		
b		0.7	
c			55%

2 Work out these amounts.

    a 15% of £68   **b** 12% of 400 kg   **c** 30% of £4.20

3 What percentage is:

    a 28 out of 50   **b** 17 out of 25   **c** 75 out of 200?

4 What is the result if:

    a 240 is increased by 15%   **b** 3600 is decreased by 11%?

5 **a** A paperboy's weekly wage went up from £10 to £12. What was the percentage increase in his wages?

    b The number of houses he has to deliver to increases from 60 to 78. What is the percentage increase in the number of houses he delivers to?

    c The newsagent then increased his new wage of £12 by 10%. What are the boy's wages now?

FM 6 The on-the-road price of a new car was £8000.

    a In the first year it depreciated in value by 20%. What was the value of the car at the end of the first year?

    b Trevor wants to buy this car when it is two years old. He has £5000 in savings. If it depreciates by a further 15% in the second year, will he be able to afford it?

AU 7
PS The members of a slimming club had a mean weight of 80 kg before they started dieting. After a month they calculated that they had lost an average of 12% in weight.

    a What was the average weight after the month?

    b One of the members realised she had misread the scale and she was 10 kg heavier than she thought. Which of these statements is true?

        i The mean weight loss will have decreased by more than 12%.

        ii The mean weight loss will have stayed at 12%.

        iii The mean weight loss will have decreased by less than 12%.

        iv There is not enough information to answer the question.

This section will show you how to:
- simplify a ratio
- express a ratio as a fraction
- divide amounts according to ratios
- complete calculations from a given ratio and partial information

Key words
cancel
common unit
ratio
simplest form

A **ratio** is a way of comparing the sizes of two or more quantities.

A ratio can be expressed in a number of ways. For example, if Joy is five years old and James is 20 years old, the ratio of their ages is:

Joy's age : James's age

which is: $\qquad$ 5 : 20

which simplifies to: $\qquad$ 1 : 4 (dividing both sides by 5)

A ratio is usually given in one of these three ways.

Joy's age : James's age	or	5 : 20	or	1 : 4
Joy's age to James's age	or	5 to 20	or	1 to 4
$\dfrac{\text{Joy's age}}{\text{James's age}}$	or	$\dfrac{5}{20}$	or	$\dfrac{1}{4}$

Common units

When working with a ratio involving different units, *always convert them to a **common unit***. A ratio can be simplified only when the units of each quantity are the *same*, because the ratio itself has no units. Once the units are the same, the ratio can be simplified or **cancelled**.

For example, the ratio of 125 g to 2 kg must be converted to the ratio of 125 g to 2000 g, so that you can simplify it.

125 : 2000

Divide both sides by 25: $\qquad$ 5 : 80

Divide both sides by 5: $\qquad$ 1 : 16 $\qquad$ The ratio 125 : 2000 can be simplified to 1 : 16.

EXAMPLE 15

Express 25 minutes : 1 hour as a ratio in its simplest form.

The units must be the same, so change 1 hour into 60 minutes.

25 minutes : 1 hour = 25 minutes : 60 minutes

$\qquad$ = $\qquad$ 25 : 60 $\qquad$ Cancel the units (minutes)

$\qquad$ = $\qquad$ 5 : 12 $\qquad$ Divide both sides by 5

So, 25 minutes : 1 hour simplifies to 5 : 12

Ratios as fractions

A ratio in its **simplest form** can be expressed as portions of a quantity by expressing the whole numbers in the ratio as fractions with the same denominator (bottom number).

EXAMPLE 16

A garden is divided into lawn and shrubs in the ratio $3 : 2$.

What fraction of the garden is covered by **a** lawn, **b** shrubs?

The denominator (bottom number) of the fraction comes from adding the number in the ratio (that is, $2 + 3 = 5$).

a the lawn covers $\frac{3}{5}$ of the garden

b and the shrubs cover $\frac{2}{5}$ of the garden.

EXERCISE 3G

1 Express each of the following ratios in its simplest form.

a $6 : 18$	**b** $15 : 20$	**c** $16 : 24$	**d** $24 : 36$
e 20 to 50	**f** 12 to 30	**g** 25 to 40	**h** 125 to 30
i $15 : 10$	**j** $32 : 12$	**k** 28 to 12	**l** 100 to 40
m 0.5 to 3	**n** 1.5 to 4	**o** 2.5 to 1.5	**p** 3.2 to 4

2 Write each of the following ratios of quantities in its simplest form. (Remember to always express both parts in a common unit before you simplify.)

a £5 to £15	**b** £24 to £16	**c** 125 g to 300 g
d 40 minutes : 5 minutes	**e** 34 kg to 30 kg	**f** £2.50 to 70p
g 3 kg to 750 g	**h** 50 minutes to 1 hour	**i** 1 hour to 1 day
j 12 cm to 2.5 mm	**k** 1.25 kg : 500 g	**l** 75p : £3.50
m 4 weeks : 14 days	**n** 600 m: 2 km	**o** 465 mm : 3 m
p 15 hours : 1 day		

3 A length of wood is cut into two pieces in the ratio $3 : 7$. What fraction of the original length is the longer piece?

4 Jack and Thomas find a bag of marbles that they share between them in the ratio of their ages. Jack is 10 years old and Thomas is 15 years old. What fraction of the marbles did Jack get?

5 Dave and Sue share a pizza in the ratio $2 : 3$. They eat it all.

a What fraction of the pizza did Dave eat?

b What fraction of the pizza did Sue eat?

6 A camp site allocates space to caravans and tents in the ratio 7 : 3. What fraction of the total space is given to:

　a the caravans

　b the tents?

7 Two sisters, Amy and Katie, share a packet of sweets in the ratio of their ages. Amy is 15 and Katie is 10. What fraction of the sweets does each sister get?

8 The recipe for a fruit punch is 1.25 litres of fruit crush to 6.75 litres of lemonade. What fraction of the punch is each ingredient?

PS 9 Three cows, Gertrude, Gladys and Henrietta, produced milk in the ratio 2 : 3 : 4. Henrietta produced $1\frac{1}{2}$ litres more than Gladys. How much milk did the three cows produce altogether?

10 In a safari park at feeding time, the elephants, the lions and the chimpanzees are given food in the ratio 10 to 7 to 3. What fraction of the total food is given to:

　a the elephants　　　　**b** the lions　　　　**c** the chimpanzees?

11 Three brothers, James, John and Joseph, share a huge block of chocolate in the ratio of their ages. James is 20, John is 12 and Joseph is 8. What fraction of the bar of chocolate does each brother get?

12 The recipe for a pudding is 125 g of sugar, 150 g of flour, 100 g of margarine and 175 g of fruit. What fraction of the pudding is each ingredient?

PS 13 June wins three-quarters of her bowls matches. She loses the rest.

What is the ratio of wins to losses?

AU 14 Three brothers share some cash.
The ratio of Mark's and David's share is 1 : 2
The ratio of David's and Paul's share is 1 : 2

What is the ratio of Mark's share to Paul's share?

Dividing amounts in a given ratio

To divide an amount in a given ratio, you first look at the ratio to see how many parts there are altogether.

For example, 4 : 3 has 4 parts and 3 parts giving 7 parts altogether.

　7 parts is the whole amount.

　1 part can then be found by dividing the whole amount by 7.

　3 parts and 4 parts can then be worked out from 1 part.

EXAMPLE 17

Divide £28 in the ratio 4 : 3

4 + 3 = 7 parts altogether

So 7 parts = £28

Dividing by 7:

1 part = £4

4 parts = 4 × £4 = £16 and 3 parts = 3 × £4 = £12

So £28 divided in the ratio 4 : 3 = £16 : £12

To divide an amount in a given ratio you can also use fractions. You first express the whole numbers in the ratio as fractions with the same common denominator. Then you multiply the amount by each fraction.

EXAMPLE 18

Divide £40 between Peter and Hitan in the ratio 2 : 3

Changing the ratio to fractions gives:

$$\text{Peter's share} = \frac{2}{(2+3)} = \frac{2}{5}$$

$$\text{Hitan's share} = \frac{3}{(2+3)} = \frac{3}{5}$$

So Peter receives £40 × $\frac{2}{5}$ = £16 and Hitan receives £40 × $\frac{3}{5}$ = £24.

EXERCISE 3H

1 Divide the following amounts according to the given ratios.

a 400 g in the ratio 2 : 3　　　　　　**b** 280 kg in the ratio 2 : 5

c 500 in the ratio 3 : 7　　　　　　　**d** 1 km in the ratio 19 : 1

e 5 hours in the ratio 7 : 5　　　　　**f** £100 in the ratio 2 : 3 : 5

g £240 in the ratio 3 : 5 : 12　　　　**h** 600 g in the ratio 1 : 5 : 6

i £5 in the ratio 7 : 10 : 8　　　　　**j** 200 kg in the ratio 15 : 9 : 1

2 The ratio of female to male members of Lakeside Gardening Club is 7 : 3. The total number of members of the group is 250.

a How many members are female?

PS **b** What percentage of members are male?

3 A supermarket aims to stock branded goods and their own goods in the ratio 2 : 3. They stock 500 kg of breakfast cereal.

a What percentage of the cereal stock is branded?

b How much of the cereal stock is their own?

4 The Illinois Department of Health reported that, for the years 1981 to 1992 when they tested a total of 357 horses for rabies, the ratio of horses with rabies to those without was 1 : 16.

How many of these horses had rabies?

5 Being overweight increases the chances of an adult suffering from heart disease. A way to test whether an adult has an increased risk is shown below:

For women, there is increased risk when $W/H > 0.8$

For men, there is increased risk when $W/H > 1.0$

| W = waist measurement |
| H = hip measurement |

a Find whether the following people have an increased risk of heart disease.

Miss Mott: waist 26 inches, hips 35 inches

Mrs Wright: waist 32 inches, hips 37 inches

Mr Brennan: waist 32 inches, hips 34 inches

Ms Smith: waist 31 inches, hips 40 inches

Mr Kaye: waist 34 inches, hips 33 inches

b Give three examples of waist and hip measurements that would suggest no risk of heart disease for a man, but would suggest a risk for a woman.

6 Rewrite the following scales as ratios as simply as possible.

a 1 cm to 4 km   **b** 4 cm to 5 km   **c** 2 cm to 5 km

d 4 cm to 1 km   **e** 5 cm to 1 km   **f** 2.5 cm to 1 km

g 8 cm to 5 km   **h** 10 cm to 1 km   **i** 5 cm to 3 km

7 A map has a scale of 1 cm to 10 km.

a Rewrite the scale as a ratio in its simplest form.

b What is the actual length of a lake that is 4.7 cm long on the map?

c How long will a road be on the map if its actual length is 8 km?

HINTS AND TIPS

1 km = 1000 m
= 100 000 cm

8 A map has a scale of 2 cm to 5 km.

a Rewrite the scale as a ratio in its simplest form.

b How long is a path that measures 0.8 cm on the map?

c How long should a 12 km road be on the map?

9 The scale of a map is 5 cm to 1 km.

    a Rewrite the scale as a ratio in its simplest form.

    b How long is a wall that is shown as 2.7 cm on the map?

    c The distance between two points is 8 km; how far will this be on the map?

10 You can simplify a ratio by changing it into the form $1 : n$. For example, $5 : 7$ can be rewritten as

$$\frac{5}{5} : \frac{7}{5} = 1 : 1.4$$

Rewrite each of the following ratios in the form $1 : n$.

    a $5 : 8$ **b** $4 : 13$ **c** $8 : 9$

    d $25 : 36$ **e** $5 : 27$ **f** $12 : 18$

    g 5 hours : 1 day   **h** 4 hours : 1 week   **i** £4 : £5

AU **11** In a triangle ABC, the angles are in the ratio A : B : C = 5 : 1 : 3.

Work out the size of the largest angle.

> **HINTS AND TIPS**
>
> Angles in a triangle add up to 180°

12 The diagram shows a quadrilateral ABCD.

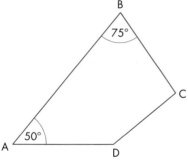

> **HINTS AND TIPS**
>
> Angles in a quadrilateral add up to 360°

The ratio of angle C to angle D is 2 : 3.

Work out the size of angle D.

Calculating with ratios when only part of the information is known

---

EXAMPLE 19

A fruit drink is made by mixing orange squash with water in the ratio 2 : 3
How much water needs to be added to 5 litres of orange squash to make the drink?

  2 parts is 5 litres

Dividing by 2:

  1 part is 2.5 litres

  3 parts = 2.5 litres × 3 = 7.5 litres

So 7.5 litres of water is needed to make the drink.

EXAMPLE 20

Two business partners, Lubna and Adama, divided their total profit in the ratio 3 : 5.
Lubna received £2100. How much did Adama get?

Lubna's £2100 was $\frac{3}{8}$ of the total profit. (Check that you know why.)
$\frac{1}{8}$ of the total profit = £2100 ÷ 3 = £700

So Adama's share, which was $\frac{5}{8}$, amounted to £700 × 5 = £3500.

EXERCISE 3I

1 Derek, aged 15, and Ricki, aged 10, shared all the conkers they found in the woods in the same ratio as their ages. Derek had 48 conkers.

    a Simplify the ratio of their ages.

    b How many conkers did Ricki have?

    c How many conkers did they find altogether?

2 A blend of tea is made by mixing Lapsang with Assam in the ratio 3 : 5. I have a lot of Assam tea but only 600 g of Lapsang. How much Assam do I need to make the blend using all the Lapsang?

3 The ratio of male to female spectators at ice hockey games is 4 : 5. At the Steelers' last match, 4500 men watched the match. What was the total attendance at the game?

4 A teacher always arranged the content of each of his lessons to Year 10 as 'teaching' and 'practising learnt skills' in the ratio 2 : 3.

    a If a lesson lasted 35 minutes, how much teaching would he do?

    b If he decided to teach for 30 minutes, how long would the lesson be?

5 A 'good' children's book is supposed to have pictures and text in the ratio 17 : 8. In a book I have just looked at, the pictures occupy 23 pages.

    a Approximately how many pages of text should this book have to be deemed a 'good' children's book?

    b What percentage of a 'good' children's book will be text?

6 Three business partners, Kevin, John and Margaret, put money into a business in the ratio 3 : 4 : 5. They shared any profits in the same ratio. Last year, Margaret made £3400 out of the profits. How much did Kevin and John make last year?

AU 7 **a** Iqra is making a drink from lemonade, orange and ginger ale in the ratio 40 : 9 : 1. If Iqra has only 4.5 litres of orange, how much of the other two ingredients does she need to make the drink?

    b Another drink made from lemonade, orange and ginger ale uses the ratio 10 : 2 : 1.

    Which drink has a larger proportion of ginger ale, Iqra's or this one? Show how you work out your answer.

PS 8 There is a group of boys and girls waiting for school buses. 25 girls get on the first bus. The ratio of boys to girls at the stop is now 3 : 2. 15 boys get on the second bus. There are now the same number of boys and girls at the bus stop. How many students altogether were originally at the bus stop?

PS 9 A jar contains 100 cc of a mixture of oil and water in the ratio 1 : 4. Enough oil is added to make the ratio of oil to water 1 : 2. How much water must be added to make the ratio of oil to water 1 : 3?

GRADE BOOSTER

G You can find equivalent fractions, decimals and percentages

E You can find simple percentages of a quantity

E You can find any percentages of a quantity

E You can convert fractions to decimals and decimals to fractions

E You can simplify ratios

D You can find a new quantity after an increase or decrease by a percentage and find one quantity as a percentage of another

C You can find a percentage change, for example percentage increase or percentage decrease

C You can solve problems using ratios

What you should know now

- How to find equivalent percentages, decimals and fractions
- How to calculate percentages, percentage increases and decreases
- How to calculate one quantity as a percentage of another
- How to divide any amount according to a given ratio
- How to complete calculations from a given ratio and partial information

1 Complete the table below.

Fraction	Decimal	Percentage
$\frac{1}{2}$	0.5	
	0.7	70%
$\frac{3}{100}$		3%

(Total 3 marks)

AQA, June 2005, Paper 1 Foundation, Question 2

2 **a** **i** Write $\frac{7}{16}$ as a decimal.

   ii Write 27% as a decimal.

   b Write these values in order of size, smallest first.

   0.7 $\frac{6}{10}$ 65%   0.095

3 Mr and Mrs Jones are buying a tumble dryer that normally costs £250. They save 12% in a sale.

   a What is 12% of £250?

   b How much do they pay for the tumble dryer?

4 Which is the larger amount?

   40% of £30 $\frac{3}{5}$ of £25

5 The price of a white fridge is £250.

   A silver fridge costs 8% more than the white fridge.

   Calculate the **extra** cost of the silver fridge.   (2)

(Total 2 marks)

AQA, June 2005, Module 5, Paper 2 Foundation, Question 8

6 Mr Shaw's bill for new tyres is £120 plus VAT. VAT is charged at $17\frac{1}{2}$%.

   What is his total bill?

7 Supermarkets often make 'Buy one, get one free' offers. What percentage saving is this?

   10%, 50%, 100% or 200%

8 A school raises £660 from a sponsored walk.

   a $\frac{1}{4}$ of the money is spent on books.

   The rest of the money is spent on sports equipment.

   How much money is spent on sports equipment?   (2)

   b The £660 was raised by teachers and pupils in the ratio 1 : 9

   The pupils raised the greater amount of money.

   How much money did the pupils raise?   (2)

(Total 4 marks)

AQA, November 2006, Module 3 Foundation, Question 8

9 In 2006 the population of a town was 68 000.

   By 2007 the population had decreased by 3.2%.

   Work out the population of the town in 2007.   (3)

(Total 3 marks)

AQA, March 2008, Module 3 Foundation, Question 8

10 Tom measured up a wall for ceramic tiles. He worked out that he needed 200 tiles, which he had seen advertised at 60p each.

   a How much did Tom expect to pay for the 200 tiles?

   b When he went to the store he decided to buy 10% more tiles than he needed, in case of breakages. Then he found that the price of the tiles had been reduced by 10%.

   i How many tiles did he buy?

   ii What was the new price per tile?

   c Calculate the percentage saving of the actual cost over the expected cost.

11 A TV originally cost £300.

   In a sale, its price was reduced by 20%, then this sale price was reduced by a further 10%.

   Show why this is not a 30% reduction of the original price.

12 **a** Write 60% as a decimal.   (1)

   b Work out 38% of £146.   (2)

   c What percentage is £108 of £150?   (2)

(Total 5 marks)

AQA, March 2006, Module 3 Foundation, Question 7

Worked Examination Questions

FM 1 The land area of a farm is 385 acres.

Two-fifths of the land is used to grow barley.

96 acres is pasture.

22% is used to keep livestock.

The rest is unused.

How many acres are unused?

$\frac{2}{5} \times 385 = 154$

> You will get 1 mark for setting up the calculation and 1 mark for the answer. You could also do this as 0.4×385.

$0.22 \times 385 = 84.7$ acres

> You will get 1 mark for setting up the calculation and 1 mark for the answer.

$385 - 96 - 154 - 84.7 = 50.3$ or 50 acres

> Work out the area by subtracting all the areas from 385. You will get 1 mark for this.

Total: 5 marks

Worked Examination Questions

PS AU **2** David is mixing compost with soil.
He mixes 2 kg of compost with 1 kg of soil.

Kayren is also mixing compost with soil.
She mixes 3 kg of compost with 2 kg of soil.

Which mixture has the greater percentage of compost?
You **must** show your working.

David uses 2 kg of compost out of
3 kg altogether.

Percentage $= \frac{2}{3} \times 100\%$

You will get 1 mark for method, used at least once.

$= 66.6\%$ or 67%

You will get 1 mark for accuracy of 66.6% (67%) or 60%.

Kayren uses 3 kg of compost out
of 5 kg altogether.

Percentage $= \frac{3}{5} \times 100\%$

$= 60\%$

You will get 1 mark for accuracy of the other percentage and stating David's mixture. Stating David's mixture without showing working scores no marks.

So David's mixture has the greater percentage
of compost.

Total: 3 marks

We use water every day: it is vital to our survival and important in making our lives more comfortable.

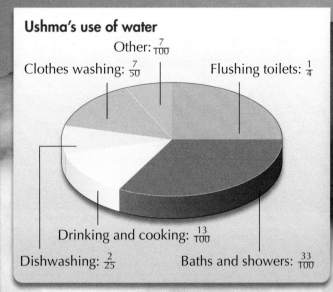

Ushma's use of water

Other: $\frac{7}{100}$

Clothes washing: $\frac{7}{50}$

Flushing toilets: $\frac{1}{4}$

Drinking and cooking: $\frac{13}{100}$

Dishwashing: $\frac{2}{25}$

Baths and showers: $\frac{33}{100}$

Your task

Ushma has noticed that her water bill has increased over the last few months. The rates charged by the water company have not changed, so Ushma knows that she is paying more money because she is using more water. She is not only concerned that she is spending too much money on water: she is also worried about the impact that her increased water use is having on the environment.

1 Using the information opposite, write a report analysing Ushma's use of water.

   In your report you should think about:

   ● the percentage of water that Ushma uses for different domestic activities

   ● the cost of each domestic activity that involves water

   ● where she could realistically reduce her water usage and by how much

   ● the impact that reducing her usage would have on her water bill

   ● what her total percentage decrease in water usage would be

   ● how best to represent your findings.

2 Seeing her water bill inspires Ushma to research how water is used on a global scale. The pie chart and table show the information she discovered.

   What can Ushma tell from this information about water supplies throughout the world? Can she work out how reducing her water usage will impact on global water supplies?

   Express your ideas as they might be seen in a newspaper.

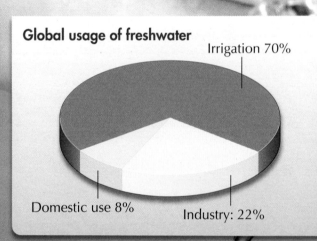

Global usage of freshwater

Irrigation 70%

Domestic use 8%

Industry: 22%

Global distribution of water	
Location	**Volume (km^3)**
Oceans	1 338 000 000
Ice	24 364 000
Groundwater	23 400 000
Lakes and reservoirs	176 400
Soil moisture	16 500
Water vapour	12 900
Rivers	2120
Swamp water	11 470

Handy hint

Use the internet to look up facts to do with water usage. How do water companies and global organisations represent their 'water facts'?

Handy hint

Water bills include a 'standing charge', which is a fixed amount paid by all customers regardless of how much water they have used. They also include a variable charge, which will change according to how much water the household has used.

Water utility bill

 South UK Water

Account number 12349876001	Date: 15-03-2010
Bill period January – March 2010	Meter No. 367X93007

Bill summary	
Water supply standing charge	£35.80
Water supply	£301.10
Surface drainage	£39.78
Sewerage standing charge	£1.68
Sewerage	£321.45
Total	**£699.81**
Payment due by: 14 April 2010	

contact us at: www.south-ukwater.co.uk

Water utility bill

 South UK Water

Account number 12349876001	Date: 15-12-2009
Bill period October – December 2009	Meter No. 367X93007

Bill summary	
Water supply standing charge	£35.80
Water supply	£258.00
Surface drainage	£39.78
Sewerage standing charge	£1.68
Sewerage	£294.48
Total	**£629.74**
Payment due by: 14 Jan 2010	

contact us at: www.south-ukwater.co.uk

Why this chapter matters

If you were asked to circle one of these to describe mathematics, which would it be?

Art   Science   Sport   Language

In fact, you could circle them all.

Art

Mathematics is important in art. The *Mona Lisa*, probably the most famous painting in the world, uses the proportions of the 'golden ratio' (approximately 1.618) as shown by the red rectangles marked on this copy of the painting. This 'golden ratio' is supposed to be particularly attractive to the human eye.

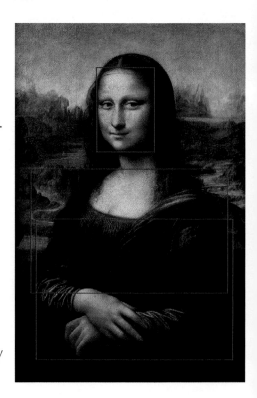

Science

Obviously, you cannot do much science without using mathematics. In 1962, a *Mariner* space probe went off course and had to be destroyed because someone had used a wrong symbol in a mathematical formula that was part of its programming.

Sport

Is mathematics a sport? There are national and international competitions each year that use mathematics. For example, there is a world Suduko championship each year and university students compete in the annual 'Mathematics Olympiad'.

Language

But perhaps the most important description in the list above is mathematics as language. As we saw in Chapter 1, mathematics is the only universal language. If you write the equation $3x = 9$, it will be understood by people in all countries.

Algebra is the way that the language of mathematics is expressed.

Algebra comes from the Arabic *al-jabr* which means something similar to 'completion'. It was used in a book written in 820AD by a Persian mathematician called al-Khwarizmi.

The use of symbols then developed until the middle of the 17th century, when René Descartes developed what is regarded as the basis of the algebra we use today.

Algebra: Basic algebra

1 The language of algebra

2 Simplifying expressions

3 Expanding brackets

4 Factorisation

5 Substitution

This chapter will show you …

- **F** how to substitute numbers into expressions and formulae
- **F** how to use letters to represent numbers
- **F** how to form simple algebraic expressions
- **E** how to simplify expressions by collecting like terms
- **D** how to factorise expressions
- **D** how to express simple rules in algebraic form

Visual overview

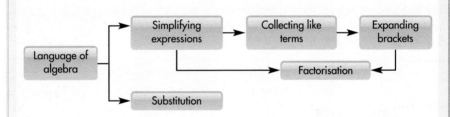

What you should already know

- The **BIDMAS/BODMAS** rule, which gives the order in which you must do the operations of arithmetic when they occur together (KS3 level 5, GCSE grade F)

Quick check

1 Write the answer to each expression.

   a $(5 - 1) \times 2$

   b $5 - (1 \times 2)$

2 Work out $(7 - 5) \times (5 + 4 - 2)$.

3 **a** Put brackets in the calculation to make the answer 40.

     $2 + 3 + 5 \times 4$

   b Put brackets in the calculation to make the answer 34.

     $2 + 3 + 5 \times 4$

The language of algebra

This section will show you how to:

- use letters, numbers and mathematical symbols to write algebraic expressions and formulae

Key words

equation
expression
formula
solve
symbol
term
variable

Algebra is based on the idea that if something works with numbers, it will work with letters. The main difference is that when you work only with numbers, the answer is also a number. When you work with letters, you get an **expression** as the answer.

Algebra follows the same rules as arithmetic, and uses the same **symbols** (+, −, × and ÷). Below are seven important algebraic rules.

- Write '4 more than x' as $4 + x$ or $x + 4$.

- Write '6 less than p' or 'p minus 6' as $p - 6$.

- Write '4 times y' as $4 \times y$ or $y \times 4$ or $4y$. The last one of these is the neatest way to write it.

- Write 'b divided by 2' as $b \div 2$ or $\dfrac{b}{2}$.

- When a number and a letter or a letter and a letter appear together, there is a hidden multiplication sign between them. So, $7x$ means $7 \times x$ and ab means $a \times b$.

- Always write '$1 \times x$' as x.

- Write 't times t' as $t \times t$ or t^2.

Here are some algebraic words that you need to know.

Variable: This is what the letters used to represent numbers are called. These letters can take on any value, so they are said to 'vary'.

Expression: This is any combination of letters and numbers. For example, $2x + 4y$ and $\dfrac{p - 6}{5}$ are expressions.

Equation: An equation contains an equals sign and at least one variable. The important fact is that a value can be found for the variable. This is called **solving** the equation. You will learn more about equations in chapter 12.

Formula: These are like equations in that they contain an equals sign, but there is more than one variable and they are rules for working out amounts such as area or the cost of taxi fares.

For example, $V = x^3$, $A = \frac{1}{2}bh$ and $C = 3 + 4m$ are all formulae.

Term: These are the separate parts of expressions, equations or formula.

For example, in $3x + 2y - 7$, there are three terms: $3x$, $+2y$ and -7.

Identity: An identity is similar to an equation, but is true for all values of the variable(s). Instead of the usual equals (=) sign, ≡ is used. For example, $2x \equiv 7x - 5x$.

FM Functional Maths   **AU** (AO2) Assessing Understanding   **PS** (AO3) Problem Solving

EXAMPLE 1

What is the area of each of these rectangles?

a 4 cm by 6 cm   **b** 4 cm by w cm   **c** l cm by w cm

The rule for working out the area of a rectangle is:

area = length × width

So, the area of rectangle **a** is $4 \times 6 = 24$ cm^2

The area of rectangle **b** is $4 \times w = 4w$ cm^2

The area of rectangle **c** is $l \times w = lw$ cm^2

Now, if A represents the area of rectangle **c**:

$A = lw$

This is an example of a rule expressed algebraically.

EXAMPLE 2

What is the perimeter of each of these rectangles?

a 6 cm by 4 cm   **b** 4 cm by w cm   **c** l cm by w cm

The rule for working out the perimeter of a rectangle is:

perimeter = twice the longer side + twice the shorter side

So, the perimeter of rectangle **a** is $2 \times 6 + 2 \times 4 = 20$ cm

The perimeter of rectangle **b** is $2 \times 4 + 2 \times w = 8 + 2w$ cm

The perimeter of rectangle **c** is $2 \times l + 2 \times w = 2l + 2w$ cm

Now, let P represent the perimeter of rectangle **c**, so:

$P = 2l + 2w$

which is another example of a rule expressed algebraically.

Expressions such as $A = lw$ and $P = 2l + 2w$ are **formulae**.

As the two examples above show, a formula states the connection between two or more quantities, each of which is represented by a different letter.

In a formula, the letters are replaced by numbers when a calculation has to be made. This is called *substitution* and is explained on page 326.

EXAMPLE 3

Say if the following are expressions (E), equations (Q) or formula (F).

a $x - 5 = 7$ **b** $P = 4x$ **c** $2x - 3y$

a is an **equation** (Q) as it can be solved to give $x = 12$.

b is a **formula** (F). This is the formula for the perimeter of a square with a side of x.

c is an **expression** (E) with two terms.

EXERCISE 4A

1 Write down the algebraic expression for:

   a 2 more than x **b** 6 less than x

   c k more than x **d** x minus t

   e x added to 3   **f** d added to m

   g y taken away from b **h** p added to t added to w

   i 8 multiplied by x **j** h multiplied by j

   k x divided by 4   **l** 2 divided by x

   m y divided by t **n** w multiplied by t

   o a multiplied by a **p** g multiplied by itself.

2 Here are four squares.

   i   **ii**   **iii**   **iv**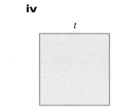

   a Work out the area and perimeter of each square.

   b Copy and complete these rules.

      i The perimeter, P, of a square of side s centimetres is $P = \ldots\ldots$

      ii The area, A, of a square of side s centimetres is $A = \ldots\ldots$

3 Asha, Bernice and Charu are three sisters. Bernice is x years old. Asha is three years older than Bernice. Charu is four years younger than Bernice.

   a How old is Asha?

   b How old is Charu?

4 An approximation method of converting from degrees Celsius to degrees Fahrenheit is given by this rule:

   Multiply by 2 and add 30.

   Using C to stand for degrees Celsius and F to stand for degrees Fahrenheit, complete this formula.

   $F = \ldots\ldots$

5 Cows have four legs. Which of these formulae connects the number of legs (L) and the number of cows (C)?

a $C = 4L$

b $L = C + 4$

c $L = 4C$

d $L + C = 4$

6 There are 3 feet in a yard. The rule $F = 3Y$ connects the number of feet (F) and the number of yards (Y). Write down rules, using the letters shown, to connect:

a   the number of centimetres (C) in metres (M)

b   the number of inches (N) in feet (F)

c   the number of wheels (W) on cars (C)

d   the number of heads (H) on people (P).

> **HINTS AND TIPS**
>
> Check your formula with a numerical example.
> In 4 yards there are 12 feet, so, if $F = 3Y$ is correct, then $12 = 3 \times 4$, which is true.

7 a   Anne has three bags of marbles. Each bag contains n marbles. How many marbles does she have altogether?

b   Bea gives her another three marbles. How many marbles does Anne have now?

c   Anne puts one of her new marbles in each bag. How many marbles are there now in each bag?

d   Anne takes two marbles out of each bag. How many marbles are there now in each bag?

8 Simon has n cubes.

- Rob has twice as many cubes as Simon.

- Tom has two more than Simon.

- Vic has three fewer than Simon.

- Will has three more than Rob.

How many cubes does each person have?

> **HINTS AND TIPS**
>
> Remember that you do not have to write down a multiplication sign between numbers and letters, or letters and letters.

9 **a** John has been drawing squares and writing down the area and the perimeter of each of them. He has drawn three squares. Finish his work by writing down the missing areas and perimeters.

b Write down the area and the perimeter of this partly-covered square.

10 **a** I go shopping with £10 and spend £6. How much do I have left?

b I go shopping with £10 and spend £x. How much do I have left?

c I go shopping with £y and spend £x. How much do I have left?

d I go shopping with £$3x$ and spend £x. How much do I have left?

11 Give the total cost of:

a five pens at 15p each

b x pens at 15p each

c four pens at Ap each

d y pens at Ap each.

12 A boy went shopping with £A. He spent £B. How much did he have left?

13 Five ties cost £A. What is the cost of one tie?

PS **14** **a** My dad is 72 and I am T years old. How old shall we each be in x years' time?

b My mum is 64 years old. In two years' time she will be twice as old as I am. What age am I now?

15 I am twice as old as my son. I am T years old.

a How old is my son?

b How old will my son be in four years' time?

c How old was I x years ago?

16 What is the perimeter of each of these figures?

a

Square

b

Equilateral triangle

c

Regular hexagon

E

17 Write down the number of marbles each student ends up with.

Student	Action	Marbles
Andrea	Start with three bags each containing n marbles and give away one marble from each bag	
Bert	Start with three bags each containing n marbles and give away one marble from one bag	
Colin	Start with three bags each containing n marbles and give away two marbles from each bag	
Davina	Start with three bags each containing n marbles and give away n marbles from each bag	
Emma	Start with three bags each containing n marbles and give away n marbles from one bag	
Florinda	Start with three bags each containing n marbles and give away m marbles from each bag	

AU 18 The answer to $3 \times 4m$ is $12m$.

Write down two **different** expressions for which the answer is $12m$.

AU 19 Three expressions for the perimeter of a rectangle with length l and width w are:

$$P = l + w + l + w$$

$$P = 2l + 2w$$

$$P = 2(l + w)$$

> **HINTS AND TIPS**
>
> Just pick some easy numbers for l and w.

Show, using a numerical example, that they all give the same result.

PS 20 My sister is three years older than I am.

The sum of our ages is 29.

How old am I?

> **HINTS AND TIPS**
>
> If I am x years old, work out how old my sister is in terms of x and use this to set up a simple equation.

PS 21 Ali has 65p and Heidi has 95p. How much should Heidi give to Ali so they both have the same amount?

22 Say if the following are expressions (E), equations (Q) or formula (F).

a $2x - 5$

b $s = \sqrt{A}$

c $2x - 3 = 1$

Simplifying expressions

This section will show you how to:

- simplify algebraic expressions by multiplying terms
- simplify algebraic expressions by collecting like terms

Key words

coefficient

like terms

simplify

Simplifying an algebraic expression means making it neater and, usually, shorter by combining its terms where possible.

Multiplying expressions

When you multiply algebraic expressions, first you multiply the numbers, then the letters.

EXAMPLE 4

Simplify:

a $2 \times t$ **b** $m \times t$ **c** $2t \times 5$ **d** $3y \times 2m$

The convention is to write the number first then the letters, but if there is no number just put the letters in alphabetical order. The number in front of the letter is called the **coefficient**.

a $2 \times t = 2t$ **b** $m \times t = mt$ **c** $2t \times 5 = 10t$ **d** $3y \times 2m = 6my$

In an examination you will not be penalised for writing $2ba$ instead of $2ab$, but you will be penalised if you write $ab2$ as this can be confused with powers, so *always* write the number first.

EXAMPLE 5

Simplify:

a $t \times t$ **b** $3t \times 2t$ **c** $3t^2 \times 4t$ **d** $2t^3 \times 4t^2$

Multiply the same variables, using powers. The indices are added together.

a $t \times t = t^2$ (Remember: $t = t^1$)   **b** $3t \times 2t = 6t^2$

c $3t^2 \times 4t = 12t^3$ **d** $2t^3 \times 4t^2 = 8t^5$

EXERCISE 4B

1 Simplify the following expressions.

HINTS AND TIPS

Remember to multiply numbers and add indices.

a $2 \times 3t$

b $5y \times 3$

c $2w \times 4$

d $5b \times b$

e $2w \times w$

f $4p \times 2p$

g $3t \times 2t$

h $5t \times 3t$

i $m \times 2t$

j $5t \times q$

k $n \times 6m$

l $3t \times 2q$

m $5h \times 2k$

n $3p \times 7r$

AU 2 a Which of the following expressions are equivalent?

$2m \times 6n$ $4m \times 3n$ $2m \times 6m$ $3m \times 4n$

b The expressions $2x$ and x^2 are the same for only two values of x. What are these values?

PS 3 A square and a rectangle have the same area.

The rectangle has sides $2x$ cm and $8x$ cm.

What is the length of a side of the square?

4 Simplify the following expressions.

a $y^2 \times y$

b $3m \times m^2$

c $4t^2 \times t$

d $3n \times 2n^2$

e $t^2 \times t^2$

f $h^3 \times h^2$

g $3n^2 \times 4n^3$

h $3a^4 \times 2a^3$

i $k^5 \times 4k^2$

j $-t^2 \times -t$

k $-4d^2 \times -3d$

l $-3p^4 \times -5p^2$

m $3mp \times p$

n $3mn \times 2m$

o $4mp \times 2mp$

FM 5 There are 2000 students at Highville school. One student starts a rumour by telling it to two other students. The next day, those two students each tell the rumour to two other students who have not heard it already. The next day, those four students each tell the rumour to two other students who have not heard it before, and so on. How many days will it be before the whole school has heard the rumour?

HINTS AND TIPS

Fill in a table like this.

Day	1	2	3	4
Number told	2	4	8	16
Total who know	3	7	15	31

In an examination there will always be space to the right of the table to draw in two columns – one for the mid-point and the other for the mid-point frequency.

Collecting like terms

Like terms are those that are multiples of the same variable or of the same combination of variables. For example, a, $3a$, $9a$, $\frac{1}{4}a$ and $-5a$ are all like terms.

So are $2xy$, $7xy$ and $-5xy$, and so are $6x^2$, x^2 and $-3x^2$.

Collecting like terms generally involves two steps.

- Collect like terms into groups.

- Then combine the like terms in each group.

Only like terms can be added or subtracted to simplify an expression. For example,

$a + 3a + 9a - 5a$	simplifies to	$8a$
$2xy + 7xy - 5xy$	simplifies to	$4xy$

Note that the variable does not change. All you have to do is combine the coefficients.

For example,

$$6x^2 + x^2 - 3x^2 = (6 + 1 - 3)x^2 = 4x^2$$

But an expression such as $4p + 8t + 5x - 9$ cannot be simplified, because $4p$, $8t$, $5x$ and 9 are *not like terms*, which *cannot* be combined.

EXAMPLE 6

Simplify the expression:

$$7x^2 + 3y - 6z + 2x^2 + 3z - y + w + 9$$

Write out the expression:
$$7x^2 + 3y - 6z + 2x^2 + 3z - y + w + 9$$

Then collect like terms:
$$\boxed{7x^2 + 2x^2} \; \boxed{+3y - y} \; \boxed{-6z + 3z} \; \boxed{+ w} \; \boxed{+ 9}$$

Then combine them:
$$9x^2 \quad + \quad 2y \quad - \quad 3z \quad + w + 9$$

So, the expression in its simplest form is:
$$9x^2 + 2y - 3z + w + 9$$

EXERCISE 4C

1 Joseph is given £t, John has £3 more than Joseph, and Joy has £$2t$.

    a How much more money has Joy than Joseph?

    b How much do the three of them have altogether?

2 Write down an expression for the perimeter of each of these shapes.

a
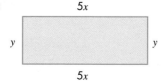

b

c
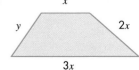

3 Write each of these expressions in a shorter form.

a $a + a + a + a + a$

b $c + c + c + c + c + c$

c $4e + 5e$

d $f + 2f + 3f$

e $5j + j - 2j$

f $9q - 3q - 3q$

g $3r - 3r$

h $2w + 4w - 7w$

i $5x^2 + 6x^2 - 7x^2 + 2x^2$

j $8y^2 + 5y^2 - 7y^2 - y^2$

k $2z^2 - 2z^2 + 3z^2 - 3z^2$

> **HINTS AND TIPS**
>
> The term a has a coefficient of 1, i.e. $a = 1a$, but you do not need to write the 1.

4 Simplify each of the following expressions.

a $3x + 4x$

b $5t - 2t$

c $-2x - 3x$

d $-k - 4k$

e $m^2 + 2m^2 - m^2$

f $2y^2 + 3y^2 - 5y^2$

> **HINTS AND TIPS**
>
> Remember that only **like** terms can be added or subtracted.
> If all the terms cancel out, just write 0 rather than $0x^2$, for example.

5 Simplify each of the following expressions.

a $5x + 8 + 2x - 3$

b $7 - 2x - 1 + 7x$

c $4p + 2t + p - 2t$

d $8 + x + 4x - 2$

e $3 + 2t + p - t + 2 + 4p$

f $5w - 2k - 2w - 3k + 5w$

g $a + b + c + d - a - b - d$

h $9k - y - 5y - k + 10$

6 Simplify these expressions. (Be careful – two of them will not simplify.)

a $c + d + d + d + c$

b $2d + 2e + 3d$

c $f + 3g + 4h$

d $5u - 4v + u + v$

e $4m - 5n + 3m - 2n$

f $3k + 2m + 5p$

g $2v - 5w + 5w$

h $2w + 4y - 7y$

i $5x^2 + 6x^2 - 7y + 2y$

j $8y^2 + 5z - 7z - 9y^2$

k $2z^2 - 2x^2 + 3x^2 - 3z^2$

7 Find the perimeter of each of these shapes, giving it in its simplest form.

a

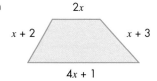

b

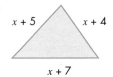

c

$x + 3$

$y + 1$

AU 8 $3x + 5y + 2x - y = 5x + 4y$

Write down two other **different** expressions which are equal to $5x + 4y$.

AU 9 Find the missing terms to make these equations true.

a $4x + 5y + \ldots\ldots\ldots - \ldots\ldots\ldots = 6x + 3y$

b $3a - 6b - \ldots\ldots\ldots + \ldots\ldots\ldots = 2a + b$

PS 10 ABCDEF is an L-shape.

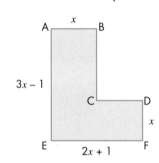

$AB = DF = x$

$AE = 3x - 1$ and $EF = 2x + 1$

> **HINTS AND TIPS**
>
> Make sure your explanation uses expressions. Do not try to explain in words alone.

a Explain why the length $BC = 2x - 1$.

b Find the perimeter of the shape in terms of x.

c If $x = 2.5$ cm, what is the perimeter of the shape?

FM 11 Sean wants to measure the size of a rectangular lawn but he does not have a tape measure. Instead he measures the length and width using his pace and his shoe length. He finds that the length is four paces plus a shoe length and the width is two paces and two shoe lengths. He writes this as $4p + s$ and $2p + 2s$.

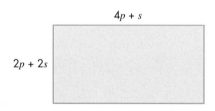

> **HINTS AND TIPS**
>
> Convert 1 m to 100 cm, then convert the answer back to metres.

a Work out the perimeter in terms of p and s.

b Later he finds that his pace is 1 m and his shoe length is 25 cm. What is the actual perimeter of the lawn? Give your answer in metres.

AU 12 A teacher asks her class to work out the perimeter of this L shape.

Tia says: 'There is information missing so you cannot work out the perimeter.'

Maria says: 'The perimeter is $4x - 1 + 4x - 1 + 3x + 2 + 3x + 2$.'

Who is correct?

Explain your answer.

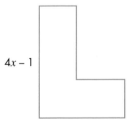

Expanding brackets

This section will show you how to:
- expand brackets such as $2(x - 3)$
- expand and simplify brackets

Key words

expand
multiply out
simplify

Expanding

In mathematics, the term '**expand**' usually means '**multiply out**'. For example, expressions such as $3(y + 2)$ and $4y^2(2y + 3)$ can be expanded by multiplying them out.

Remember that there is an invisible multiplication sign between the outside number and the opening bracket. So $3(y + 2)$ is really $3 \times (y + 2)$, and $4y^2(2y + 3)$ is really $4y^2 \times (2y + 3)$.

You expand by multiplying *everything inside* the brackets by what is outside the brackets.

EXAMPLE 7

Expand $3(y + 2)$.

$3(y + 2) = 3 \times (y + 2) = 3 \times y + 3 \times 2 = 3y + 6$

EXAMPLE 8

Expand $4y^2(2y + 3)$.

$4y^2(2y + 3) = 4y^2 \times (2y + 3) = 4y^2 \times 2y + 4y^2 \times 3 = 8y^3 + 12y^2$

Look at these next examples of expansion, which show how each term inside the brackets has been multiplied by the term outside the brackets.

$2(m + 3) = 2m + 6$ $y(y^2 - 4x) = y^3 - 4xy$

$3(2t + 5) = 6t + 15$ $3x^2(4x + 5) = 12x^3 + 15x^2$

$m(p + 7) = mp + 7m$ $3(2 + 3x) = 6 + 9x$

$x(x - 6) = x^2 - 6x$ $2x(3 - 4x) = 6x - 8x^2$

$4t(t + 2) = 4t^2 + 8t$ $3t(2 + 5t - p) = 6t + 15t^2 - 3pt$

D

C

1 Expand these expressions.

   a $2(3 + m)$ **b** $5(2 + l)$ **c** $3(4 - y)$ **d** $4(5 + 2k)$

   e $4(3d - 2n)$ **f** $t(t + 3)$ **g** $m(m + 5)$ **h** $k(k - 3)$

   i $g(3g + 2)$ **j** $y(5y - 1)$ **k** $p(5 - 3p)$ **l** $3m(m + 4)$

   m $3t(5 - 4t)$ **n** $3d(2d + 4e)$ **o** $2y(3y + 4k)$ **p** $5m(3m - 2p)$

2 Expand these expressions.

   a $y(y^2 + 5)$ **b** $h(h^3 + 7)$ **c** $k(k^2 - 5)$ **d** $3t(t^2 + 4)$

   e $3d(5d^2 - d^3)$ **f** $3w(2w^2 + t)$ **g** $5a(3a^2 - 2b)$ **h** $3p(4p^3 - 5m)$

   i $m^2(5 + 4m)$ **j** $t^3(t + 2t^2)$ **k** $g^2(5t - 4g^2)$ **l** $3t^2(5t + m)$

FM 3 The local supermarket is offering £1 off a large tin of biscuits. Morris wants five tins.

   a If the original price of one tin is £t, which of the expressions below represents how much it will cost Morris to buy five tins?

      $5(t - 1)$ $5t - 1$ $t - 5$ $5t - 5$

   b Morris has £20 to spend. If each tin is £4.50, will he have enough money for five tins? Show your working to justify your answer.

AU 4 Dylan wrote the following.

   $3(5x - 4) = 8x - 4$

   Dylan has made two mistakes.

   Explain the mistakes that Dylan has made.

> **HINTS AND TIPS**
>
> It is not enough just to give the right answer. You must try to explain, for example, why Dylan wrote $8x$ and what he should really have written if this is wrong.

PS 5 The expansion $2(x + 3) = 2x + 6$ can be shown by this diagram.

	x	3
2	$2x$	6

   a What expansion is shown by this diagram?

3	$6y$	9

   b Write down an expansion that can be shown by this diagram.

$12z$	8

Expand and simplify

This usually means that you need to expand more than one set of brackets and **simplify** the resulting expressions.

You will often be asked to expand and simplify expressions.

EXAMPLE 9

Expand and simplify $3(4 + m) + 2(5 + 2m)$.

$3(4 + m) + 2(5 + 2m) = 12 + 3m + 10 + 4m = 22 + 7m$

EXAMPLE 10

Expand and simplify $3t(5t + 4) - 2t(3t - 5)$.

$3t(5t + 4) - 2t(3t - 5) = 15t^2 + 12t - 6t^2 + 10t = 9t^2 + 22t$

Notice that multiplying $-2t$ by -5 gives an answer of $+10t$.

EXAMPLE 11

Expand and simplify $4a(2b - 3f) - 3b(a + 2f)$.

$4a(2b - 3f) - 3b(a + 2f) = 8ab - 12af - 3ab - 6bf = 5ab - 12af - 6bf$

EXERCISE 4E

1 Simplify these expressions.

  a $4t + 3t$ **b** $2y + y$ **c** $3d + 2d + 4d$

  d $5e - 2e$ **e** $4p - p$ **f** $3t - t$

  g $2t^2 + 3t^2$ **h** $3ab + 2ab$ **i** $7a^2d - 4a^2d$

2 Expand and simplify these expressions.

  a $3(4 + t) + 2(5 + t)$ **b** $5(3 + 2k) + 3(2 + 3k)$

  c $4(1 + 3m) + 2(3 + 2m)$ **d** $2(5 + 4y) + 3(2 + 3y)$

  e $4(3 + 2f) + 2(5 - 3f)$ **f** $5(1 + 3g) + 3(3 - 4g)$

  g $3(2 + 5t) + 4(1 - t)$ **h** $4(3 + 3w) + 2(5 - 4w)$

> **HINTS AND TIPS**
>
> Expand the expression before trying to collect like terms. If you try to expand and collect at the same time you will probably make a mistake.

3 Expand and simplify these expressions.

a $4(3 + 2h) - 2(5 + 3h)$ **b** $5(3g + 4) - 3(2g + 5)$

c $3(4y + 5) - 2(3y + 2)$ **d** $3(5t + 2) - 2(4t + 5)$

e $5(5k + 2) - 2(4k - 3)$ **f** $4(4e + 3) - 2(5e - 4)$

g $3(5m - 2) - 2(4m - 5)$ **h** $2(6t - 1) - 3(3t - 4)$

HINTS AND TIPS

Be careful with minus signs. They are causes of the most common errors students make in examinations. Remember $-2 \times -4 = 8$ but $-2 \times 5 = -10$. You will learn more about multiplying and dividing with negative numbers in Chapter 3.

4 Expand and simplify these expressions.

a $m(4 + p) + p(3 + m)$ **b** $k(3 + 2h) + h(4 + 3k)$

c $t(2 + 3n) + n(3 + 4t)$ **d** $p(2q + 3) + q(4p + 7)$

e $3h(2 + 3j) + 2j(2h + 3)$ **f** $2y(3t + 4) + 3t(2 + 5y)$

g $4r(3 + 4p) + 3p(8 - r)$ **h** $5k(3m + 4) - 2m(3 - 2k)$

FM 5 A two-carriage train has f first-class seats and $2s$ standard-class seats.

A three-carriage train has $2f$ first-class seats and $3s$ standard-class seats.

On a weekday, five two-carriage trains and two three-carriage trains travel from Hull to Liverpool.

a Write down an expression for the total number of first-class and standard-class seats available during the day.

b On average on any day, half of the first-class seats are used. Each first-class seat costs £60.

On average on any day, three-quarters of the standard-class seats are used. Each standard-class seat costs £40.

How much money does the rail company earn in an average day on this route? Give your answer in terms of f and s.

c $f = 15$ and $s = 80$. It costs the rail company £30 000 per day to operate this route. How much profit do they make on an average day?

AU 6 Fill in whole-number values so that the following expansion is true.

$$3(\ldots\ldots x + \ldots\ldots y) + 2(\ldots\ldots x + \ldots\ldots y) = 11x + 17y$$

HINTS AND TIPS

There is more than one answer. You don't have to give them all.

PS 7 A rectangle with sides 5 and $3x + 2$ has a smaller rectangle with sides 3 and $2x - 1$ cut from it.

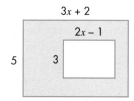

HINTS AND TIPS

Write out the expression for the difference of the two answers and then work it out.

Work out the remaining area.

Factorisation

This section will show you how to:
- 'reverse' the process of expanding brackets by taking out a common factor from each term in an expression

Key words
factor
factorisation

Factorisation is the opposite of expansion. It puts an expression into brackets.

To factorise an expression, look for the common **factors** in every term of the expression. Follow through the examples below to see how this works.

EXAMPLE 12

Factorise each expression.   **a** $6t + 9m$ **b** $6my + 4py$
c $8kp + 4k - 12km$ **d** $8kp + 4kt - 12km$

a The common factor is 3, so $6t + 9m = 3(2t + 3m)$

b The common factor is $2y$, so $6my + 4py = 2y(3m + 2p)$

c The common factor is $4k$, so $8kp + 4k - 12km = 4k(2p + 1 - 3m)$

d The common factor is $4k$, so $8kp + 4kt - 12km = 4k(2p + t - 3m)$

Notice that if you multiply out each answer you will get the expressions you started with.

This diagram may help you to see the difference and the connection between expansion and factorisation.

Note: When the whole term is the common factor, as in part **c**, then you are left with 1, not 0, inside the brackets.

Expanding

$3(2 + 3) = 6 + 9$

Factorising

EXERCISE 4F

1 Factorise the following expressions. The first three have been started for you.

a $6m + 12t = 6(\quad)$ **b** $9t + 3p = 3(\quad)$ **c** $8m + 12k = 4(\quad)$

d $4r + 8t$ **e** $mn + 3m$ **f** $5g^2 + 3g$

g $4w - 6t$ **h** $8p - 6k$ **i** $16h - 10k$

j $2mp + 2mk$ **k** $4bc + 2bk$ **l** $6ab + 4ac$

m $3y^2 + 2y$ **n** $4t^2 - 3t$ **o** $4d^2 - 2d$

p $3m^2 - 3mp$

HINTS AND TIPS

First look for a common factor of the numbers and then look for common factors of the letters.

D

2 Factorise the following expressions.

a $6p^2 + 9pt$

b $8pt + 6mp$

c $8ab - 4bc$

d $12a^2 - 8ab$

e $9mt - 6pt$

f $16at^2 + 12at$

g $5b^2c - 10bc$

h $8abc + 6bed$

i $4a^2 + 6a + 8$

j $6ab + 9bc + 3bd$

k $5t^2 + 4t + at$

l $6mt^2 - 3mt + 9m^2t$

m $8ab^2 + 2ab - 4a^2b$

n $10pt^2 + 15pt + 5p^2t$

3 Factorise the following expressions where possible. List those that cannot be factorised.

a $7m - 6t$

b $5m + 2mp$

c $t^2 - 7t$

d $8pt + 5ab$

e $4m^2 - 6mp$

f $a^2 + b$

g $4a^2 - 5ab$

h $3ab + 4cd$

i $5ab - 3b^2c$

FM 4 Three friends have a meal together. They each have a main meal costing £6.75 and a dessert costing £3.25.

Chris says that the bill in pounds, will be $3 \times 6.75 + 3 \times 3.25$.

Mary says that she has an easier way to work out the bill as $3 \times (6.75 + 3.25)$.

a Explain why Chris and Mary's methods both give the correct answer.

b Explain why Mary's method is better.

c What is the total bill?

AU 5 Three students are asked to factorise the expression $12m - 8$.

These are their answers.

Aidan	Bella	Craig
$2(6m - 4)$	$4(3m - 2)$	$4m\left(3 - \dfrac{2}{m}\right)$

All the answers are factorised correctly, but only one is the normally accepted answer.

a Which student gave the answer that is normally accepted as correct?

b Explain why the other two students' answers are not normally accepted as correct answers.

PS 6 Explain why $5m + 6p$ cannot be factorised.

Algebra dominoes

This is an activity for two people.

You need some card to make a set of algebra dominoes like those below.

| $4 \times n$ | t^2 | | $2b$ | $2 - t$ | | $\dfrac{12n}{2}$ | $0.5n$ | | $5w$ | $b + b$ |

| $3t - 2$ | $3 \times 2y$ | | $5 + y$ | $n + 2 + n + 3$ | | $\dfrac{4t + 2n}{2}$ | $t \times t$ | | $5b - 3b$ | $6n$ |

| $6y$ | $t - 2$ | | $2a + 2$ | $2t + 2 - 3t$ | | $y + 5$ | $7n - n$ | | $3n + 3n$ | $4t - 2 - t$ |

| b^2 | $\dfrac{1}{2}n$ | | $t + 3 - 2$ | $2n - 1$ | | $t + 5$ | $b \times 2$ | | $10w \div 2$ | $n + n + n + n$ |

| $2t + n$ | $4n$ | | $n + 2 + n - 3$ | $2n + 5$ | | $\dfrac{n}{2}$ | $2(a + 1)$ | | $n \div 2$ | $n \times 4$ |

Turn the dominoes over and shuffle them. Deal five dominoes to each player.

One player starts by putting down a domino.

The other player may put down a domino that matches either end of the domino on the table. For example, if one player has put down the domino with $6n$ at one end, the other player could put down a domino that has an expression equal in value, such as the domino with $3n + 3n$ because $3n + 3n = 6n$. Otherwise, this player must pick up a domino from the spares.

The first player follows, playing a domino or picking one up, and so on, in turn.

The winner is the first player who has no dominoes.

Make up your own set of algebra dominoes.

Substitution

This section will show you how to:
- substitute numbers for letters in formulae and evaluate the resulting numerical expression
- use a calculator to evaluate numerical expressions

One of the most important features of algebra is the use of expressions and **formulae**, and the **substitution** of real numbers into them.

The value of an expression, such as $3x + 2$, changes when different values of x are substituted into it. For example, the expression $3x + 2$ has the value:

5 when $x = 1$ 14 when $x = 4$

and so on. A formula expresses the value of one variable as the others in the formula change. For example, the formula for the area, A, of a triangle of base b and height h is:

$$A = \frac{b \times h}{2}$$

When $b = 4$ and $h = 8$:

$$A = \frac{4 \times 8}{2} = 16$$

EXAMPLE 13

The formula for the area of a trapezium is:

$$A = \frac{(a + b)h}{2}$$

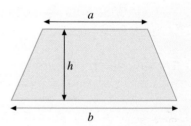

Find the area of the trapezium when $a = 5$, $b = 9$ and $h = 3$.

$$A = \frac{(5 + 9) \times 3}{2} = \frac{14 \times 3}{2} = 21$$

Always substitute the numbers for the letters before trying to work out the value of the expression. You are less likely to make a mistake this way. It is also useful to write **brackets** around each number, especially with negative numbers.

EXERCISE 4G

1 Find the value of $3x + 2$ when:

   a $x = 2$ **b** $x = 5$ **c** $x = 10$

2 Find the value of $4k - 1$ when:

   a $k = 1$ **b** $k = 3$ **c** $k = 11$

3 Find the value of $5 + 2t$ when:

   a $t = 2$ **b** $t = 5$ **c** $t = 12$

4 Evaluate $15 - 2f$ when: **a** $f = 3$ **b** $f = 5$ **c** $f = 8$

5 Evaluate $5m + 3$ when: **a** $m = 2$ **b** $m = 6$ **c** $m = 15$

6 Evaluate $3d - 2$ when: **a** $d = 4$ **b** $d = 5$ **c** $d = 20$

> **HINTS AND TIPS**
>
> It helps to put the numbers in brackets.
> $3(2) + 2 = 6 + 2 = 8$
> $3(5) + 2 = 15 + 2 = 17$
> etc …

FM 7 A taxi company uses the following rule to calculate their fares.

      Fare = £2.50 plus 50p per kilometre.

   a How much is the fare for a journey of 3 km?

   b Farook pays £9.00 for a taxi ride. How far was the journey?

   c Maisy knows that her house is 5 miles from town. She has £5.50 left in her purse after a night out. Has she got enough for a taxi ride home?

AU 8 Kaz knows that x, y and z have the values 2, 8 and 11, but she does not know which variable has which value.

   a What is the maximum value that the expression $2x + 6y - 3z$ could be?

   b What is the minimum value that the expression $5x - 2y + 3z$ could be?

> **HINTS AND TIPS**
>
> You could just try all combinations, but if you think for a moment you will find that the $6y$ term must give the largest number. This will give you a clue to the other terms.

PS 9 The formula for the area, A, of a rectangle with length l and width w is $A = lw$.

The formula for the area, T, of a triangle with base b and height h is $T = \frac{1}{2}bh$.

Find values of l, w, b and h so that $A = T$.

10 Find the value of $\dfrac{8 \times 4h}{5}$ when: **a** $h = 5$ **b** $h = 10$ **c** $h = 25$

11 Find the value of $\dfrac{25 - 3p}{2}$ when: **a** $p = 4$ **b** $p = 8$ **c** $p = 10$

12 Evaluate $\dfrac{x}{3}$ when: **a** $x = 6$ **b** $x = 24$ **c** $x = -30$

D

13 Evaluate $\dfrac{A}{4}$ when: **a** $A = 12$ **b** $A = 10$ **c** $A = -20$

14 Find the value of $\dfrac{12}{y}$ when: **a** $y = 2$ **b** $y = 4$ **c** $y = 6$

15 Find the value of $\dfrac{24}{x}$ when: **a** $x = 2$ **b** $x = 3$ **c** $x = 16$

FM 16 A holiday cottage costs £150 per day to rent.

A group of friends decide to rent the cottage for seven days.

a Which formula represents the cost of the rental for each person if there are n people in the group? Assume that they share the cost equally.

$$\dfrac{150}{n} \qquad \dfrac{150}{7n} \qquad \dfrac{1050}{n} \qquad \dfrac{150n}{n}$$

> **HINTS AND TIPS**
>
> To check your choice in part **a**, make up some numbers and try them in the formula. For example, take $n = 5$.

b Eventually 10 people go on the holiday. When they get the bill, they find that there is a discount for a seven-day rental.

After the discount, they each find it cost them £12.50 less than they expected.

How much does a 7-day rental cost?

C

AU 17 **a** p is an odd number and q is an even number.

Say if each of these expressions is odd or even.

  i $p + q$ **ii** $p^2 + q$ **iii** $2p + q$ **iv** $p^2 + q^2$

b x, y and z are all odd numbers.

Write an expression, using x, y and z, so that the value of the expression is always even.

> **HINTS AND TIPS**
>
> There are many answers for **b** and **a** should give you a clue.

PS 18 A formula for the cost of delivery, in pounds, of orders from a do-it-yourself warehouse is:

$$D = 2M - \dfrac{C}{5}$$

where D is the cost of the delivery, M is the distance in miles from the store and C is the cost of the goods to be delivered.

> **HINTS AND TIPS**
>
> Note: a rebate is a refund of some of the money that someone has already paid for goods or services.

a How much does the delivery cost when $M = 30$ and $C = 200$?

b Bob buys goods worth £300 and lives 10 miles from the store.

  i The formula gives the cost of delivery as a negative value. What is this value?

  ii Explain why Bob will not get a rebate from the store.

c Maya buys goods worth £400. She calculates that her cost of delivery will be zero. What is the greatest distance that Maya could live from the store?

GRADE BOOSTER

F You can use a formula expressed in words

F You can substitute numbers into expressions and use letters to write a simple algebraic expression

E You can simplify expressions by collecting like terms

D You can use letters to write more complicated expressions, expand expressions with brackets and factorise simple expressions

C You can expand and simplify expressions with brackets and factorise expressions

What you should know now

- How to simplify a variety of algebraic expressions by multiplying, collecting like terms and expanding brackets

- How to factorise expressions by removing common factors

- How to substitute into expressions, using positive or negative whole numbers and decimals

1 Simplify each expression.

   a $p + 5p - 2p$

   b $3q \times 4r$

   c $7t - 10t$

2 **a** Simplify: $5a + 2b - a + 5b$

   b Expand: $5(p + 2q - 3r)$

3 **a** Matt buys 10 boxes of apple juice at 24 pence each.

     i Calculate the total cost.

     ii He pays with a £10 note. How much change will he receive?

4 Graham is y years old.

   Harriet is 5 years older than Graham.

   a Write down an expression for Harriet's age.

   b Jane is half as old as Harriet. Write down an expression for Jane's age.

5 **a** Simplify $7x + 8x - 2x$ (1)

   b Use the formula $H = 5P + 3L$

     to find P when $H = 26$ and $L = 2$ (3)

               (Total 4 marks)

AQA, November 2008, Paper 2 Foundation, Question 13

6 A golf ball is travelling towards a hole.

   The distance of the ball from the hole, s feet, after time t seconds, is given by

   $s = t^2 - 6t + 9$

   a The ball drops into the hole after 3 seconds. By working out s when $t = 3$, show that this is correct.   (3)

               (Total 3 marks)

AQA, June 2008, Paper 1 Foundation, Question 19(a)

7 **a** Find the value of a^3 when $a = 4$.

   b Find the value of $5x + 3y$ when $x = -2$ and $y = 4$.

   c There are p seats in a standard class coach and q seats in a first class coach. A train has five standard class coaches and two first class coaches.

     Write down an expression in terms of p and q for the total number of seats in the train.

8 $d = 3e + 2h^2$

   Calculate the value of d when $e = 3.7$ and $h = 2$.

9 **a** Expand and simplify this expression.

     $2(x + 3) + 5(x + 2)$

   b Expand and simplify this expression.

     $(4x + y) - (2x - y)$

10 **a** Multiply out $4(x - 3)$ (1)

   b Factorise $x^2 + 5x$ (1)

               (Total 2 marks)

AQA, June 2007, Paper 2 Foundation, Question 22

11 Shapes are made from quarter circles and rectangles. For example

   The area of a quarter circle is Q cm^2.
   The area of a rectangle is R cm^2.
   This shape  has an area of $2Q + R$ cm^2.

   a Write down the area of this shape in terms of Q and R.   (1)

   b This shape has an area of $R - Q$ cm^2.

     Write down the area of this shape in terms of Q and R.   (2)

               (Total 3 marks)

AQA, June 2009, Paper 2 Foundation, Question 20

12 **a** Expand $6(x - 7)$ (1)

   b Expand and simplify

     $x(2x + 3) - 4(x^2 - 1)$ (2)

               (Total 3 marks)

AQA, November 2008, Paper 2 Foundation, Question 24

   C   **D**   **E**   **F**

Worked Examination Questions

1 Factorise completely:

$4x^2 - 8xy$

$$4x^2 - 8xy = 4x \times x - 4x \times 2y$$
$$= 4x(x - 2y)$$

(Total: 2 marks)

Note the words 'Factorise completely'. This is a clue that there is more than one common factor. Look for a common factor of 4 and 8, e.g. 4. Look for a common factor of x^2 and xy, e.g. x.

Split up the terms, using the common factors.

Write as a factorised expression.

The correct answer gets 2 marks (1 for accuracy and 1 for method). A partial factorisation such as $4(x^2 - 2xy)$ or $2x(2x - 4y)$ would get 1 mark.

(AU) **2** A rectangle has a length of $2x + 4$ and width of $x + 2$.

a Show that the perimeter can be written as $6(x + 2)$.

b Mark says that the perimeter must always be an even number.

Find a value of x that proves that Mark is wrong.

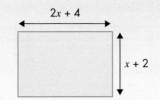

a Perimeter $= 2 \times (2x + 4) + 2 \times (x + 2)$
$= 2 \times 2(x + 2) + 2(x + 2)$
$= 4(x + 2) + 2(x + 2)$
$= 6(x + 2)$

Write down an expression for the perimeter. This will get you 1 method mark.

Take out the common factor.

Collect the like terms. These final two steps will get you 1 mark.

An alternative is
$2x + 4 + 2x + 4 + x + 2 + x + 2 = 6x + 12$
$6x + 12 = 6(x + 2)$

(2 marks)

b $6(x + 2) = 15$
$x + 2 = 2.5$
$x = 0.5$

Pick an odd value for the perimeter, say 15. Any value that works will be worth 1 mark.

(1 mark)

There are many answers. Any value that makes the value inside the brackets 'something and a half', such as $x = 1.5$ or $x = 2.5$ will work.

(Total: 3 marks)

Many people go walking each weekend. It is good exercise and can be a very enjoyable pastime.

When walkers set out they often try to estimate the length of time the walk will take. There are many factors that could influence this, but one rule that can help in estimating how long the walk will take is Naismith's Rule.

Naismith's rule

Naismith's rule is a rule of thumb that you can use when planning a walk, by calculating how long it will take. The rule was devised by William Naismith, a Scottish mountaineer, in 1892.

The basic rule is:

Allow 1 hour for every 3 miles (5 km) forward, plus $\frac{1}{2}$ hour for every 1000 feet (300 m) of ascent.

Getting started

Before you begin your main task, you may find it useful to fill in the following table to practise using Naismith's Rule.

Can you use algebra to display the rule?

Day	Distance (km)	Height (m)	Time (m)
1	16	250	
2	18	0	
3	11	340	
4	13	100	
5	14	120	

Now, in small groups think about:

- What kind of things influence the speed at which you walk?
- Do different types of routes make people walk at different rates?
- If there is a large group of people will they all walk at the same rate?

Use all the ideas you have just discussed as you move on to your main task.

Your task

You are going to compare data to see if Naismith's rule is still a useful way to work out how much time to allow for different walks.

The table on the right shows the actual times taken by a school group as they did five different walks in five days. Use this information to work out the following:

1 If the group had started at the same times and had the same breaks, how long would the group have taken each day, according to Naismith's rule?

2 Do you think Naismith's rule is still valid today? Explain your reasons.

3 If your friend was going to climb Ben Nevis, setting out at 11.30 am, would you advise them to do the walk? You will need to research the distance and climb details of the pathway up Ben Nevis, in order to advise them fully.

Day	Distance (km)	Height (m)	Time (minutes)	Time (hours/ minutes)	Start	Breaks	Finish
1	16	250	255	4 h 15 m	10.00 am	2 h	4.15 pm
2	18	0	270	4 h 30 m	10.00 am	1 h 30 m	4.00 pm
3	11	340	199	3 h 19 m	09.30 am	2 h 30 m	3.19 pm
4	13	100	195	3 h 15 m	10.30 am	2 h 30 m	4.15 pm
5	14	120	222	3 h 42 m	10.30 am	2 h 30 m	4.42 pm

Today, people always seem to be asking "Am I average?" But, how do we answer this question? The idea of an 'average' is not a modern one. History shows us that people have always been concerned with averages as the examples below demonstrate.

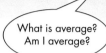

What is average? Am I average?

Mean in Ancient India

There is a story about Rtuparna who was born in India around 5000BC. He wanted to estimate the amount of fruit on a single tree. He counted how much fruit was on one branch, then estimated the number of branches on the tree. He multiplied the estimated number of branches by the counted fruit on one branch and was amazed that the total was very close to the actual counted number of fruit when it was picked.

This is seen as a first attempt at an arithmetic **mean**, because the one branch he chose would have been an average one representing all the branches. So the number of fruit on that branch would have been in the middle of the smallest and largest number of fruit on other branches on the tree.

Rtuparna may have been estimating the amount of fruit on a mango tree. These trees are common in India.

Mode in Ancient Greece

The Athenian army would have needed regular-sized bricks, as on this wall, to make their calculations accurate and useful.

This story comes from the Peloponnesian War in Ancient Greece (431–404BC). It is about a battle between the Peloponnesian League (led by Sparta) and the Delian League (led by Athens).

The Athenians had to get over the Peloponnesian Wall so they needed to work out the height of the wall. They did this by looking at the wall and counting the layers of bricks. This was done by hundreds of soldiers at the same time because many of them would get it wrong – but the majority would get it about right.

They then had to guess the height of one brick and so calculate the total height of the wall. They could then make ladders long enough to reach the top of the wall.

This is seen as an early use of the **mode**: the number of layers that occurred the most was deemed as the one most likely to be correct.

The other average that we use is the **median**, and there is no record of any use of this (which finds the middle value) being used until the early 17th century.

These ancient examples demonstrate that we do not always work out the average in the same way – we must choose a method that is appropriate to the situation. Bearing this in mind, how will you seek to answer the question "Am I average?"?

Statistics: Averages

This chapter will show you ...

- **F** how to calculate the mode, median, mean and range of small sets of discrete data
- **E** how to decide which is the best average for different types of data
- **D** how to calculate the mode, median, mean and range from frequency tables of discrete data
- **D** how to draw frequency polygons
- **C** how to use and recognise the modal class and calculate an estimate of the mean from frequency tables of grouped data

Visual overview

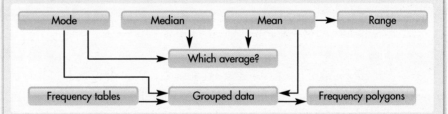

What you should already know

- How to collect and organise data **(KS3 level 4, GCSE grade F)**
- How to draw frequency tables **(KS3 level 4, GCSE grade F)**
- How to extract information from tables and diagrams **(KS3 level 4, GCSE grade F)**

Quick check

The marks for 15 students in a mathematics test are:

2, 3, 4, 5, 5, 6, 6, 6, 7, 7, 7, 7, 7, 8, 10

a What is the most common mark?

b What is the middle value in the list?

c Find the difference between the highest mark and the lowest mark.

d Find the total of all 15 marks.

Average is a term often used when describing or comparing sets of data, for example, the average rainfall in Britain, the average score of a batsman, an average weekly wage or the average mark in an examination.

In each of the above examples, you are representing the whole set of many values by just a single, 'typical' value, which is called the average.

The idea of an average is extremely useful, because it enables you to compare one set of data with another set by comparing just two values – their averages.

There are several ways of expressing an average, but the most commonly used averages are the **mode**, the **median** and the **mean**.

5.1 The mode

This section will show you how to:
- find the mode from lists of data and from frequency tables

Key words
frequency
modal class
modal value
mode

The **mode** is the value that occurs the most in a set of data. That is, it is the value with the highest **frequency**.

The mode is a useful average because it is very easy to find and it can be applied to non-numerical data (qualitative data). For example, you could find the modal style of skirts sold in a particular month.

EXAMPLE 1

Suhail scored the following number of goals in 12 school football matches:

1 2 1 0 1 0 0 1 2 1 0 2

What is the mode of his scores?

The number which occurs most often in this list is 1. So, the mode is 1.

You can also say that the modal score or **modal value** is 1.

EXAMPLE 2

Barbara asked her friends how many books they had each taken out of the school library during the previous month. Their responses were:

2 1 3 4 6 4 1 3 0 2 6 0

Find the mode.

Here, there is *no mode*, because no number occurs more than any of the others.

FM Functional Maths **AU** (AO2) Assessing Understanding **PS** (AO3) Problem Solving

EXERCISE 5A

1 Find the mode for each set of data.

a 3, 4, 7, 3, 2, 4, 5, 3, 4, 6, 8, 4, 2, 7

b 47, 49, 45, 50, 47, 48, 51, 48, 51, 48, 52, 48

c −1, 1, 0, −1, 2, −2, −2, −1, 0, 1, −1, 1, 0, −1, 2, −1, 2

d $\frac{1}{2}$, $\frac{1}{4}$, 1, $\frac{1}{2}$, $\frac{3}{4}$, $\frac{1}{4}$, 0, 1, $\frac{3}{4}$, $\frac{1}{4}$, 1, $\frac{1}{4}$, $\frac{3}{4}$, $\frac{1}{4}$, $\frac{1}{2}$

e 100, 10, 1000, 10, 100, 1000, 10, 1000, 100, 1000, 100, 10

f 1.23, 3.21, 2.31, 3.21, 1.23, 3.12, 2.31, 1.32, 3.21, 2.31, 3.21

> **HINTS AND TIPS**
>
> It helps to put the data in order or group all the same things together.

2 Find the modal category for each set of data.

a red, green, red, amber, green, red, amber, green, red, amber

b rain, sun, cloud, sun, rain, fog, snow, rain, fog, sun, snow, sun

c α, γ, α, β, γ, α, α, γ, β, α, β, γ, β, β, α, β, γ, β

d ❄, ☆, ★, ★, ☆, ❄, ★, ☆, ★, ☆, ★, ❄, ✪, ☆, ★, ★, ☆

FM 3 Joan did a survey to find the shoe sizes of students in her class. The bar chart illustrates her data.

a How many students are in Joan's class?

b What is the modal shoe size?

c Can you tell from the bar chart which are the boys or which are the girls in her class?

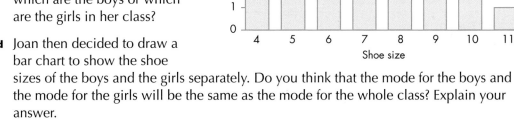

d Joan then decided to draw a bar chart to show the shoe sizes of the boys and the girls separately. Do you think that the mode for the boys and the mode for the girls will be the same as the mode for the whole class? Explain your answer.

4 The frequency table shows the marks that Form 10MP obtained in a spelling test.

Mark	3	4	5	6	7	8	9	10
Frequency	1	2	6	5	5	4	3	4

a Write down the mode for their marks.

AU **b** Do you think this is a typical mark for the form? Explain your answer.

5 The grouped frequency table shows the number of e-mails each household in Orchard Street received during one day.

No. of e-mails	0–4	5–9	10–14	15–19	20–24	25–29	30–34	35–39
Frequency	9	12	14	11	10	8	4	2

a Draw a bar chart to illustrate the data.

b How many households are there in Orchard Street?

c How many households received 20 or more e-mails?

AU **d** How many households did not receive any e-mails during the week? Explain your answer.

e Write down the modal class for the data in the table.

> **HINTS AND TIPS**
>
> You cannot find the mode of the data in a grouped frequency table. So, instead, you need to find the **modal class**, which is the class interval with the highest frequency.

6 Explain why the mode is often referred to as the 'shopkeeper's average'.

7 This table shows the colours of eyes of the students in form 11P.

	Blue	Brown	Green
Boys	4	8	1
Girls	8	5	2

a How many students are in form 11P?

b What is the modal eye colour for:

   i boys   **ii** girls   **iii** the whole form?

AU **c** After two students join the form the modal eye colour for the whole form is blue. Which of the following statements is true?

   ● Both students had green eyes.

   ● Both students had brown eyes.

   ● Both students had blue eyes.

   ● You cannot tell what their eye colours were.

AU **8** Here is a large set of raw data.

   5 6 8 2 4 8 9 8 1 3 4 2 7 2 4 6 7 5 3 8

   9 1 3 1 5 6 2 5 7 9 4 1 4 3 3 5 6 8 6 9

   8 4 8 9 3 4 6 7 7 4 5 4 2 3 4 6 7 6 5 5

a What problems may occur if you attempted to find the mode by counting individual numbers?

b Explain a method that would make finding the mode more efficient and accurate.

c Use your method to find the mode of the data.

The median

This section will show you how to:

● find the median from a list of data, a table of data and a stem-and-leaf diagram

Key words

median
middle value

The **median** is the **middle value** of a list of values when they are put in *order* of size, from lowest to highest.

The advantage of using the median as an average is that half the data-values are below the median value and half are above it. Therefore, the average is only slightly affected by the presence of any particularly high or low values that are not typical of the data as a whole.

EXAMPLE 3

Find the median for the following list of numbers:

2, 3, 5, 6, 1, 2, 3, 4, 5, 4, 6

Putting the list in numerical order gives:

1, 2, 2, 3, 3, **4**, 4, 5, 5, 6, 6

There are 11 numbers in the list, so the middle of the list is the 6th number. Therefore, the median is 4.

EXAMPLE 4

Find the median of the data shown in the frequency table.

Value	2	3	4	5	6	7
Frequency	2	4	6	7	8	3

First, add up the frequencies to find out how many pieces of data there are.

The total is 30 so the median value will be between the 15th and 16th values.

Now, add up the frequencies to give a running total, to find out where the 15th and 16th values are.

Value	2	3	4	5	6	7
Frequency	2	4	6	7	8	3
Total frequency	2	6	12	19	27	30

There are 12 data-values up to the value 4 and 19 up to the value 5.

Both the 15th and 16th values are 5, so the median is 5.

To find the median in a list of *n* values, written in order, use the rule:

$$\text{median} = \frac{n+1}{2}\text{th value}$$

For a set of data that has a lot of values, it is sometimes more convenient and quicker to draw a stem-and-leaf diagram. Example 5 shows you how to do this.

EXAMPLE 5

The ages of 20 people attending a conference were as follows:

28, 34, 46, 23, 28, 34, 52, 61, 45, 34, 39, 50, 26, 44, 60, 53, 31, 25, 37, 48

Find the modal age and median age of the group.

Taking the tens to be the 'stem' and the units to be the 'leaves', draw the stem-and-leaf diagram as shown below.

```
2 | 3  5  6  8  8
3 | 1  4  4  4  7  9
4 | 4  5  6  8
5 | 0  2  3
6 | 0  1
```
Key 2 | 3 represents 23 people

The most common value is 34, so the mode is 34.

There is an even number of values in this list, so the middle of the list is between the two central values, which are the 10th and 11th values. To find the central values count up 10 from the lowest value, 23, 25, 26, 28, 28, 31 … or *down* 10 from the highest value 61, 60, 53, 52, 50, 48 …

Therefore, the median is exactly midway between 37 and 39.

Hence, the median is 38.

EXERCISE 5B

1 Find the median for each set of data.

a 7, 6, 2, 3, 1, 9, 5, 4, 8

b 26, 34, 45, 28, 27, 38, 40, 24, 27, 33, 32, 41, 38

c 4, 12, 7, 6, 10, 5, 11, 8, 14, 3, 2, 9

d 12, 16, 12, 32, 28, 24, 20, 28, 24, 32, 36, 16

e 10, 6, 0, 5, 7, 13, 11, 14, 6, 13, 15, 1, 4, 15

f −1, −8, 5, −3, 0, 1, −2, 4, 0, 2, −4, −3, 2

g 5.5, 5.05, 5.15, 5.2, 5.3, 5.35, 5.08, 5.9, 5.25

HINTS AND TIPS

Remember to put the data in order before finding the median.

HINTS AND TIPS

If there is an even number of pieces of data, the median will be halfway between the two middle values.

2 A group of 15 sixth-formers had lunch in the school's cafeteria. Given below are the amounts that they spent.

£2.30, £2.20, £2, £2.50, £2.20, £3.50, £2.20, £2.25, £2.20, £2.30, £2.40, £2.20, £2.30, £2, £2.35

a Find the mode for the data.

b Find the median for the data.

AU **c** Which is the better average to use? Explain your answer.

3 **a** Find the median of 7, 4, 3, 8, 2, 6, 5, 2, 9, 8, 3.

b Without putting them in numerical order, write down the median for each of these sets.

i 17, 14, 13, 18, 12, 16, 15, 12, 19, 18, 13

ii 217, 214, 213, 218, 212, 216, 215, 212, 219, 218, 213

iii 12, 9, 8, 13, 7, 11, 10, 7, 14, 13, 8

iv 14, 8, 6, 16, 4, 12, 10, 4, 18, 16, 6

4 Given below are the age, height and weight of each of the seven players in a netball team.

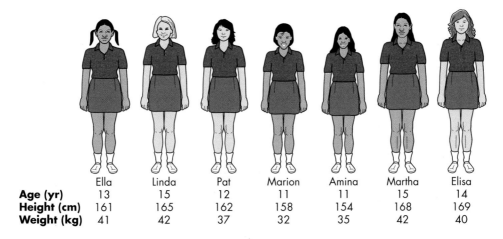

	Ella	Linda	Pat	Marion	Amina	Martha	Elisa
Age (yr)	13	15	12	11	11	15	14
Height (cm)	161	165	162	158	154	168	169
Weight (kg)	41	42	37	32	35	42	40

a Find the median age of the team. Which player has the median age?

b Find the median height of the team. Which player has the median height?

c Find the median weight of the team. Which player has the median weight?

AU **d** Who would you choose as the average player in the team? Give a reason for your answer.

5 The table shows the number of sandwiches sold in a corner shop over 25 days.

Sandwiches sold	10	11	12	13	14	15	16
Frequency	2	3	6	4	3	4	3

a What is the modal number of sandwiches sold?

b What is the median number of sandwiches sold?

6 The bar chart shows the marks that Mrs Woodhead gave her students for their first Functional Mathematics task.

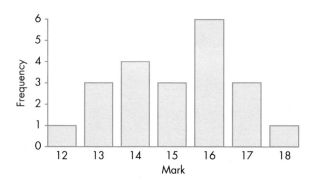

a How many students are there in Mrs Woodhead's class?

b What is the modal mark?

c Copy and complete this frequency table.

Mark	12	13	14	15	16	17	18
Frequency	1	3					

d What is the median mark?

PS 7 **a** Write down a list of nine numbers that has a median of 12.

b Write down a list of 10 numbers that has a median of 12.

c Write down a list of nine numbers that has a median of 12 and a mode of 8.

d Write down a list of 10 numbers that has a median of 12 and a mode of 8.

8 The following stem-and-leaf diagram shows the times taken for 15 students to complete a mathematical puzzle.

Key 1 | 7 represents 17 seconds

```
1 | 7  8  8  9
2 | 2  2  2  5  6  9
3 | 3  4  5  5  8
```

a What is the modal time taken to complete the puzzle?

b What is the median time taken to complete the puzzle?

9 The stem-and-leaf diagram shows the marks for 13 boys and 12 girls in form 7E in a science test.

HINTS AND TIPS

Read the boys' marks from right to left.

Key 2 | 3 represents 32 marks for boys
3 | 5 represents 35 marks for girls

			Boys			Girls					
		6	4	2	3	5	7	9			
	9	9	6	2	4	2	2	3	8	8	8
7	6	6	6	5	3	5	1	1	5		

a What was the modal mark for the boys?

b What was the modal mark for the girls?

c What was the median mark for the boys?

d What was the median mark for the girls?

FM e Who did better in the test, the boys or the girls? Give a reason for your answer.

HINTS AND TIPS

To find the middle value of two numbers, add them together and divide the result by 2. For example, for 43 and 48, 43 + 48 = 91, 91 ÷ 2 = 45.5.

PS 10 A list contains seven even numbers. The largest number is 24. The smallest number is half the largest. The mode is 14 and the median is 16. Two of the numbers add up to 42. What are the seven numbers?

11 The marks of 25 students in an English examination were as follows.

55, 63, 24, 47, 60, 45, 50, 89, 39, 47, 38, 42, 69, 73, 38, 47, 53, 64, 58, 71, 41, 48, 68, 64, 75

Draw a stem-and-leaf diagram to find the median.

PS 12 Look at this list of numbers.

4, 4, 5, 8, 10, 11, 12, 15, 15, 16, 20

a Add four numbers to make the median 12.

b Add six numbers to make the median 12.

c What is the least number of numbers to add that will make the median 4?

AU 13 Here are five payments.

£3, £5, £8, £100, £3000

Explain why the median is not a good average to use in this set of payments.

The mean

This section will show you how to:
- calculate the mean of a set of data

Key words
average
mean

The **mean** of a set of data is the sum of all the values in the set divided by the total number of values in the set. That is:

$$\text{mean} = \frac{\text{sum of all values}}{\text{total number of values}}$$

This is what most people mean when they use the term '**average**'.

Another name for this average is the arithmetic **mean**.

The advantage of using the mean as an average is that it takes into account all the values in the set of data.

EXAMPLE 6

Find the mean of 4, 8, 7, 5, 9, 4, 8, 3.

Sum of all the values = 4 + 8 + 7 + 5 + 9 + 4 + 8 + 3 = 48

Total number of values = 8

Therefore, mean = $\dfrac{48}{8} = 6$

EXAMPLE 7

The ages of 11 players in a football squad are:

21, 23, 20, 27, 25, 24, 25, 30, 21, 22, 28

What is the mean age of the squad?

Sum of all the ages = 266

Total number in squad = 11

Therefore, mean age = $\dfrac{266}{11} = 24.1818… = 24.2$ (1 decimal place)

When the answer is not exact, it is usual to round the mean to 1 decimal place.

Using a calculator

If your calculator has a statistical mode, the mean of a set of numbers can be found by simply entering the numbers and then pressing the $\boxed{\overline{x}}$ key. On some calculators, the statistical mode is represented by $\boxed{\text{SD}}$.

EXAMPLE 8

The mean weight of eight members of a rowing crew is 89 kg. When the cox is included, the mean weight is 85 kg.

What is the weight of the cox?

The eight crew members have a total weight of 8 × 89 = 712 kg.

With the cox the total weight is 9 × 85 = 765 kg.

So the cox weighs 765 − 712 = 53 kg.

EXAMPLE 9

Find the mean of 2, 3, 7, 8 and 10.

First put your calculator into statistical mode.

Then press the following keys:

   $\boxed{2}$ $\boxed{\text{DATA}}$ $\boxed{3}$ $\boxed{\text{DATA}}$ $\boxed{7}$ $\boxed{\text{DATA}}$ $\boxed{8}$ $\boxed{\text{DATA}}$ $\boxed{1}$ $\boxed{0}$ $\boxed{\text{DATA}}$ $\boxed{\overline{x}}$

You should find that the mean is given by $\boxed{\overline{x}}$ = 6.

You can also find the number of data-values by pressing the $\boxed{n}$ key.

Your calculator may use the following process for finding the mean.

Set up into statistics mode by keying:

   $\boxed{\text{MODE}}$ $\boxed{2}$ (stat) $\boxed{1}$ (1-var)

Enter the data 12, 16, 17 by keying:

   $\boxed{1}$ $\boxed{2}$ $\boxed{=}$ $\boxed{1}$ $\boxed{6}$ $\boxed{=}$ $\boxed{1}$ $\boxed{7}$

Calculate the mean by keying:

   $\boxed{\text{AC}}$ $\boxed{\text{SHIFT}}$ $\boxed{1}$ (stat) $\boxed{5}$ (var) $\boxed{2}$ $(\overline{x})$ $\boxed{=}$

This gives: mean = 15

> **HINTS AND TIPS**
>
> It is generally more difficult to find the mean on a calculator by this method than simply adding up the data values and dividing. So don't use this method to find the mean unless you are very confident about using it.

F

EXERCISE 5C

1 Find, without the help of a calculator, the mean for each set of data.

  a 7, 8, 3, 6, 7, 3, 8, 5, 4, 9

  b 47, 3, 23, 19, 30, 22

  c 42, 53, 47, 41, 37, 55, 40, 39, 44, 52

  d 1.53, 1.51, 1.64, 1.55, 1.48, 1.62, 1.58, 1.65

  e 1, 2, 0, 2, 5, 3, 1, 0, 1, 2, 3, 4

2 Calculate the mean for each set of data, giving your answer correct to 1 decimal place. You may use your calculator.

  a 34, 56, 89, 34, 37, 56, 72, 60, 35, 66, 67

  b 235, 256, 345, 267, 398, 456, 376, 307, 282

  c 50, 70, 60, 50, 40, 80, 70, 60, 80, 40, 50, 40, 70

  d 43.2, 56.5, 40.5, 37.9, 44.8, 49.7, 38.1, 41.6, 51.4

  e 2, 3, 1, 0, 2, 5, 4, 3, 2, 0, 1, 3, 4, 5, 0, 3, 1, 2

FM **3** The table shows the marks that 10 students obtained in Mathematics, English and Science in their Year 10 examinations.

Student	Abigail	Brian	Chloe	David	Eric	Frances	Graham	Howard	Ingrid	Jane
Maths	45	56	47	77	82	39	78	32	92	62
English	54	55	59	69	66	49	60	56	88	44
Science	62	58	48	41	80	56	72	40	81	52

  a Work out the mean mark for Mathematics.

  b Work out the mean mark for English.

  c Work out the mean mark for Science.

  d Which student obtained marks closest to the mean in all three subjects?

  e How many students were above the average mark in all three subjects?

4 Heather kept a record of the amount of time she spent on her homework over 10 days:

  $\frac{1}{2}$ h, 20 min, 35 min, $\frac{1}{4}$ h, 1 h, $\frac{1}{2}$ h, $1\frac{1}{2}$ h, 40 min, $\frac{3}{4}$ h, 55 min

Calculate the mean time, in minutes, that Heather spent on her homework.

> **HINTS AND TIPS**
>
> Convert all times to minutes, for example, $\frac{1}{4}$h = 15 minutes.

5 The weekly wages of 10 people working in an office are:

£350 £200 £180 £200 £350 £200 £240 £480 £300 £280

a Find the modal wage.

b Find the median wage.

c Calculate the mean wage.

d Which of the three averages best represents the office staff's wages? Give a reason for your answer.

> **HINTS AND TIPS**
>
> Remember that the mean can be distorted by extreme values.

6 The ages of five people in a group of walkers are 38, 28, 30, 42 and 37.

a Calculate the mean age of the group.

b Steve, who is 41, joins the group. Calculate the new mean age of the group.

7 a Calculate the mean of 3, 7, 5, 8, 4, 6, 7, 8, 9 and 3.

b Calculate the mean of 13, 17, 15, 18, 14, 16, 17, 18, 19 and 13. What do you notice?

c Write down, without calculating, the mean for each of the following sets of data.

   i 53, 57, 55, 58, 54, 56, 57, 58, 59, 53

   ii 103, 107, 105, 108, 104, 106, 107, 108, 109, 103

   iii 4, 8, 6, 9, 5, 7, 8, 9, 10, 4

> **HINTS AND TIPS**
>
> Look for a connection between the original data and the new data. For example in **i** the numbers are 50 more.

PS 8 Two families were in a competition.

Speed family		Roberts family	
Brian	aged 59	Frank	aged 64
Kath	aged 54	Marylin	aged 62
James	aged 34	David	aged 34
Helen	aged 34	James	aged 32
John	aged 30	Tom	aged 30
Joseph	aged 24	Helen	aged 30
Joy	aged 19	Evie	aged 16

Each family had to choose four members with a mean age of between 35 and 36.

Choose two teams, one from each family, that have this mean age between 35 and 36.

AU 9 Asif had an average batting score of 35 runs.
He had scored 315 runs in nine games of cricket.

What is the least number of runs he needs to score in the next match if he is to get a higher average score?

10 The mean age of a group of eight walkers is 42. Joanne joins the group and the mean age changes to 40. How old is Joanne?

The range

This section will show you how to:

- find the range of a set of data and compare different sets of data, using the mean and the range

Key words

consistency

range

spread

The **range** for a set of data is the highest value of the set minus the lowest value.

The range is *not* an average. It shows the **spread** of the data. It is, therefore, used when comparing two or more sets of similar data. You can also use it to comment on the **consistency** of two or more sets of data.

EXAMPLE 10

Rachel's marks in 10 mental arithmetic tests were 4, 4, 7, 6, 6, 5, 7, 6, 9 and 6.

Therefore, her mean mark is $60 \div 10 = 6$ and the range is $9 - 4 = 5$.

Adil's marks in the same tests were 6, 7, 6, 8, 5, 6, 5, 6, 5 and 6.

Therefore, his mean mark is $60 \div 10 = 6$ and the range is $8 - 5 = 3$.

Although the means are the same, Adil has a smaller range. This shows that Adil's results are more consistent.

EXERCISE 5D

G

1 Find the range for each set of data.

a 3, 8, 7, 4, 5, 9, 10, 6, 7, 4

b 62, 59, 81, 56, 70, 66, 82, 78, 62, 75

c 1, 0, 4, 5, 3, 2, 5, 4, 2, 1, 0, 1, 4, 4

d 3.5, 4.2, 5.5, 3.7, 3.2, 4.8, 5.6, 3.9, 5.5, 3.8

e 2, –1, 0, 3, –1, –2, 1, –4, 2, 3, 0, 2, –2, 0, –3

2 The table shows the maximum and minimum temperatures at midday for five cities in England during a week in August.

	Birmingham	Leeds	London	Newcastle	Sheffield
Maximum temperature (°C)	28	25	26	27	24
Minimum temperature (°C)	23	22	24	20	21

    a Write down the range of the temperatures for each city.

    b What do the ranges tell you about the weather for England during the week?

FM 3 Over a three-week period, the school tuck shop took the following amounts.

	Monday	Tuesday	Wednesday	Thursday	Friday
Week 1	£32	£29	£36	£30	£28
Week 2	£34	£33	£25	£28	£20
Week 3	£35	£34	£31	£33	£32

    a Calculate the mean amount taken each week.

    b Find the range for each week.

    c What can you say about the total amounts taken for each of the three weeks?

4 In a womens' golf tournament, the club chairperson had to choose either Sheila or Fay to play in the first round. In the previous eight rounds, their scores were as follows.

    Sheila's scores: 75, 92, 80, 73, 72, 88, 86, 90

    Fay's scores: 80, 87, 85, 76, 85, 79, 84, 88

> **HINTS AND TIPS**
>
> The best person to choose may not be the one with the biggest mean but could be the most consistent player.

    a Calculate the mean score for each golfer.

    b Find the range for each golfer.

AU   **c** Which golfer would you choose to play in the tournament? Explain why.

5 Dan has a choice of two buses to get to school: Number 50 or Number 63. Over a month, he kept a record of the number of minutes each bus was late when it set off from his home bus stop.

    No. 50: 4, 2, 0, 6, 4, 8, 8, 6, 3, 9

    No. 63: 3, 4, 0, 10, 3, 5, 13, 1, 0, 1

    a For each bus, calculate the mean number of minutes late.

    b Find the range for each bus.

AU   **c** Which bus would you advise Dan to catch? Give a reason for your answer.

PS 6 The table gives the ages and heights of 10 children.

Name	Age (years)	Height (cm)
Billy	9	121
Isaac	4	73
Lilla	8	93
Lewis	10	118
Evie	3	66
Andrew	6	82
Oliver	4	78
Beatrice	2	69
Isambard	9	87
Chloe	7	82

a Chloe is having a party. She wants to invite as many children as possible but does not want the range of ages to be more than 5. Who will she invite?

b This is a sign at a theme park:

You have to be taller than … cm

and

shorter than … cm to go on this ride

Isaac is the shortest person who can go on the ride and Isambard is the tallest.

What are the smallest and largest missing values on the sign?

AU 7 a The age range of a school quiz team is 20 years and the mean age is 34. Who would you expect to be in this team? Explain your answer.

b Another team has an average age of $15\frac{1}{2}$ and a range of 1. Who would you expect to be in this team? Explain your answer.

ACTIVITY

Your time is up

You are going to find out how good you are at estimating 1 minute.

You need a stopwatch and a calculator.

This is a group activity. One person in the group acts as a timekeeper, says 'Start' and starts the stopwatch.

When someone thinks 1 minute has passed, they say 'Stop', and the timekeeper writes down the actual time, in seconds, that has passed. The timekeeper should try to record everyone's estimate.

Repeat the activity, with every member of the group taking a turn as the timekeeper.

Collate all the times and, from the data, find the mean (to the nearest second) and the range.

- How close is the mean to 1 minute?
- Why is the range useful?
- What strategies did people use to estimate 1 minute?

Repeat the activity for estimating different times, for example, 30 seconds or 2 minutes.

Write a brief report on what you find out about people's ability to estimate time.

Which average to use

This section will show you how to:
- understand the advantages and disadvantages of each type of average and decide which one to use in different situations

Key words
appropriate
extreme values
representative

An average must be truly **representative** of a set of data. So, when you have to find an average, it is crucial to choose the **appropriate** type of average for this particular set of data.

If you use the wrong average, your results will be distorted and give misleading information.

This table, which compares the advantages and disadvantages of each type of average, will help you to make the correct decision.

	Mode	Median	Mean
Advantages	Very easy to find Not affected by **extreme values** Can be used for non-numerical data	Easy to find for ungrouped data Not affected by extreme values	Easy to find Uses all the values The total for a given number of values can be calculated from it
Disadvantages	Does not use all the values May not exist	Does not use all the values Often not understood	Extreme values can distort it Has to be calculated
Use for	Non-numerical data Finding the most likely value	Data with extreme values	Data with values that are spread in a balanced way

EXERCISE 5E

1 The ages of the members of a hockey team were:

29 26 21 24 26 28 35 23 29 28 29

a Give:
  i the modal age   **ii** the median age   **iii** the mean age.

b What is the range of the ages?

2 a For each set of data, find the mode, the median and the mean.

  i 6, 10, 3, 4, 3, 6, 2, 9, 3, 4   **ii** 6, 8, 6, 10, 6, 9, 6, 10, 6, 8

  iii 7, 4, 5, 3, 28, 8, 2, 4, 10, 9

AU **b** For each set of data, decide which average is the best one to use and give a reason.

E

3 A newsagent sold the following numbers of copies of *The Evening Star* on 12 consecutive evenings during a promotion exercise organised by the newspaper's publisher.

65 73 75 86 90 112 92 87 77 73 68 62

a Find the mode, the median and the mean for the sales.

AU **b** The newsagent had to report the average sale to the publisher after the promotion. Which of the three averages would you advise the newsagent to use? Explain why.

4 The mean age of a group of 10 young people was 15.

a What do all their ages add up to?

b What will be their mean age in five years' time?

5 **a** Find the median of each list below.

i 2, 4, 6, 7, 9   **ii** 12, 14, 16, 17, 19

iii 22, 24, 26, 27, 29   **iv** 52, 54, 56, 57, 59

v 92, 94, 96, 97, 99

b What do you notice about the lists and your answers?

c Use your answer above to help find the medians of the following lists.

i 132, 134, 136, 137, 139   **ii** 577, 576, 572, 574, 579

iii 431, 438, 439, 432, 435   **iv** 855, 859, 856, 851, 857

d Find the mean of each of the sets of numbers in part **a**.

D

AU **6** Decide which average you would use for each of the following. Give a reason for your answer.

a The average mark in an examination

b The average pocket money for a group of 16-year-old students

c The average shoe size for all the girls in Year 10

d The average height for all the artistes on tour with a circus

e The average hair colour for students in your school

f The average weight of all newborn babies in a hospital's maternity ward.

7 A pack of matches consisted of 12 boxes. The contents of each box were counted as:

34 31 29 35 33 30 31 28 29 35 32 31

On the box it stated 'Average contents 32 matches'. Is this correct?

PS **8** Here are three expressions, where x represents a number.

$x + 5$ $3x$ $2x - 2$

a Work out the median when $x = 4$ **b** Work out the mean when $x = 3$

9 A firm showed the annual salaries for its employees as:

Chairman	£83 000
Managing director	£65 000
Floor manager	£34 000
Skilled worker 1	£28 000
Skilled worker 2	£28 000
Machinist	£20 000
Computer engineer	£20 000
Secretary	£20 000
Office junior	£8 000

a Give:

    i the modal salary   **ii** the median salary   **iii** the mean salary.

b The management suggested a pay rise of 6% for all employees. The shopfloor workers suggested a pay rise of £1500 for all employees.

    i One of the suggestions would cause problems for the firm. Which one is that and why?

    ii What difference would each suggestion make to the modal, median and mean salaries?

AU 10 Mr Brennan, a caring maths teacher, told each student their test mark and only gave the test statistics to the whole class. He gave the class the modal mark, the median mark and the mean mark.

a Which average would tell a student whether they were in the top half or the bottom half of the class?

b Which average tells the students nothing really?

c Which average allows a student to gauge how well they have done compared with everyone else?

FM 11 Three players were hoping to be chosen for the basketball team.

The following table shows their scores in the last few games they played.

Tom	16, 10, 12, 10, 13, 8, 10
David	16, 8, 15, 25, 8
Mohaned	15, 2, 15, 3, 5

The teacher said they would be chosen by their best average score.

Which average would each boy want to be chosen by?

PS 12 These expressions represent five numbers.

    $3x$ $4x$ $4x$ $5x$ $9x$

a Write down the range in terms of x.

b Work out the value of x when the mean = 40.

D

C

PS 13 **a** Find five numbers that have **both** the properties below:

- a range of 5
- a mean of 5.

b Find five numbers that have **all** the properties below:

- a range of 5
- a median of 5
- a mean of 5.

14 These expressions represent four numbers.

$$x + 1 \qquad x + 5 \qquad 2x - 1 \qquad 4x + 7$$

Work out the mean in terms of x.

Give your answer in its simplest form.

AU 15 What is the average pay at a factory with 10 employees?

The boss said: "£43 295"
A worker said: "£18 210"

They were both correct.
Explain how this can be.

PS 16 A list of nine numbers has a mean of 7.6. What number must be added to the list to give a new mean of 8?

PS 17 A dance group of 17 teenagers had a mean weight of 44.5 kg. To enter a competition there needed to be 18 teenagers with an average weight of 44.4 kg or less. What is the maximum weight that the eighteenth person must be?

5.6 ## Frequency tables

This section will show you how to:
- revise finding the mode and median from a frequency table
- learn how to calculate the mean from a frequency table

Key word
frequency table

When a lot of information has been gathered, it is often convenient to put it together in a **frequency table**. From this table you can then find the values of the three averages and the range.

EXAMPLE 11

A survey was done on the number of people in each car leaving the Meadowhall Shopping Centre, in Sheffield. The results are summarised in the table below.

Number of people in each car	1	2	3	4	5	6
Frequency	45	198	121	76	52	13

For the number of people in a car, calculate:

a the mode   **b** the median   **c** the mean.

a The modal number of people in a car is easy to spot. It is the number with the largest frequency, which is 198. Hence, the modal number of people in a car is 2.

b The median number of people in a car is found by working out where the middle of the set of numbers is located. First, add up frequencies to get the total number of cars surveyed, which comes to 505. Next, calculate the middle position.

$(505 + 1) \div 2 = 253$

Now add the frequencies across the table to find which group contains the 253rd item. The 243rd item is the end of the group with 2 in a car. Therefore, the 253rd item must be in the group with 3 in a car. Hence, the median number of people in a car is 3.

c To calculate the mean number of people in a car, multiply the number of people in the car by the frequency. This is best done in an extra column. Add these to find the total number of people and divide by the total frequency (the number of cars surveyed).

Number in car	Frequency	Number in these cars
1	45	$1 \times 45 = 45$
2	198	$2 \times 198 = 396$
3	121	$3 \times 121 = 363$
4	76	$4 \times 76 = 304$
5	52	$5 \times 52 = 260$
6	13	$6 \times 13 = 78$
Totals	505	1446

Hence, the mean number of people in a car is $1446 \div 505 = 2.9$ (to 1 decimal place).

Using your calculator

The previous example can also be done by using the statistical mode that is available on some calculators. However, not all calculators are the same, so you will have either to read your instruction manual or experiment with the statistical keys on your calculator.

You may find one labelled:

[DATA] or [M+] or [Σ+] or [$\bar{x}$] , where $\bar{x}$ is printed in blue.

Try the following key strokes:

[1] [×] [4] [5] [DATA]   [2] [×] [1] [9] [8] [DATA] ... [6] [×] [1] [3] [DATA] [$\bar{x}$]

D

1 Find **i** the mode, **ii** the median and **iii** the mean from each frequency table below.

a A survey of the shoe sizes of all the Year 10 boys in a school gave these results.

Shoe size	4	5	6	7	8	9	10
Number of students	12	30	34	35	23	8	3

b A survey of the number of eggs laid by hens over a period of one week gave these results.

Number of eggs	0	1	2	3	4	5	6
Frequency	6	8	15	35	48	37	12

c This is a record of the number of babies born each week over one year in a small maternity unit.

Number of babies	0	1	2	3	4	5	6	7	8	9	10	11	12	13	14
Frequency	1	1	1	2	2	2	3	5	9	8	6	4	5	2	1

d A school did a survey on how many times in a week students arrived late at school. These are the findings.

Number of times late	0	1	2	3	4	5
Frequency	481	34	23	15	3	4

2 A survey of the number of children in each family of a school's intake gave these results.

Number of children	1	2	3	4	5
Frequency	214	328	97	26	3

a Assuming each child at the school is shown in the data, how many children are at the school?

b Calculate the mean number of children in a family.

c How many families have this mean number of children?

FM **d** How many families would consider themselves average from this survey?

FM **3** A dentist kept records of how many teeth he extracted from his patients.

In 1989 he extracted 598 teeth from 271 patients.

In 1999 he extracted 332 teeth from 196 patients.

In 2009 he extracted 374 teeth from 288 patients.

a Calculate the average number of teeth taken from each patient in each year.

AU **b** Explain why you think the average number of teeth extracted falls each year.

4 One hundred cases of apples delivered to a supermarket were inspected and the numbers of bad apples were recorded.

Bad apples	0	1	2	3	4	5	6	7	8	9
Frequency	52	29	9	3	2	1	3	0	0	1

Give:

a the modal number of bad apples per case

b the mean number of bad apples per case.

5 Two dice are thrown together 60 times. The sums of the scores are shown below.

Score	2	3	4	5	6	7	8	9	10	11	12
Frequency	1	2	6	9	12	15	6	5	2	1	1

Find: **a** the modal score **b** the median score **c** the mean score.

6 During a one-month period, the number of days off taken by 100 workers in a factory were noted as follows.

Number of days off	0	1	2	3	4
Number of workers	35	42	16	4	3

Calculate:

a the modal number of days off

b the median number of days off

c the mean number of days off.

7 Two friends often played golf together. They recorded their scores for each hole over the last five games to compare who was more consistent and who was the better player. Their results were summarised in the following table.

No. of shots to hole ball	1	2	3	4	5	6	7	8	9
Roger	0	0	0	14	37	27	12	0	0
Brian	5	12	15	18	14	8	8	8	2

a What is the modal score for each player?

b What is the range of scores for each player?

c What is the median score for each player?

d What is the mean score for each player?

AU **e** Which player is the more consistent and why?

AU **f** Who would you say is the better player and why?

D

PS **8** A tea stain on a newspaper removed four numbers from the following frequency table of goals scored in 40 league football matches one weekend.

Goals	0	1	2		5
Frequency	4	6	9		3

The mean number of goals scored is 2.4.

What could the missing four numbers be?

AU **9** Talera made day trips to Manchester frequently during a year.

The table shows how many days in a week she travelled.

Days	0	1	2	3	4	5
Frequency	17	2	4	13	15	1

Explain how you would find the median number of days Talera travelled in a week to Manchester.

5.7 Grouped data

This section will show you how to:
- identify the modal class
- calculate an estimate of the mean from a grouped table

Key words
estimated
grouped data
mean
modal class

Sometimes the information you are given is grouped in some way (called **grouped data**), as in Example 12, which shows the range of weekly pocket money given to Year 12 students in a particular class.

Normally, grouped tables use continuous data, which is data that can have any value within a range of values, for example, height, weight, time, area and capacity. In these situations, the **mean** can only be **estimated** as you do not have all the information.

Discrete data is data that consists of separate numbers, for example, goals scored, marks in a test, number of children and shoe sizes.

In both cases, when using a grouped table to estimate the mean, first find the midpoint of the interval by adding the two end-values and then dividing by two.

EXAMPLE 12

Pocket money, p (£)	$0 < p \leq 1$	$1 < p \leq 2$	$2 < p \leq 3$	$3 < p \leq 4$	$4 < p \leq 5$
No. of students	2	5	5	9	15

a Write down the **modal class**.

b Calculate an estimate of the mean weekly pocket money.

a The modal class is easy to pick out, since it is simply the one with the largest frequency. Here the modal class is £4 to £5.

b To estimate the mean, assume that each person in each class has the 'midpoint' amount, then build up the following table.

To find the midpoint value, the two end-values are added together and then divided by two.

Pocket money, p (£)	Frequency (f)	Midpoint (m)	$f \times m$
$0 < p \leq 1$	2	0.50	1.00
$1 < p \leq 2$	5	1.50	7.50
$2 < p \leq 3$	5	2.50	12.50
$3 < p \leq 4$	9	3.50	31.50
$4 < p \leq 5$	15	4.50	67.50
Totals	36		120

The estimated mean will be £120 ÷ 36 = £3.33 (rounded to the nearest penny).

Note the notation for the classes:

$0 < p \leq 1$ means any amount above 0p up to and including £1.

$1 < p \leq 2$ means any amount above £1 up to and including £2, and so on.

If you had written 0.01–1.00, 1.01–2.00 and so on for the groups, then the midpoints would have been 0.505, 1.505 and so on. This would not have had a significant effect on the final answer as it is only an estimate.

Note that you **cannot** find the **median** from a grouped table as you do not know the actual values.

You also **cannot** find the **range** but you can say what limits there are. In the table above the smallest possible value for pocket money in the first group is 1p (this is unlikely but it cannot be 0 as the range is $0 < p \leq 1$) and the largest is £1. In the last group the smallest possible value is £4.01 and the largest is £5. This means the range must be between £5 – 1p = £4.99 and £4.01 – £1 = £3.01.

EXERCISE 5G

1 For each table of values given below, find:

   i the modal group

   ii an estimate for the mean.

> **HINTS AND TIPS**
>
> When you copy the tables, draw them vertically as in Example 12.

a

x	$0 < x \leqslant 10$	$10 < x \leqslant 20$	$20 < x \leqslant 30$	$30 < x \leqslant 40$	$40 < x \leqslant 50$
Frequency	4	6	11	17	9

b

y	$0 < y \leqslant 100$	$100 < y \leqslant 200$	$200 < y \leqslant 300$	$300 < y \leqslant 400$	$400 < y \leqslant 500$	$500 < x \leqslant 600$
Frequency	95	56	32	21	9	3

c

z	$0 < z \leqslant 5$	$5 < z \leqslant 10$	$10 < z \leqslant 15$	$15 < z \leqslant 20$
Frequency	16	27	19	13

d

Weeks	1–3	4–6	7–9	10–12	13–15
Frequency	5	8	14	10	7

2 Jason brought 100 pebbles back from the beach and weighed them all, recording each weight to the nearest gram. His results are summarised in the table below.

Weight, w (g)	$40 < w \leqslant 60$	$60 < w \leqslant 80$	$80 < w \leqslant 100$
Frequency	5	9	22

Weight, w (g)	$100 < w \leqslant 120$	$120 < w \leqslant 140$	$140 < w \leqslant 160$
Frequency	27	26	11

Find:

  a the modal weight of the pebbles

  b an estimate of the total weight of all the pebbles

  c an estimate of the mean weight of the pebbles.

3 A gardener measured the heights of all his daffodils to the nearest centimetre and summarised his results as follows.

Height (cm)	10–14	15–18	19–22	23–26	27–40
Frequency	21	57	65	52	12

  a How many daffodils did the gardener have?

  b What is the modal height of the daffodils?

  c What is the estimated mean height of the daffodils?

FM A survey was created to see how quickly the AA attended calls that were not on a motorway. The following table summarises the results.

Time (min)	1–15	16–30	31–45	46–60	61–75	76–90	91–105
Frequency	2	23	48	31	27	18	11

a How many calls were used in the survey?

b Estimate the mean time taken per call.

c Which average would the AA use for the average call-out time?

d What percentage of calls do the AA get to within the hour?

5 One hundred light bulbs were tested by their manufacturer to see whether the average life-span of the manufacturer's bulbs was over 200 hours. The following table summarises the results.

Life span, h (hours)	$150 < h \leqslant 175$	$175 < h \leqslant 200$	$200 < h \leqslant 225$	$225 < h \leqslant 250$	$250 < h \leqslant 275$
Frequency	24	45	18	10	3

a What is the modal length of time a bulb lasts?

b What percentage of bulbs last longer than 200 hours?

c Estimate the mean life-span of the light bulbs.

d Do you think the test shows that the average life-span is over 200 hours? Fully explain your answer.

FM AU Three supermarkets each claimed to have the lowest average price increase over the year. The following table summarises their price increases.

Price increase (p)	1–5	6–10	11–15	16–20	21–25	26–30	31–35
Soundbuy	4	10	14	23	19	8	2
Springfields	5	11	12	19	25	9	6
Setco	3	8	15	31	21	7	3

Using their average price increases, make a comparison of the supermarkets and write a report on which supermarket, in your opinion, has the lowest price increases over the year. Do not forget to justify your answers.

FM 7
AU The table shows the distances run, over a month, by an athlete who is training for a marathon.

Distance, d (miles)	$0 < d \leqslant 5$	$5 < d \leqslant 10$	$10 < d \leqslant 15$	$15 < d \leqslant 20$	$20 < d \leqslant 25$
Frequency	3	8	13	5	2

a A marathon is 26.2 miles. It is recommended that an athlete's daily average mileage should be at least a third of the distance of the race for which they are training. Is this athlete doing enough training?

b The athlete records the times of some runs and calculates that her average pace for all runs is $6\frac{1}{2}$ minutes to a mile. Explain why she is wrong to expect a finishing time for the marathon of $26.2 \times 6\frac{1}{2}$ minutes ≈ 170 minutes.

c The runner claims that the difference in length between her shortest and longest run is 21 miles. Could this be correct? Explain your answer.

PS 8 The table shows the points scored in a general-knowledge competition by all the players.

Points	0–9	10–19	20–29	30–39	40–49
Frequency	8	5	10	5	2

Helen noticed that two numbers were the wrong way round and that this made a difference of 1.7 to the arithmetic mean.

Which two numbers were the wrong way round?

AU 9 The profit made each week by a charity shop is shown in the table below.

Profit	£0–£500	£501–£1000	£1001–£1500	£1501–£2000
Frequency	15	26	8	3

Explain how you would estimate the mean profit made each week.

AU 10 The table shows the number of members of 100 football clubs.

Members	20–29	30–39	40–49	50–59	60–69
Frequency	16	34	27	18	5

a Roger claims that the median number of members is 39.5.

Is he correct? Explain your answer.

b He also says that the range of the number of members is 34.

Could he be correct? Explain your answer.

Frequency polygons

This section will show you how to:
- draw frequency polygons for discrete and continuous data

Key words
continuous data
discrete data
frequency polygon

To help people understand it, statistical information is often presented in pictorial or diagrammatic form, which includes the pie chart, the line graph, the bar chart and the stem-and-leaf diagram. These were covered in Chapter 5. Another method of showing data is by **frequency polygons**.

Frequency polygons can be used to represent both ungrouped data and grouped data, as shown in Example 13 and Example 14 respectively and are appropriate for both **discrete data** and **continuous data**.

Frequency polygons show the shapes of distributions and can be used to compare distributions.

EXAMPLE 13

No. of children	0	1	2	3	4	5
Frequency	12	23	36	28	16	11

This is the frequency polygon for the ungrouped data in the table.

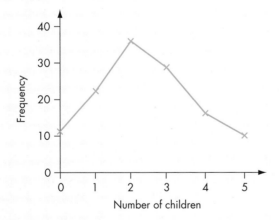

Note:
- The coordinates are plotted from the ordered pairs in the table.
- The polygon is completed by joining up the plotted points with straight lines.

EXAMPLE 14

Weight, w (kg)	$0 < w \leqslant 5$	$5 < w \leqslant 10$	$10 < w \leqslant 15$
Frequency	4	13	25

Weight, w (kg)	$15 < w \leqslant 20$	$20 < w \leqslant 25$	$25 < w \leqslant 30$
Frequency	32	17	9

This is the frequency polygon for the grouped data in the table.

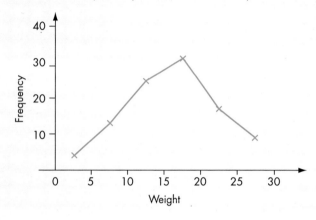

Note:

• The midpoint of each group is used, just as it was in estimating the mean.

• The ordered pairs of midpoints with frequency are plotted, namely:

   (2.5, 4), (7.5, 13), (12.5, 25), (17.5, 32), (22.5, 17), (27.5, 9)

• The polygon should be left like this. Any lines you draw before and after this have no meaning.

If you only have a frequency polygon you can work out the mean from the information on the graph.

EXAMPLE 15

The frequency polygon shows the lengths of 50 courgettes.

Work out the mean length.

The points are plotted at the midpoints so the mean is

$(7 \times 92.5 + 10 \times 97.5 + 16 \times 102.5 + 12 \times 107.5 + 5 \times 112.5) \div 50$

$= 102.3$ mm

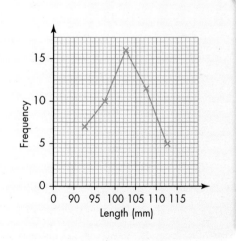

EXERCISE 5H

1. The following table shows how many students were absent from one particular class throughout the year.

Students absent	1	2	3	4	5
Frequency	48	32	12	3	1

   a Draw a frequency polygon to illustrate the data.

   b Estimate the mean number of absences each lesson.

2. The table below shows the number of goals scored by a hockey team in one season.

Goals	1	2	3	4	5
Frequency	3	9	7	5	2

   a Draw the frequency polygon for this data.

   b Estimate the mean number of goals scored per game this season.

3. After a spelling test, all the results were collated for girls and boys as below.

Number correct	1–4	5–8	9–12	13–16	17–20
Boys	3	7	21	26	15
Girls	4	8	17	23	20

   a Draw frequency polygons to illustrate the differences between the boys' scores and the girls' scores.

   b Estimate the mean score for boys and girls separately, and comment on the results.

> **HINTS AND TIPS**
>
> The highest point of the frequency polygon is the modal value.

4. A doctor was concerned at the length of time her patients had to wait to see her when they came to the morning surgery. The survey she did gave her the following results.

Time, m (min)	$0 < m \leqslant 10$	$10 < m \leqslant 20$	$20 < m \leqslant 30$
Monday	5	8	17
Tuesday	9	8	16
Wednesday	7	6	18

Time, m (min)	$30 < m \leqslant 40$	$40 < m \leqslant 50$	$50 < m \leqslant 60$
Monday	9	7	4
Tuesday	3	2	1
Wednesday	2	1	1

   a Using the same pair of axes, draw a frequency polygon for each day.

   b What is the average amount of time spent waiting each day?

   c Why might the average times for each day be different?

5 The frequency polygon shows the amounts of money spent in a corner shop by the first 40 customers one morning.

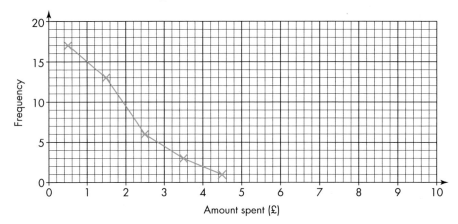

Amount spent (£)

a i Use the frequency polygon to complete the table for the amounts spent by the first 40 customers.

Amount spent, m (£)	$0 < m \leqslant 1$	$1 < m \leqslant 2$	$2 < m \leqslant 3$	$3 < m \leqslant 4$	$4 < m \leqslant 5$
Frequency					

ii Work out the mean amount of money spent by these 40 customers.

b Mid-morning another 40 customers visit the shop and the shopkeeper records the amounts they spend. The table below shows the data.

Amount spent, m (£)	$0 < m \leqslant 2$	$2 < m \leqslant 4$	$4 < m \leqslant 6$	$6 < m \leqslant 8$	$8 < m \leqslant 10$
Frequency	3	5	18	10	4

i Copy the graph above and draw the frequency polygon to show this data.

ii Calculate the mean amount spent by the 40 mid-morning customers.

c Comment on the differences between the frequency polygons and the average amounts spent by the different groups of customers.

6 The frequency polygon shows the ages of 50 staff in a school.

a Draw up a grouped table to show the data.

b Calculate an estimate of the mean age.

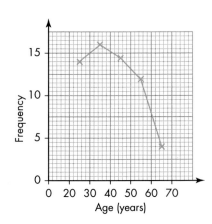

Age (years)

PS 7 The frequency polygon shows the lengths of time that students spent on homework one weekend.

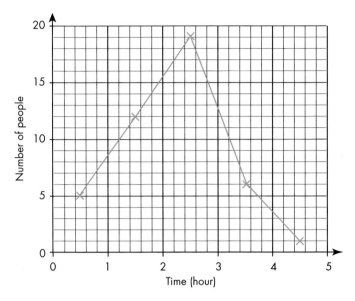

Calculate an estimate of the mean time spent on homework by the students.

AU 8 The frequency polygon shows the times that a number of people waited at a Post Office before being served one morning.

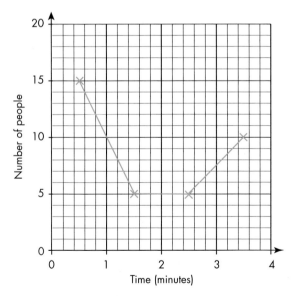

Julie said: "Most people spent 30 seconds waiting."

Explain why this might be wrong.

GRADE BOOSTER

G You can find the mode and median from a list of data

F You can find the range of a set of data and find the mean of a small set of data

E You can find the median and range from a stem-and-leaf diagram

D You can find the mean from a frequency table of discrete data and also draw a frequency polygon for such data

C You can find an estimate of the mean from a grouped table of continuous data and draw a frequency polygon for continuous data

What you should know now

- How to find the range, mode, median and mean of sets of discrete data
- Which average to use in different situations
- How to find the modal class and an estimated mean for continuous data
- How to draw frequency polygons for discrete and continuous data

1 The table shows how many children there were in each family of form 7J.

Number of children	Frequency
1	5
2	11
3	5
4	2
5	2

a How many children were in the form?

b What is the modal number of children per family?

c What is the median number of children per family?

d What is the mean number of children per family?

2 Find: **a** the mode **b** the median of:

6, 6, 6, 8, 9, 10, 11, 12, 13

3 The marks of ten people in a maths test are shown.

10 15 12 13 9 20 16 11 4 15

Work out the range of these marks.

4 The temperature, in degrees Celsius, at midday at a holiday camp on six summer days was recorded:

20 16 24 29 20 17

Work out the mean temperature at midday for these six days.

5 The number of points scored by the Tigers in the last 10 rugby matches is listed.

38 16 18 76 32
16 16 40 60 42

a Write down the mode of these scores. (1)

b Calculate the mean of these scores. (3)

c Calculate the range of these scores. (1)

(Total 5 marks)

AQA, June 2007, Module 1 Foundation, Question 2

6 Calculate the mean of these numbers.

34 27 38 27 45 17 (3)

(Total 3 marks)

AQA, June 2008, Paper 2 Foundation, Question 12

7 The table shows the ages of 40 people in a village.

Age, x (years)	Frequency
$0 < x \leqslant 20$	4
$20 < x \leqslant 40$	12
$40 < x \leqslant 60$	16
$60 < x \leqslant 80$	6
$80 < x \leqslant 100$	2

a How many people are more than 60 years old? (1)

b Write down the modal class for the ages of the people. (1)

c Draw a frequency polygon to represent this data. (2)

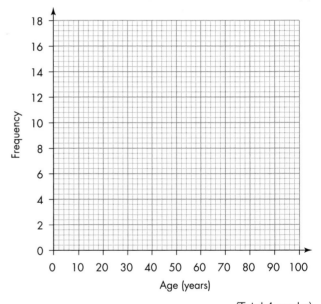

(Total 4 marks)

AQA, November 2008, Paper 1 Foundation, Question 22

8 **a** Work out:

i the mean **ii** the range of:

61, 63, 61, 86, 78, 75, 80, 68, 84 and 84.

b Fred wants to plant a conifer hedge. At the local garden centre he looks at 10 plants from two different varieties of conifer.

All the plants have been growing for six months.

The Sprucy Pine plants have a mean height of 74 cm and a range of 25 cm.

The Evergreen plants have a mean height of 52 cm and a range of 5 cm.

  i Give one reason why Fred might decide to plant a hedge of Sprucy Pine trees.

  ii Give one reason why Fred might decide to plant a hedge of Evergreen trees.

9 The stem-and leaf-diagram shows the number of packages 15 drivers delivered.

Key 3 | 5 means 35 packages

```
3 | 5 7
4 | 1 3 8 8
5 | 0 2 5 6 7 9
6 | 6 9
7 | 2
```

a What is the range of the packets delivered?

b What is the median of the packets delivered?

c What is the mode of the packets delivered?

10 A rounders coach records the number of rounders the players in her squad scored in a season.

All the players scored at least once.

She shows the data in a stem-and-leaf diagram.

Key | 2 | 7 represents 27 rounders scored

```
0 | 1 1 2 7
1 | 2 5 5
2 | 3 7
3 | 6
4 | 0
5 | 0 9
```

a What is the range of the data? (1)

b How many players are there in the squad? (1)

c What is the median number of rounders scored? (1)

d Calculate the mean number of rounders scored. (3)

(Total 6 marks)

AQA, June 2006, Paper 2 Foundation, Question 21

11 The stem-and-leaf diagram shows the marks of 15 students in a Science test.

Key | 2 | 8 represents a mark of 28

```
1 | 0
2 | 8 8 9 9
3 | 1 2 3 7 8 9
4 | 5 8 9
5 | 4
```

a What is the range of the marks? (1)

b What is the median mark? (1)

c Explain why the median is a suitable average to use. (1)

(Total 3 marks)

AQA, May 2008, Paper 1 Foundation, Question 22

12 The number of goals scored in 15 hockey matches is shown in the table.

Number of goals	Number of matches
1	2
3	1
5	5
6	3
9	4

Calculate the mean number of goals scored. (3)

(Total 3 marks)

AQA, June 2005, Paper 2 Foundation, Question 23

13 The numbers of people in 50 cars are recorded.

Number of people	Frequency
1	24
2	13
3	8
4	4
5	1

Calculate the mean number of people per car.

14 The table shows the distances travelled to work by 40 office workers.

Distance travelled, d (km)	Frequency
$0 < d \leqslant 2$	10
$2 < d \leqslant 4$	16
$4 < d \leqslant 6$	8
$6 < d \leqslant 8$	5
$8 < d \leqslant 10$	1

Calculate an estimate of the mean distance travelled to work by these office workers.

Worked Examination Questions

1 The weights, in kilograms, of a rowing boat crew are:

91, 81, 89, 91, 91, 85, 89, 38

    a Work out:

        i the modal weight   **ii** the median weight   **iii** the range of the weights   **iv** their mean weight.

AU **b** Which of the averages best describes the data? Explain your answer.

a **i** The modal weight is 91 kg.

> Look for the weight that occurs most frequently. This is worth 1 mark.

    ii The median weight is 89 kg.

> Arrange the weights in order of size: 38, 81, 85, 89, 89, 91, 91, 91. The middle two are 89. This worth 1 mark.

    iii The range is 53 kg.

> Highest value – lowest value = 91 – 38. This is worth 1 mark.

    iv The mean weight =
655 ÷ 8 = 81.875

> Add up all the weights and divide the total by the number of weights (8). You get 1 mark for method and 1 mark for accuracy.

(5 marks)

b The mean as it is the only average that takes into account all the weights.

> The correct answer is worth 1 mark.

(1 mark)   (**Total:** 6 marks)

2 The mean speed of each member of a cycling club over a long-distance race was recorded and a frequency polygon was drawn.

PS Work out an estimate of the mean speed for the whole club.

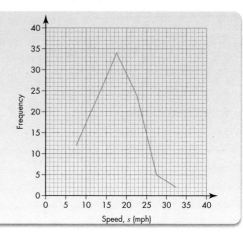

Create a grouped frequency table.

Speed, s (mph)	Frequency, f	Midpoint, m	f × m
5–10	12	7.5	90
10–15	23	12.5	287.5
15–20	34	17.5	595
20–25	24	22.5	540
25–30	5	27.5	137.5
30–35	2	32.5	65
	100		1715

> The table must include the midpoint values, the frequency and f × m.
>
> You get 1 mark for giving the frequencies and showing a total.
>
> You get 1 mark for attempting to work out f × m and showing a total.

An estimate of the mean is 1715 ÷ 100 = 17.15 mph

> You get 1 mark for dividing the total f × m by the total f and 1 mark for the correct answer.

(**Total:** 4 marks)

Averages are used to compare data and make statements about sets of data. They are used every day for a wide variety of purposes, from describing the weather to analysing the economy. In this activity averages will be applied to a sporting event.

A fishing competition

Kath's dad ran a fishing competition on the river Avon during the summer.

Kath kept an accurate record of the data collected during the competition.

This table shows a summary of Kath's records, collected from all of the anglers in the competition, for the first four weeks of July.

	Week 1	Week 2	Week 3	Week 4	Week 5
Mean number of fish caught	12.1	12.3	12.2	11.8	
Mean time spent fishing (hrs)	5.6	6.1	5.8	5.4	
Mean weight of fish caught (g)	1576	1728	1635	1437	
Mean length of longest fish caught (cm)	21.7	20.6	21.6	21.9	

In the last week of July, Kath again collected data from the anglers. She did not have time to add this data to her table.

The following tables show all the data that she collected in Week 5.

Number of fish caught	Frequency
0 – 5	6
6 – 10	11
11 – 15	8
16 – 20	5

Time spent fishing (hrs)	Frequency
0 – 4	2
4 hrs 1 min – 5	14
5 hrs 1 min – 6	6
6 hrs 1 min – 7	8

Weight of fish caught (g)	Frequency
0 – 500	1
501 – 1000	8
1001 – 1500	18
1501 – 2000	3

Longest fish caught (cm)	Frequency
0 – 10	2
11 – 15	6
16 – 20	12
21 – 25	10

Your task

Kath must do a presentation at the end of the month to summarise the fishing competition. Help her to write her presentation.

The presentation must include the following:

- The data for Week 5 inserted into the main table
- A graph to represent the data for all five weeks
- Statements to compare the five weeks of the competition
- A description of the 'average angler'.

Getting started

Start by thinking about how averages are calculated. What do the mean, median, mode and range show when applied to sets of data? How do they apply to frequency tables?

What are the mean, median, mode and range for this set of data?

3, 4, 8, 3, 2, 4, 5, 3, 4, 6, 8, 4, 2, 9, 1

Now think about the averages that you will need in your presentation and how these averages are best represented. Use your ideas in your presentation.

Why this chapter matters

Perimeter and area both require an understanding of basic measurements. It is useful to go back in history and find out about the different measurements that have been used over time. Look back to the beginning of Chapter 1 in Book 1 to remind yourself of some of the ways that ancient people recorded numbers using simple symbols.

Measures of length and area in ancient times

When humans first recognised the need to measure length they used what was available. Most commonly, they used parts of the human body, such as the width of a man's thumb, which became known as an *inch*. Other measures included the *palm*, the width of a man's hand, and the *span*, the width of a man's spread fingers. Similarly, measures of area were based upon everyday units, such as the area of a strip of ground farmed by a peasant.

Over the centuries, recognising the need for standard units, people adopted common measures. In the UK, the imperial system was used. More recently, the metric system has been accepted, almost worldwide.

Ancient Greek units of length and area

Greek measures of length were also based on the relative lengths of body parts, such as the foot (*pous*) and the finger (*dactylos*). An area called a *plethron* was the amount of land a yoke (a pair) of oxen could plough in one day.

	Greek unit	Metric equivalent
Length	*dactylos*	1.8 cm
	pous	29.6 cm
Area	*plethron*	12 140 m²

Ancient Roman units of length and area

The Romans used a system based on the Greek units, with some influence from the Egyptian, Hebrew and Mesopotamian systems. They would have used these units when they built the Colosseum in Rome in the first century, although they used their own words. For finger (*dactylos*) they used *uncia*, for foot (*pous*) they used *pes*. Other measures they used included pace (*passus*), the length of a stadium (*stadium*) and what is now known as a mile (*milliarium*).

For area, the Romans used the square foot (*pes quadratus*), the acre (*acnua*) and the *clima*, which was a quarter of an *acnua*.

Even today, Olympic stadiums are a similar size to some of the ancient Roman buildings.

	Roman unit	Metric equivalent
Length	*uncia*	2.46 cm
	pes	29.6 cm
	passum	1.48 m
	stadium	185 m
	milliarium	1.48 km
Area	*pes quadratus*	876 cm²
	clima	315 m²
	acnua	1260 m²

The Greeks would have used the units given in the table above when they built the Parthenon in Athens in the fifth century BC.

The Colosseum has a perimeter of approximately 1835 Roman *pes*. The perimeter was important to the building's designers as they wanted to created equally sized entrance arches – a perimeter of 1835 *pes* allowed them to create 80 grand entrance arches.

Geometry: Perimeter and area

This chapter will show you ...

- **G** how to find the perimeters and areas of shapes by counting squares
- **F** how to work out the perimeters and areas of rectangles
- **D** how to work out the areas of triangles, parallelograms, trapeziums and compound shapes

Visual overview

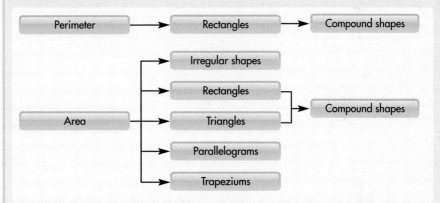

What you should already know

- The common units of length (KS3 level 3, GCSE grade G)
- What is meant by 'area'.
  The common units of area (KS3 level 3, GCSE grade G)

Quick check

This rectangle has sides of length 8 cm and 2 cm.

a What is the total length of all four sides?

b How many centimetre squares are there in the rectangle?

8 cm

2 cm

This section will show you how to:
- find the perimeter of a rectangle and compound shapes

Key words
compound shape
perimeter
rectangle

The **perimeter** of a rectangle is the sum of the lengths of all its sides.

ACTIVITY

Round about

Using centimetre-squared paper, draw this **rectangle**.

Measure its perimeter. You should get:

   3 cm + 2 cm + 3 cm + 2 cm = 10 cm

Draw a different rectangle that also has a perimeter of 10 cm.

There are only three different rectangles that each have a perimeter of 14 cm and whole numbers of centimetres for their length and width. Can you draw all three?

Can you draw a rectangle that has a perimeter of 7 cm?

Can you do it using only whole squares?

If not, why not? If you can, what is different about it?

Try drawing a rectangle that has a perimeter of 13 cm.

EXAMPLE 1

Find the perimeter of this rectangle.

7 cm

3 cm

Perimeter = 7 + 3 + 7 + 3 = 20 cm

FM Functional Maths **AU** (AO2) Assessing Understanding **PS** (AO3) Problem Solving

A **compound shape** is any 2D shape that is made up of other simple shapes such as rectangles and triangles.

EXAMPLE 2

Find the perimeter of this compound shape.

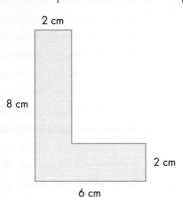

The lengths of the two missing sides are 6 cm and 4 cm.

So, the perimeter = 2 + 6 + 4 + 2 + 6 + 8 = 28 cm

EXERCISE 6A

1 Find the perimeter of each of the following shapes. Draw them first on squared paper if it helps you.

a
4 cm

1 cm

b
2 cm

2 cm

c
4 cm

3 cm

d
3 cm

3 cm

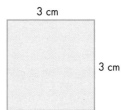

e
6 cm

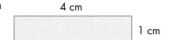

2 cm

f
2 cm

1 cm

F

2 Find the perimeter of each of the following shapes.

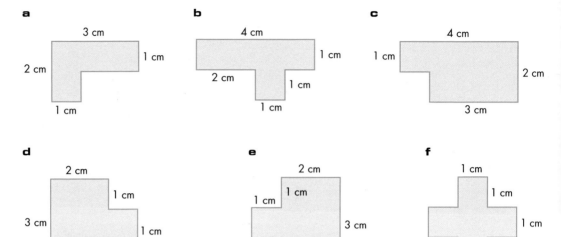

a
3 cm
1 cm
2 cm
1 cm

b
4 cm
1 cm
2 cm
1 cm
1 cm

c
4 cm
1 cm
2 cm
3 cm

d
2 cm
1 cm
3 cm
1 cm
1 cm
4 cm

e
2 cm
1 cm
1 cm
3 cm
1 cm

f
1 cm
1 cm
1 cm
1 cm

FM **3** Joe is putting new skirting board in a room.
This is the plan of the room.
The width of each door is 1 m.

What is the minimum length of skirting board
he needs to buy?

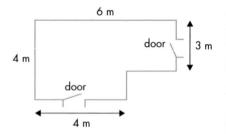

6 m
door   3 m
4 m
door
4 m

PS **4** The sides of this square are 5 cm.

Katie puts two of these squares together to make a rectangle.
She says that the perimeter of the rectangle is 40 cm.
Is she correct?

Give a reason for your answer.

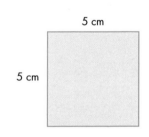

5 cm
5 cm

AU **5** Is this statement true or false?

Explain your decision …

The perimeter of this
shape is 24 cm.

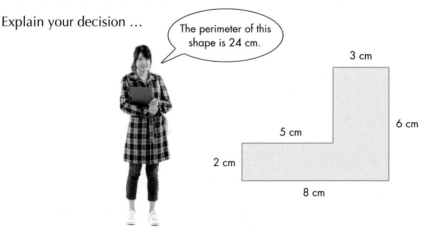

3 cm
5 cm
6 cm
2 cm
8 cm

Area of an irregular shape

This section will show you how to:

- estimate the area of an irregular 2D shape by counting squares

Key words

area

estimate

ACTIVITY

A different area

Using centimetre-squared paper, draw a rectangle 2 cm by 6 cm.

Check that it has a perimeter of 16 cm.

Count the number of squares inside the rectangle. This should come to 12.

This means that the **area** of this shape is 12 square centimetres.

Draw a different rectangle that has an area of 12 square centimetres but a perimeter that is smaller than 16 cm.

Draw another different rectangle that also has an area of 12 square centimetres, but a perimeter that is larger than 16 cm.

Using whole squares only, how many rectangles can you draw that have *different* perimeters but the *same* area of 16 square centimetres?

To find the area of an irregular shape, you can put a square grid over the shape and **estimate** the number of complete squares that are covered.

The most efficient way to do this is:

- First, count all the whole squares.

- Second, put together parts of squares to make whole and almost whole squares.

- Finally, add together the two results.

EXAMPLE 3

Below is a map of a lake. Each square represents 1 km². Estimate the area of the lake.

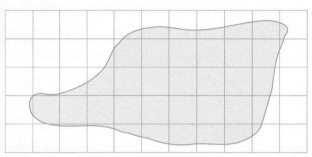

First, count all the whole squares. You should count 16.

Next, put together the parts of squares around the edge of the lake.

This should make up about 10 squares.

Finally, add together the 16 and the 10 to get an area of 26 km².

Note: This is only an *estimate*. Someone else may get a slightly different answer. However, provided the answer is close to 26, it is acceptable.

EXERCISE 6B

1 These shapes were drawn on centimetre-squared paper. By counting squares, estimate the area of each of them, giving your answers in square centimetres.

a

b

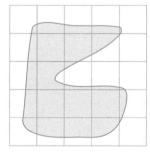

c

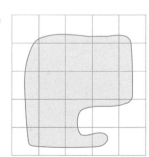

d

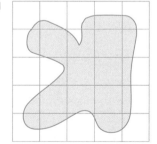

2 On a piece of 1-centimetre squared paper, draw round each of your hands to find its area. Do both hands have the same area?

3 Draw an irregular shape of your own, on centimetre-squared paper. First, guess the area of the shape. Then by counting squares, estimate the area of the shape. How close was your guess to your estimate?

AU 4 Estimate the area of this oval shape.

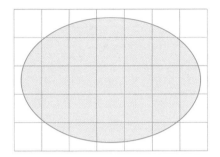

FM 5 Mr Ahmed needs to estimate the area of his lawn.

He draws a sketch of the lawn on squared paper, so that the length of each square on his sketch represents 1 m.

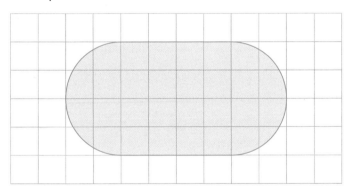

Find an estimate for the area of the lawn.

PS 6 This shape is drawn on centimetre-squared paper.

The area of the shape must be less than 24 cm². Explain how you know this.

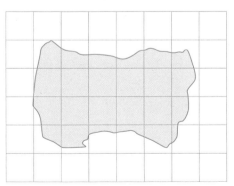

This section will show you how to:
- find the area of a rectangle
- use the formula for the area of a rectangle

Key words

area
length
width

Look at these rectangles and their areas.

Area 6 cm²

Area 9 cm²

Area 15 cm²

Notice that the area of each rectangle is given by its length multiplied by its width.

So, the formula to find the area of a rectangle is:

area = length × width

As an algebraic formula, this is written as:

$A = lw$

EXAMPLE 4

Calculate the area of this rectangle.

Area of rectangle = length × width
= 11 cm × 4 cm
= 44 cm²

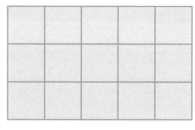

EXERCISE 6C

1 Calculate the area and the perimeter for each of the rectangles.

a 7 cm, 5 cm

b 11 cm, 3 cm

c 15 cm, 3 cm

d 10 cm, 7 cm

e 8 cm, 7 cm

f 5 cm, 2 cm

2 Calculate the area and the perimeter for each of the rectangles.

a 8.2 cm, 6.5 cm

b 11.8 cm, 7.2 cm

3 Copy and complete the table on the right for rectangles **a** to **h**.

	Length	Width	Perimeter	Area
a	7 cm	3 cm		
b	5 cm	4 cm		
c	4 cm		12 cm	
d	5 cm		16 cm	
e	6 mm			18 mm^2
f	7 mm			28 mm^2
g		2 m	14 m	
h		5 m		35 m^2

FM 4 A rectangular field is 150 m long and 45 m wide.

Fencing is needed to go all the way around the field.

The fencing is sold in 10-metre long pieces.

How many pieces will Kevin need to buy?

FM 5 A rugby pitch is 160 m long and 70 m wide.

a Before a game, the players have to run about 1500 m to help them loosen up. How many times will they need to run round the perimeter of the pitch to do this?

b The groundsman waters the pitch at the rate of 100 m^2 per minute. How long will it take him to water the whole pitch?

FM 6 How much will it cost to buy enough carpet for a rectangular room 12 m by 5 m, if the carpet costs £13.99 per square metre?

7 What is the perimeter of a square with an area of 100 cm^2?

AU 8 Jim is tiling this wall.

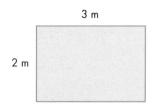

3 m

2 m

> **HINTS AND TIPS**
>
> Find how many tiles fit across the length and the height of the wall.

Each tile measures 25 cm by 25 cm. How many tiles does he need?

AU 9 Which rectangle has the largest area?

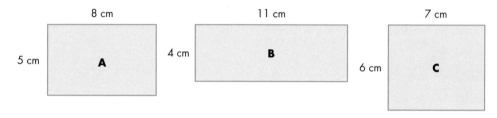

8 cm

5 cm   **A**

11 cm

4 cm   **B**

7 cm

6 cm   **C**

Explain your answer.

PS 10 Doubling the length and width of a rectangle doubles the area of the rectangle.
Is this statement:

• always true

• sometimes true

• never true?

> **HINTS AND TIPS**
>
> Draw some diagrams with different lengths and widths.

Explain your answer.

11 a The two squares on the right have the same area.
Calculate the areas of square A and square B.
Copy and complete: 1 cm^2 = …… mm^2

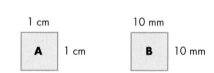

1 cm

A   1 cm

10 mm

B   10 mm

b Change the following into square millimetres.

i 3 cm^2 **ii** 5 cm^2 **iii** 6.3 cm^2

12 a The two squares on the right have the same area.
Calculate the areas of square A and square B.
Copy and complete: 1 m^2 = …… cm^2

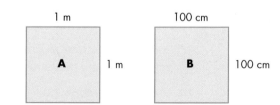

1 m

A   1 m

100 cm

B   100 cm

b Change the following into square centimetres.

i 2 m^2 **ii** 4 m^2 **iii** 5.6 m^2

Area of a compound shape

Some 2D shapes are made up of two or more rectangles or triangles.

These **compound shapes** can be split into simpler shapes, which makes it easy to calculate the **areas** of these shapes.

EXAMPLE 5

Find the area of the shape below.

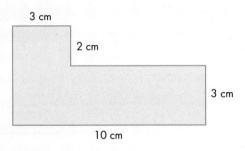

First, split the shape into two rectangles, A and B.

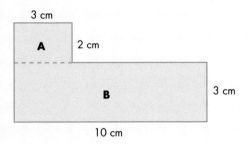

Then, calculate the area of each one.

area of A = 2 × 3 = 6 cm^2

area of B = 10 × 3 = 30 cm^2

The area of the shape is given by:

area of A + area of B = 6 + 30 = 36 cm^2

EXERCISE 6D

1 Calculate the area of each of the compound shapes below.

a
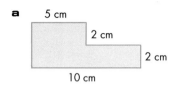

b
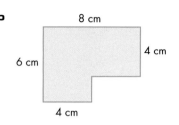

c
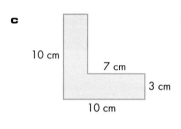

d
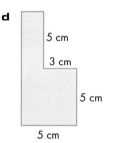

e
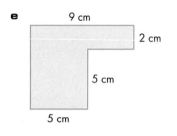

f
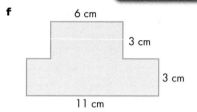

g
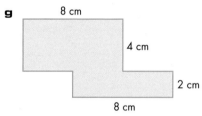

h
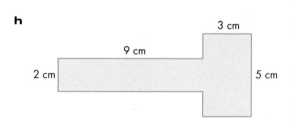

FM **2** A square lawn of side 5 m has a rectangular path, 1 m wide, running all the way round the outside of it. Carlos is laying paving stones on the path. The area of each one is 1m². How many paving stones does he need?

AU **3** Tom is working out the area of this shape.

This is what Tom wrote down.

> $8 \times 4 = 32$
> $10 \times 2 = 20$
> So the area is $32 + 20 = 52$ cm²

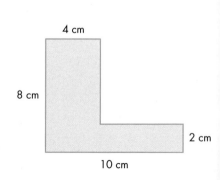

Explain why Tom is wrong.

PS **4** This compound shape is made from four rectangles that are all the same size.

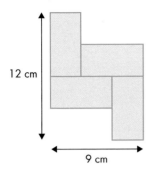

Work out the area of the compound shape.

PS **5** This shape is made from five squares that are all the same size.

It has an area of 80 cm^2.

Work out the perimeter of the shape.

FM **6** Dave is painting a wall.

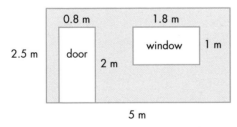

He buys a tin of paint that will cover 10 m^2.

Will he have enough paint?

> **HINTS AND TIPS**
>
> Find the total area of the door and window, then subtract this from the area of the large rectangle.

Area of a triangle

This section will show you how to:
- find the area of a triangle
- use the formula for the area of a triangle

Key words

area
base
height
perpendicular
  height
triangle

Area of a right-angled triangle

It is easy to see that the **area** of a right-angled **triangle** is half the area of the rectangle with the same **base** and **height**. Hence:

$$\text{area} = \tfrac{1}{2} \times \text{base} \times \text{height}$$

As an algebraic formula, this is written as:

$$A = \tfrac{1}{2}bh$$

Length

Width

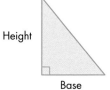

Height

Base

EXAMPLE 6

Find the area of this right-angled triangle.

4 cm

7 cm

$$\begin{aligned}\text{Area} &= \tfrac{1}{2} \times 7\text{ cm} \times 4\text{ cm} \\ &= \tfrac{1}{2} \times 28\text{ cm}^2 \\ &= 14\text{ cm}^2\end{aligned}$$

EXERCISE 6E

D

1 Write down the area and the perimeter of each triangle.

a

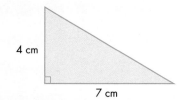

4 cm   5 cm

3 cm

b

10 cm   26 cm

24 cm

c

5 cm

13 cm   12 cm

2 Find the area of the shaded triangle RST.

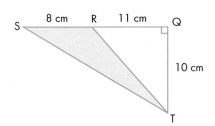

FM **3** A tree is in the middle of a garden.
Around the tree there is a square region where nothing will be planted. The dimensions of the garden are shown in the diagram.

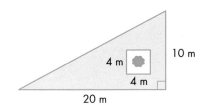

How much area can be planted?

4 Find the area of the shaded part of each triangle.

a

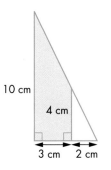

b

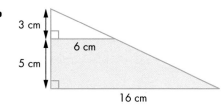

c

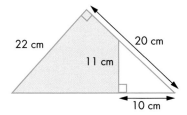

5 Which of these three triangles has the largest area?

a

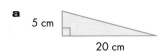

b

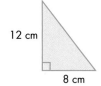

c

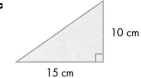

D

AU 6 This shape is made from two right-angled triangles.

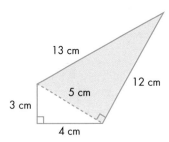

Work out the perimeter and the area of the shape.

PS 7 This compound shape is made from a rectangle and two right-angled triangles that are the same size.

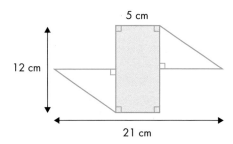

Work out the area of the shape.

FM 8 Chris is working out the area of one of the walls of his garden shed.

The diagram shows the dimensions.

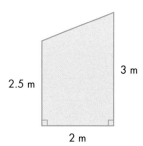

a Calculate the area of the wall.

b Chris wants to repaint the shed.

The length of the shed is 4 m and Chris knows that a tin of paint covers 10 m^2.

How many tins of paint does he need?

> **HINTS AND TIPS**
>
> Find the area of the rectangle and the right-angled triangle.

Area of any triangle

The area of any triangle is given by the formula:

area $= \frac{1}{2} \times$ base $\times$ **perpendicular height**

As an algebraic formula, this is written as:

$A = \frac{1}{2}bh$

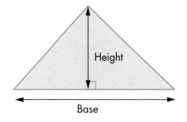

EXAMPLE 7

Calculate the area of this triangle.

Area $= \frac{1}{2} \times 9$ cm $\times 4$ cm

$= \frac{1}{2} \times 36$ cm^2

$= 18$ cm^2

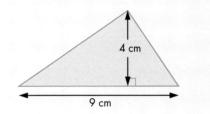

EXAMPLE 8

Calculate the area of the shape shown below.

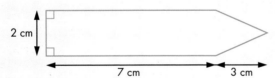

This is a compound shape that can be split into a rectangle (R) and a triangle (T).

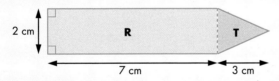

Area of the shape $=$ area of R + area of T
$= 7 \times 2 + \frac{1}{2} \times 2 \times 3$
$= 14 + 3$
$= 17$ cm^2

EXERCISE 6F

D

1 Calculate the area of each of these triangles.

a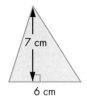
7 cm
6 cm

b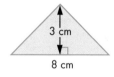
3 cm
8 cm

c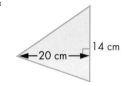
7 cm
4 cm

d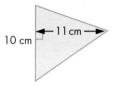
10 cm
11 cm

e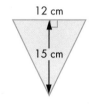
12 cm
15 cm

f
20 cm
14 cm

2 Copy and complete the following table for triangles **a** to **f**.

	Base	Perpendicular height	Area
a	8 cm	7 cm	
b		9 cm	36 cm²
c		5 cm	10 cm²
d	4 cm		6 cm²
e	6 cm		21 cm²
f	8 cm	11 cm	

PS 3 This regular hexagon has an area of 48 cm².

What is the area of the square that surrounds the hexagon?

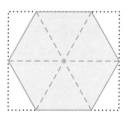

4 Find the area of each of these shapes.

a
6 cm
5 cm
10 cm

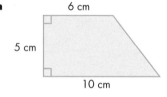

HINTS AND TIPS

Refer to Example 8 on how to find the area of a compound shape.

b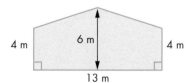
4 m
6 m
4 m
13 m

c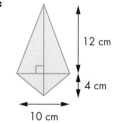
12 cm
4 cm
10 cm

5 Find the area of each shaded shape.

a

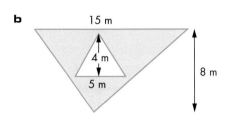

HINTS AND TIPS

Find the area of the outer shape and subtract the area of the inner shape.

b

AU **6** Write down the dimensions of two different-sized triangles that have the same area of 50 cm^2.

FM **7** Lee is making a kite. He cuts the kite from a rectangular piece of material measuring 60 cm by 40 cm.

   a What is the area of the material that is left?

   b Work out the area of the material he will need, to cover both sides of the kite.

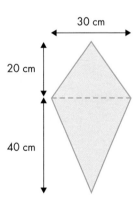

AU **8** Which triangle is the odd one out?
Give a reason for your answer.

a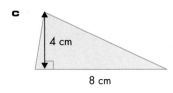

b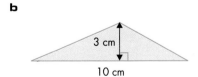

c

Area of a parallelogram

This section will show you how to:
- find the area of a parallelogram
- use the formula for the area of a parallelogram

Key words
area
base
height
parallelogram

A **parallelogram** can be changed into a rectangle by moving a triangle.

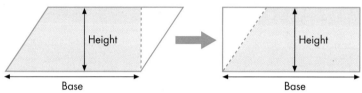

This shows that the **area** of the parallelogram is the area of a rectangle with the same **base** and **height**. The formula is:

area of a parallelogram = base × height

As an algebraic formula, this is written as:

$A = bh$

EXAMPLE 9

Find the area of this parallelogram.

Area = 8 cm × 6 cm
     = 48 cm²

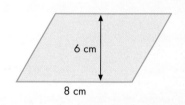

EXERCISE 6G

D

1 Calculate the area of each parallelogram below.

a

8 cm

12 cm

b

10 cm

7 cm

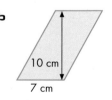

c

5 m

4 m

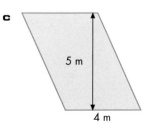

d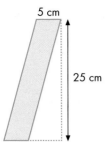

5 cm

25 cm

e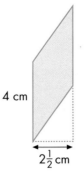

4 cm

$2\frac{1}{2}$ cm

f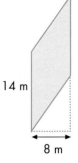

14 m

8 m

D

AU **2** Sandeep says that the area of this parallelogram is 30 cm².

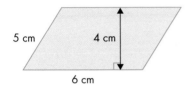

5 cm   4 cm

6 cm

Is she correct? Give a reason for your answer.

PS **3** This shape is made from four parallelograms that are all the same size.
The area of the shape is 120 cm².

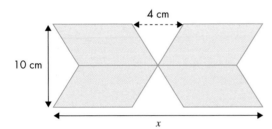

4 cm

10 cm

x

Work out the length marked x on the diagram.

FM **4** This logo, made from two identical parallelograms, is cut from a sheet of card.

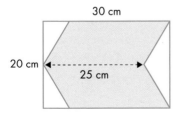

30 cm

20 cm

25 cm

a Calculate the area of the logo.

b How many logos can be cut from a sheet of card that measures 1 m by 1 m?

Area of a trapezium

This section will show you how to:
- find the area of a trapezium
- use the formula for the area of a trapezium

Key words

area
height
trapezium

The **area** of a **trapezium** is calculated by finding the average of the lengths of its parallel sides and multiplying this by the perpendicular **height** between them.

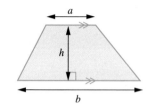

The area of a trapezium is given by this formula:

$$A = \tfrac{1}{2}(a + b)h$$

EXAMPLE 10

Find the area of the trapezium ABCD.

$$\text{Area} = \tfrac{1}{2}(4 + 7) \times 3$$

$$= \tfrac{1}{2} \times 11 \times 3$$

$$= 16.5 \text{ cm}^2$$

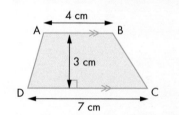

EXERCISE 6H

1 Copy and complete the following table for each trapezium.

	Parallel side 1	Parallel side 2	Perpendicular height	Area
a	8 cm	4 cm	5 cm	
b	10 cm	12 cm	7 cm	
c	7 cm	5 cm	4 cm	
d	5 cm	9 cm	6 cm	
e	3 cm	13 cm	5 cm	
f	4 cm	10 cm		42 cm^2
g	7 cm	8 cm		22.5 cm^2

2 Calculate the perimeter and the area of each trapezium.

a

6.5 cm
7 cm
5 cm
6 cm
8 cm

b

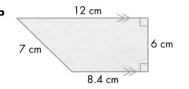

12 cm
7 cm
6 cm
8.4 cm

c

7.6 m
10 m
9 m
12 m

AU 3 A trapezium has an area of 25 cm². Its vertical height is 5 cm. Work out a possible pair of lengths for the two parallel sides.

4 Which of the following shapes has the largest area?

a

6 cm
4 cm

b

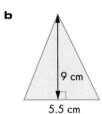

9 cm
5.5 cm

c

7 cm
3 cm
10 cm

5 Which of the following shapes has the smallest area?

a

7 cm
8 cm

b

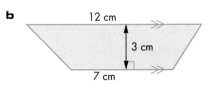

12 cm
3 cm
7 cm

c

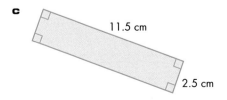

11.5 cm
2.5 cm

D

AU **6** Which of the following is the area of this trapezium?

    a 45 cm^2 **b** 65 cm^2 **c** 70 cm^2

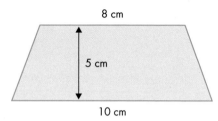

You must show your workings.

PS **7** Work out the value of a so that the square and the trapezium have the same area.

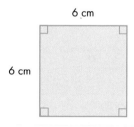 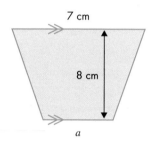

8 The side of a ramp is a trapezium, as shown in the diagram.
Calculate its area, giving your answer in square metres.

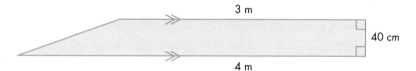

HINTS AND TIPS

Change the height into metres first.

GRADE BOOSTER

G You can find the perimeter of a 2D shape

G You can find the area of a 2D shape by counting squares

F You can find the area of a rectangle using the formula $A = lw$

D You can find the area of a triangle using the formula $A = \frac{1}{2}bh$

D You can find the area of a parallelogram using the formula $A = bh$

D You can find the area of a trapezium using the formula $A = \frac{1}{2}(a + b)h$

D You can find the area of a compound shape

What you should know now

- How to find the perimeter and area of 2D shapes by counting squares
- How to find the area of a rectangle
- How to find the area of a triangle
- How to find the area of a parallelogram and a trapezium
- How to find the area of a compound shape

1 The grids for this question are made of squares of side 1 cm.

a Find the area of this shape. (1)

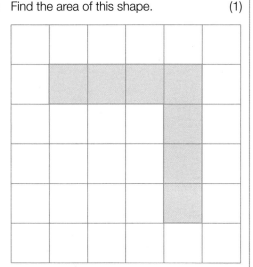

b On an centimetre-squared grid, draw a rectangle with area 8 cm². (2)

c Estimate the area of this shape. (2)

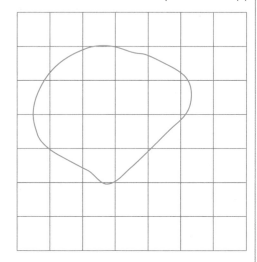

(Total 5 marks)

AQA, June 2008, Module 5, Paper 2 Foundation, Question 2

2 The diagram shows the measurements of a rectangle.

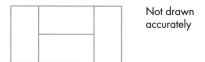

2 cm   Not drawn accurately

4 cm

Four of the rectangles are arranged to form a larger rectangle.

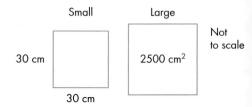

Not drawn accurately

a Work out the perimeter of the larger rectangle. (2)

b Work out the area of the larger rectangle. (2)

(Total 4 marks)

AQA, November 2008, Paper 1 Foundation, Question 11

3 A shop sells square carpet tiles in two different sizes.

Small   Large

Not to scale

30 cm   2500 cm²

30 cm

a What is the area of a small carpet tile? (2)

b What is the length of a side of a large carpet tile? (1)

c The floor of a rectangular room is 300 cm long and 180 cm wide.

How many **small** tiles are needed to carpet the floor? (3)

(Total 6 marks)

AQA, June 2005, Paper 1 Foundation, Question 15

C   D   E   F

4 Find the area of each of these shapes.

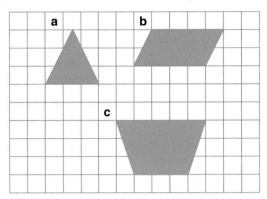

(Total 6 marks)

5 Find the area of each shape.

a (2)

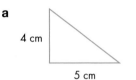

4 cm

5 cm

b (2)

7 cm

9 cm

(Total 4 marks)

6 Calculate the area of this shape.

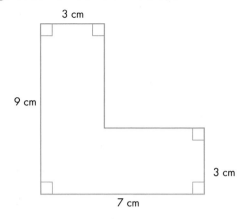

3 cm

9 cm

3 cm

7 cm

(3 marks)

AQA, Question 22, Paper 1 Foundation, June 2003

7 The diagram shows a trapezium ABCD.
AB = 7 cm, AD = 7 cm, DC = 10 cm

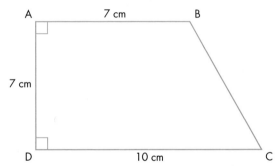

A   7 cm   B

7 cm

D   10 cm   C

Find the area of the trapezium ABCD.
Remember to state the units of your answer.

(2 marks)

AQA, Question 14b, Paper 2 Intermediate, November 2000

Worked Examination Questions

FM **1** Sam is using these square tiles to tile a wall.

25 cm

25 cm ☐

4 m

2.5 m

The tiles come in boxes of 30.

How many boxes does Sam need?

4 m = 400 cm and 2.5 m = 250 cm — First change the length and height of the wall into centimetres.

400 ÷ 25 = 16 —

250 ÷ 25 = 10 — This shows how many tiles fit across the wall and gains 1 method mark.

16 × 10 = 160 —

160 ÷ 30 = 5.33 — This shows how many tiles fit up the wall and gains 1 method mark.

Number of full boxes = 6 —

Total: 4 marks — This is the number of tiles needed and is worth 1 mark.

This is an important part of the real-life problem. We need to find the number of **full** boxes needed.

The correct answer is worth 1 mark.

2 Work out the area of this trapezium.
State the units of your answer.

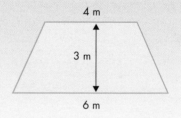

4 m

3 m

6 m

The formula is $A = \frac{1}{2}(a + b)h$ — This is on the formula sheet at the front of the examination paper.

$A = \frac{1}{2}(4 + 6) \times 3$ —

$A = \frac{1}{2} \times 10 \times 3$ — Substitute the numbers into the formula. This is worth 1 mark.

$A = 15$ —

$A = 15\ m^2$ — Make sure you work out the brackets first.

Total: 3 marks — The correct answer is worth 1 mark.

1 mark for accuracy is given for the unit m^2, even if you did not get 15.

Worked Examination Questions

PS **3** A shape is made up of two squares and an equilateral triangle as shown.

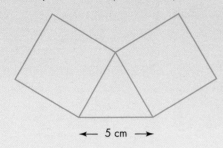

Rebecca says,

The **area** of the shape is greater than 50 cm² and less than 75 cm².

← 5 cm →

Is Rebecca correct?

Area of one square = 5 × 5 = 25 cm²

The area of two squares = 50 cm²

So, the area of the shape must be greater than 50 cm². The area of the triangle must be less than the area of a square.

The area of the shape must be less than the area of three squares, which is 75 cm².

So, Rebecca is correct.

The side of each square is 5 cm, since the triangle is equilateral.

Remember that there are two squares. Doubling your previous answer gains you 1 mark.

You do not need to find the area of the triangle.

You gain 1 mark for showing how you found 75 cm².

You gain 1 mark for making this statement and showing how you got the values 50 and 75.

Total: 3 marks

Steve is building an extension in his house to accommodate an extra family room. He will want to lay a carpet in the new room.
The carpet for the sitting room does not have to cover the whole room and can only be bought in whole metre quantities. Your task is to look at the information below and advice Steve about what size the carpet he should buy.

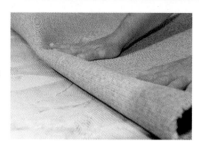

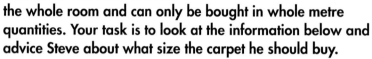

Your task:

Using all the information on these pages, advise Steve on the size of carpet he should buy.

Begin by thinking about the following points:

- What shape Steve's new room will be – think of at least **two** potential shapes for Steve's new room and plan the carpet size for each
- The dimensions of each room shape
- The area of each room that could be covered by the carpet

Remember: Keep it simple to start with and use realistic measurements.

Discuss your idea with a partner, making sure that you have taken into account all the criteria that Steve has give you. Write down the assumptions that you have made and then find the total carpet size that each of your potential rooms will require.

Steve's criteria

Steve has given you the following criteria:

- The carpet must cover at least half of the floor area but not more than three-quarters of the floor area.
- Due to roll sizes, the carpet can only be purchased in whole-metre widths and lengths, for example 2 m by 3 m.
- Steve can only tell you that his room has a total floor area of 30m².

Extension

Steve will need to add skirting board to his room once the carpet is laid. How much skirting board will he need for each of your room designs?

We use proportion and speed in our everyday lives to help us to deal with facts, or to compare two or more pieces of information.

Proportions are often used to compare sizes; speed is used to compare distances with the time taken to travel them.

A 100-m sprinter

Speed

When is a speed fast?

On 16 August 2009 Usain Bolt set a new world record for the 100-m sprint of 9.58 seconds. This is an average speed of 23.3 mph.

The sailfish is the fastest fish and can swim at 68 mph.

Sailfish

The cheetah is the fastest land animal and can travel at 75 mph.

The fastest bird is the swift which can travel at 106 mph.

Swift Cheetah

Proportion facts

Russia is the largest country. Vatican City is the smallest country. The area of Russia is nearly 39 million times the area of Vatican City.

Monaco has the most people per square mile. Mongolia has the least people per square mile. The number of people per square mile in Monaco to the number of people in Mongolia is in the ratio 10 800 : 1.

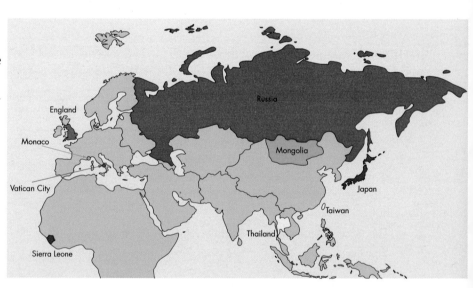

Japan has the highest life expectancy. Sierra Leone has the lowest life expectancy. On average people in Japan live over twice as long as people in Sierra Leone.

Taiwan has the most mobile phones per 100 people (106.5). This is approximately four times that of Thailand (26.04).

About one-seventh of England is green-belt land.

This chapter is about comparing pieces of information. You can compare the speeds of Usain Bolt, the sailfish, the cheetah and the swift by answering questions such as: How much faster is a sailfish than Usain Bolt? Who is fastest on land?

Now, consider what questions you would ask to compare countries, using the information given on the left.

Number: Speed and proportion

1 Speed, time and distance

2 Direct proportion problems

3 Best buys

This chapter will show you ...

- **D** how to solve problems involving direct proportion
- **D** how to compare prices of products
- **D** how to calculate speed

Visual overview

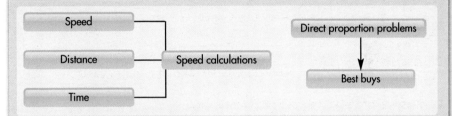

What you should already know

- Multiplication tables up to 10×10 (**KS3 level 4, GCSE grade G**)
- How to simplify fractions (**KS3 level 5, GCSE grade G**)
- How to find a fraction of a quantity (**KS3 level 5, GCSE grade F**)
- How to multiply and divide, with and without a calculator (**KS3 level 4, GCSE grade G**)

Quick check

1 Cancel the following fractions.

   a $\dfrac{6}{10}$ **b** $\dfrac{4}{20}$ **c** $\dfrac{4}{12}$ **d** $\dfrac{32}{50}$ **e** $\dfrac{36}{90}$ **f** $\dfrac{18}{24}$ **g** $\dfrac{16}{48}$

2 Find the following quantities.

   a $\dfrac{2}{5}$ of £30   **b** $\dfrac{3}{4}$ of £88   **c** $\dfrac{7}{10}$ of 250 litres   **d** $\dfrac{5}{8}$ of 24 kg

   e $\dfrac{2}{3}$ of 60 m   **f** $\dfrac{5}{6}$ of £42   **g** $\dfrac{9}{20}$ of 300 g   **h** $\dfrac{3}{10}$ of 3.5 litres

This section will show you how to:
- recognise the relationship between speed, distance and time
- calculate average speed from distance and time
- calculate distance travelled from the speed and the time taken
- calculate the time taken on a journey from the speed and the distance

Key words
average
distance
speed
time

The relationship between **speed**, **time** and **distance** can be expressed in three ways:

$$\text{speed} = \frac{\text{distance}}{\text{time}} \qquad \text{distance} = \text{speed} \times \text{time} \qquad \text{time} = \frac{\text{distance}}{\text{speed}}$$

In problems relating to speed, you usually mean **average** speed, as it would be unusual to maintain one exact speed for the whole of a journey.

This diagram will help you remember the relationships between distance (D), time (T) and speed (S).

$$D = S \times T \qquad S = \frac{D}{T} \qquad T = \frac{D}{S}$$

EXAMPLE 1

Paula drove a distance of 270 miles in 5 hours. What was her average speed?

$$\text{Paula's average speed} = \frac{\text{distance she drove}}{\text{time she took}} = \frac{270}{5} = 54 \text{ miles per hour (mph)}$$

EXAMPLE 2

Sarah drove from Sheffield to Peebles in $3\frac{1}{2}$ hours at an average speed of 60 mph. How far is it from Sheffield to Peebles?

Since:

distance = speed × time

the distance from Sheffield to Peebles is given by:

60 × 3.5 = 210 miles

Note: You need to change the time to a decimal number and use 3.5 (*not* 3.30).

FM Functional Maths   **AU** (AO2) Assessing Understanding   **PS** (AO3) Problem Solving

EXAMPLE 3

Sean is going to drive from Newcastle upon Tyne to Nottingham, a distance of 190 miles. He estimates that he will drive at an average speed of 50 mph. How long will it take him?

$$\text{Sean's time} = \frac{\text{distance he covers}}{\text{his average speed}} = \frac{190}{50} = 3.8 \text{ hours}$$

Change the 0.8 hour to minutes by multiplying by 60, to give 48 minutes.

So, the time for Sean's journey will be 3 hours 48 minutes.

Remember: When you calculate a time and get a decimal answer, as in Example 3, *do not mistake* the decimal part for minutes. You must either:

- leave the time as a decimal number and give the unit as hours, or

- change the decimal part to minutes by multiplying it by 60 (1 hour = 60 minutes) and give the answer in hours and minutes.

EXERCISE 7A

1 A cyclist travels a distance of 90 miles in 5 hours. What was her average speed?

2 How far along a motorway would you travel if you drove at 70 mph for 4 hours?

3 I drive to Bude in Cornwall from Sheffield in about 6 hours. The distance from Sheffield to Bude is 315 miles. What is my average speed?

4 The distance from Leeds to London is 210 miles. The train travels at an average speed of 90 mph. If I catch the 9.30 am train in London, at what time should I expect to arrive in Leeds?

5 How long will an athlete take to run 2000 m at an average speed of 4 metres per second?

HINTS AND TIPS

Remember to convert time to a decimal if you are using a calculator, for example, 8 hours 30 minutes is 8.5 hours.

HINTS AND TIPS

km/h means kilometres per hour.
m/s means metres per second.

6 Copy and complete the following table.

	Distance travelled	Time taken	Average speed
a	150 miles	2 hours	
b	260 miles		40 mph
c		5 hours	35 mph
d		3 hours	80 km/h
e	544 km	8 hours 30 minutes	
f		3 hours 15 minutes	100 km/h
g	215 km		50 km/h

7 Eliot drove from Sheffield to Inverness, a distance of 410 miles, in 7 hours 45 minutes.

 a Change the time 7 hours 45 minutes to a decimal.

 b What was the average speed of the journey? Round your answer to 1 decimal place.

8 Colin drives home from his son's house in 2 hours 15 minutes. He says that he drives at an average speed of 44 mph.

 a Change the 2 hours 15 minutes to a decimal.

 b How far is it from Colin's home to his son's house?

9 The distance between Paris and Le Mans is 200 km. The express train between Paris and Le Mans travels at an average speed of 160 km/h.

 a Calculate the time taken for the journey from Paris to Le Mans, giving your answer as a decimal number of hours.

 b Change your answer to part **a** to hours and minutes.

FM 10 The distance between Sheffield and Land's End is 420 miles.

 a What is the average speed of a journey from Sheffield to Land's End that takes 8 hours 45 minutes?

 b If Sam covered the distance at an average speed of 63 mph, how long would it take him?

FM 11 A train travels at 50 km/h for 2 hours, then slows down to do the last 30 minutes of its journey at 40 km/h.

 a What is the total distance of this journey?

 b What is the average speed of the train over the whole journey?

FM 12 Jade runs and walks the 3 miles from home to work each day. She runs the first 2 miles at a speed of 8 mph, then walks the next mile at a steady 4 mph.

    a How long does it take Jade to get to work?

    b What is her average speed?

13 Change the following speeds to metres per second.

    a 36 km/h   **b** 12 km/h   **c** 60 km/h

    d 150 km/h   **e** 75 km/h

14 Change the following speeds to kilometres per hour.

    a 25 m/s   **b** 12 m/s   **c** 4 m/s

    d 30 m/s   **e** 0.5 m/s

PS 15 A train travels at an average speed of 18 m/s.

    a Express its average speed in km/h.

    b Find the approximate time the train would take to travel 500 m.

    c The train set off at 7.30 on a 40 km journey. At approximately what time will it reach its destination?

16 A cyclist is travelling at an average speed of 24 km/h.

    a What is this speed in metres per second?

    b What distance does he travel in 2 hours 45 minutes?

    c How long does it take him to travel 2 km?

    d How far does he travel in 20 seconds?

AU 17 How much longer does it take to travel 100 miles at 65 mph than at 70 mph?

HINTS AND TIPS

Remember that there are 3600 seconds in an hour and 1000 metres in a kilometre. So to change from km/h to m/s multiply by 1000 and divide by 3600.

HINTS AND TIPS

To change from m/s to km/h multiply by 3600 and divide by 1000.

HINTS AND TIPS

To convert a decimal fraction of an hour to minutes, just multiply by 60.

Direct proportion problems

This section will show you how to:
- recognise and solve problems using direct proportion

Key words
direct proportion
unit cost
unitary method

Suppose you buy 12 items which each cost the *same*. The total amount you spend is 12 times the cost of one item.

That is, the total cost is said to be in **direct proportion** to the number of items bought. The cost of a single item (the **unit cost**) is the constant factor that links the two quantities.

Direct proportion is not only concerned with costs. Any two related quantities can be in direct proportion to each other.

The best way to solve all problems involving direct proportion is to start by finding the single unit value. This method is called the **unitary method**, because it involves referring to a single unit value. Work through Examples 4 and 5 to see how it is done.

Remember: Before solving a direct proportion problem, think about it carefully to make sure that you know how to find the required single unit value.

EXAMPLE 4

If eight pens cost £2.64, what is the cost of five pens?

First, find the cost of one pen. This is £2.64 ÷ 8 = £0.33

So, the cost of five pens is £0.33 × 5 = £1.65

EXAMPLE 5

Eight loaves of bread will make packed lunches for 18 people. How many packed lunches can be made from 20 loaves?

First, find how many lunches one loaf will make.

One loaf will make 18 ÷ 8 = 2.25 lunches.

So, 20 loaves will make 2.25 × 20 = 45 lunches.

EXERCISE 7B

1 If 30 matches weigh 45 g, what would 40 matches weigh?

2 Five bars of chocolate cost £2.90. Find the cost of nine bars.

3 Eight men can chop down 18 trees in a day. How many trees can 20 men chop down in a day?

4 Find the cost of 48 eggs when 15 eggs can be bought for £2.10.

5 Seventy maths textbooks cost £875.

    a How much will 25 maths textbooks cost?

    b How many maths textbooks can you buy for £100?

> **HINTS AND TIPS**
>
> **Remember** to work out the value of one unit each time. Always check that answers are sensible.

FM 6 A lorry uses 80 litres of diesel fuel on a trip of 280 miles.

    a How much diesel would the same lorry use on a trip of 196 miles?

    b How far would the lorry get on a full tank of 100 litres of diesel?

FM 7 During the winter, I find that 200 kg of coal keeps my open fire burning for 12 weeks.

    a If I want an open fire all through the winter (18 weeks), how much coal will I need to buy?

    b Last year I bought 150 kg of coal. For how many weeks did I have an open fire?

8 It takes a photocopier 16 seconds to produce 12 copies. How long will it take to produce 30 copies?

9 A recipe for 12 biscuits uses:

    200 g margarine   400 g sugar

    500 g flour   300 g ground rice

    a What quantities are needed for:

        i 6 biscuits   **ii** 9 biscuits   **iii** 15 biscuits?

PS **b** What is the maximum number of biscuits I could make if I had just 1 kg of each ingredient?

AU 10 Peter the butcher sells sausages in pack of 6 for £2.30.
Paul the butcher sells sausages in packs of 10 for £3.50.

I have £10 to spend on sausages.

If I want to buy as many sausages as possible from one shop, which shop should I use? Show your working.

Best buys

This section will show you how to:
- find the cost per unit weight
- find the weight per unit cost
- use the above to find which product is the cheaper

Key words

best buy

better value

value for money

When you wander around a supermarket and see all the different prices for the many different-sized packets, it is rarely obvious which are the '**best buys**'. However, with a calculator you can easily compare **value for money** by finding either:

the cost per unit weight   **or**   the weight per unit cost

To find:

- *cost per unit weight*, divide *cost by weight*

- *weight per unit cost*, divide *weight by cost.*

The next two examples show you how to do this.

EXAMPLE 6

A 300 g tin of cocoa costs £1.20. Find the cost per unit weight and the weight per unit cost.

First change £1.20 to 120p. Then divide, using a calculator, to get:

Cost per unit weight $120 \div 300 = 0.4$p per gram

Weight per unit cost $300 \div 120 = 2.5$ g per penny

EXAMPLE 7

A supermarket sells two different-sized packets of Whito soap powder. The medium size contains 800 g and costs £1.60 and the large size contains 2.5 kg and costs £4.75. Which is the better buy?

Find the weight per unit cost for both packets.

Medium: $800 \div 160 = 5$ g per penny

Large: $2500 \div 475 = 5.26$ g per penny

From these it is clear that there is more weight per penny with the large size, which means that the large size is the better buy.

Sometimes it is easier to use a scaling method to compare prices and find **better value**.

EXAMPLE 8

Which of these boxes of fish fingers is better value?

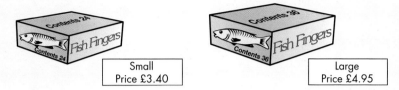

Small
Price £3.40

Large
Price £4.95

12 is a common factor of 24 and 36 so work out the cost of 12 fish fingers.

For the small box, 12 fish fingers cost £3.40 ÷ 2 = £1.70
For the large box, 12 fish fingers cost £4.95 ÷ 3 = £1.65

So the large box is better value.

EXAMPLE 9

Which of these packs of yoghurt is better value?

Price £1.45

Price £1.20

30 is the least common multiple of 5 and 6 so work out the cost of 30 yoghurts.

For the six-pack the cost of 30 yoghurts is £1.45 × 5 = £7.25
For the five-pack the cost of 30 yoghurts is £1.20 × 6 = £7.20

So the five-pack is better value.

EXERCISE 7C

1 Compare the prices of the following pairs of products and state which, if any, is the better buy.

  a Chocolate bars: £2.50 for a 5-pack, £4.50 for a 10-pack

  b Eggs: £1.08 for 6, £2.25 for 12

  c Car shampoo: £4.99 for 2 litres, £2.45 for 1 litre

  d Dishwasher tablets: £7.80 for 24, £3.90 for 12

  e Carrots: 29p for 250 grams, 95p for 750 grams

  f Bread rolls: £1.39 for a pack of 6, £5.60 for a pack of 24

  g Juice: £1.49 for 1 carton, £4 for 3 cartons

 2 Compare the following pairs of products and state which is the better buy. Explain why.

  a Coffee: a medium jar which contains 140 g for £1.10 or a large jar which contains 300 g for £2.18

  b Beans: a 125 g tin at 16p or a 600 g tin at 59p

  c Flour: a 3 kg bag at 75p or a 5 kg bag at £1.20

  d Toothpaste: a large tube containing 110 ml for £1.79 or a medium tube containing 75 ml for £1.15

  e Frosted Flakes: a large box which contains 750 g for £1.64 or a medium box which contains 500 g for £1.10

  f Rice Crisp: a medium box which contains 440 g for £1.64 or a large box which contains 600 g for £2.13

  g Hair shampoo: a bottle containing 400 ml for £1.15 or a bottle containing 550 ml for £1.60

FM 3 Julie wants to respray her car with yellow paint. In the local automart, she sees the following tins:

Small tin	350 ml at a cost of £1.79
Medium tin	500 ml at a cost of £2.40
Large tin	1.5 litres at a cost of £6.70

  a What is the cost per litre of paint in the small tin?

  b Which tin is offered at the lowest price per litre?

FM 4 Tisco's sells bottled water in three sizes.

Handy size 40 cl   Family size 2 litres   Giant size 5 litres
£0.38   £0.98   £2.50

a Work out the cost per litre of the 'handy' size.

b Which bottle is the best value for money?

PS 5 Two drivers are comparing the petrol consumption of their cars.

Ahmed says, 'I get 320 miles on a tank of 45 litres.'
Bashir says, 'I get 230 miles on a tank of 32 litres.'

Whose car is the more economical?

PS 6 Mary and Jane are arguing about which of them is better at mathematics.

Mary scored 49 out of 80 on a test.
Jane scored 60 out of 100 on a test of the same standard.

Who is better at mathematics?

PS 7
AU Paula and Kelly are comparing their running times.

Paula completed a 10-mile run in 65 minutes.
Kelly completed a 10-kilometre run in 40 minutes.

Given that 8 kilometres are equal to 5 miles, which girl has the greater average speed?

GRADE BOOSTER

D You can calculate average speeds from data

D You can calculate distance from speed and time

D You can calculate time from speed and distance

D You can compare prices of products to find the 'best buy'

D You can solve direct proportion problems

What you should know now

- The relationships between speed, time and distance
- How to solve problems involving direct proportion
- How to compare the prices of products

1 Jim's class and Rosie's class go on a trip to the zoo.

Each pupil pays the same amount.

There are 30 pupils from my class on the trip. The total cost for my class is £90

There are 25 pupils from my class on the trip.

Jim

Rosie

What is the total cost for Rosie's class?

(Total 3 marks)

AQA, June 2005, Paper 1, Question 17

2 The ingredients needed to make 500 millilitres (ml) of a fruit drink are

orange juice 300 ml
mango juice 60 ml
lemonade 140 ml

a What percentage of the fruit drink is orange juice? (2)

b Robert wants to make 750 ml of the fruit drink.

How much lemonade will he need? (2)

(Total 4 marks)

AQA, November 2006, Module 3 Foundation, Question 18

3 The same type of crystal glasses is sold in two different packs.

Small pack	Large pack
Contents **4 glasses**	Contents **12 glasses**
£3.20	**£10.20**

Which size is the better value for money? You **must** show your working.

(Total 2 marks)

AQA, November 2005, Module 3 Foundation, Question 17

4 a Sue took a holiday in Scotland.
She arrived on 26 April 2007.
She departed on 9 May 2007.
How many days did the holiday last?
Include the arrival and departure days. (2)

b Sue travelled to Scotland by car.
Her average speed was 36 miles per hour.
Her journey time was 4 hours 15 minutes.
How many miles did she travel? (3)

(Total 5 marks)

AQA, March 2008, Module 3 Foundation, Question 5

5 Here is part of a railway timetable.

	Departure Times			
Newcastle	0840	0935	1040	1122
York	0943	1034	1144	1225
Leeds	1010	–	1210	–
Derby	1124	1157	1324	1355
Birmingham	1215	1315	1415	1515

a A train leaves Newcastle at 1040.

How long is the journey to Birmingham for this train?

Give your answer in hours and minutes. (3)

b The 1225 train from York takes 1 hour 30 minutes to reach Derby.

The distance from York to Derby is 96 miles. (3)

Calculate the average speed of the train in miles per hour.

(Total 6 marks)

AQA, June 2007, Paper 2 Foundation, Question 18

6 Two advertisements for the same type of sun oil are shown.

The sun oil is usually sold in 100 ml bottles which cost £4 each.

Which offer gives the better value for money?

You **must** show all your working.

(Total 5 marks)

AQA, March 2005, Module 3 Foundation, Question 9

7 Gudrun boards a train at 10.15 in the morning and arrives at her destination at 12.45 the same day.

a How long did the journey take? (2)

She had travelled 225 miles.

b What was the average speed of the train? (2)

c She picked up her car from her father's home, 15 miles from the station, and then drove herself home, at total distance of 240 miles, at an average speed of 60 miles per hour. How long did the journey home take her? (2)

(Total 6 marks)

8 Tom saw a leaflet that advertised forest walks and cycle rides.

He went for a cycle ride that was 24 miles long. The leaflet said it should take 2 hours.

a Calculate the speed he needed to maintain, to do the ride in the stated time. (2)

b Tom cycled at 8 mph for the first quarter of the journey. How fast did he need to travel, to finish in the stated time? (2)

(Total 4 marks)

Worked Examination Questions

PS **1** Jonathan is comparing two ways to travel from his flat in London to his parents' house.

Tube, train and taxi

It takes 35 minutes to get to the train station by tube in London.

A train journey from London to Doncaster takes 1 hour 40 minutes.

From Doncaster it is 15 miles by taxi at an average speed of 20 mph.

Car

The car journey is 160 miles at an average speed of 50 mph.

Which is the slower journey: tube, train and taxi or car?

Time = Distance ÷ Speed = $\frac{15}{20}$

= 0.75 hour (or 45 minutes)

> Work out the time taken by taxi.
> You get 1 method mark for using the correct formula.

> You get 1 mark for accuracy for arriving at the correct time taken.

Total time = 35 minutes + 1 hour 40 minutes + 45 minutes

= 3 hours

> Work out the **total time** for tube, train and taxi.

Time = Distance ÷ Speed = 160 ÷ 50

= 3.2 hours (or 3 hours 12 minutes)

> This is required to compare with the car.
> This is worth 1 mark.

Car is 12 minutes slower.

Total: 6 marks

> Work out the time taken by car. You get 1 method mark for using the correct formula.
> You get 1 mark for accuracy for arriving at the correct time taken.

> State the conclusion following from your results to get 1 mark.

You are planning a dinner party but have not finalised your guest list or menu.

You have decided to make your dessert from scratch because it will be cheaper than buying it from the local shop. Your Aunt Mildred has helped you by giving you four recipes for desserts that all your friends will like: crème caramel, blueberry and lime cheesecake, mango sorbet and chocolate brownies. You have already researched the prices of the ingredients required to make these desserts and are now ready to plan which ones to make!

Getting started

Consider the questions below to get you started.

- How many 150-g portions can I get from 1 kg?
- How many 75-g chocolate bars would I need to buy if I wanted 0.5 kg?
- Which is cheaper, three 330-ml cans of lemonade at 59p per can or a litre bottle of lemonade at £1.80?
- In a school meeting, one packet of biscuits was provided for every three people attending. How many packets would be needed if 20 students attend the meeting?

Your task

You can work on your own or in pairs for this activity.

1 Imagine that 8 people confirm that they can come to your dinner party.

   Work out the ingredients you would need for 8 people for each recipe.

   Work out the total cost for each recipe for 8; then work out the cost of one portion.

2 Now suppose three more people have accepted. Repeat the task, this time catering for 11 people, and yourself.

3 Work out the cost if all invited arrived. (There would be 20 altogether, including yourself.)

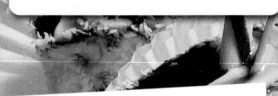

Ingredient costs

Item	Amount	Cost
Butter	500 g	£2.40
Eggs	6	£1.10
Lime	1	20p
Mango	1	£1.50
Milk	2.272 litres	£1.53
Plain chocolate	200 g	£1.09
Plain flour	1.5 kg	75p
Caster sugar	1 kg	£1.25
Granulated sugar	2 kg	£1.90
Walnuts	200 g	£2.50
Vanilla extract	30 ml	£1.55
Baking powder	50 g	£1.25
Biscuits	275 g	89p
Blueberries	500 g	£3.99
Quark	10 oz	£2.49
Double cream	284 ml	£1.89
Sour cream	284 ml	£1.39
Gelatine sachet	25 g	99p

(Ingredients can only be bought in these amounts or multiples of these amounts.)

Recipes

Crème Caramel – serves 6

500 ml milk
2 eggs
4 egg yolks
225 g caster sugar
1 teaspoon of vanilla extract

Blueberry and Lime Cheesecake – serves 8

10 oz sweet oaty biscuits
4 oz butter
1 lb 2 oz blueberries
$8\frac{1}{2}$ oz caster sugar
Grated zest and juice of two limes
20 oz of quark
$\frac{1}{2}$ pint of double cream
4 tsp powdered gelatine
$\frac{1}{2}$ pint sour cream

Mango Sorbet – serves 8

4 large ripe mangos
Juice of 2 large limes
450 ml sugar syrup
(150 g granulated sugar and
300 ml water)

Chocolate Brownies – serves 15

50 g plain flour
110 g butter
2 eggs
225 g granulated sugar
175 g walnuts
1 level teaspoon of baking powder

Extension

1 You have a maximum of £50 to spend.

   Assume that everyone will come to the party, so you are catering for 20 people, including yourself.

   Work out what combinations of dessert you could provide.

   Try to provide a variety of desserts to offer everyone a choice.

2 After the party, you decide to take Aunt Mildred a box of chocolates.

   She lives 15 miles away but you know she will provide an excellent lunch when you get there.

   You can cycle at 12 miles per hour. How long should you allow for the journey?

Why this chapter matters

The first equation ever written, using a modern equals sign, was:

$$14.\not{z}e.\ +\ .15.\not{\varphi}====71.\not{\varphi}.$$

It was written by Robert Recorde in 1557.
In today's notation this is $14\sqrt{x} + 15 = 71$.
The solution is $x = 16$.
Three of the most important equations in the world are shown on this page.

Why does the Moon keep orbiting the Earth and not fly off into space?

This is explained by Newton's law of universal gravitation, which describes the gravitational attraction between two bodies:

$$F = G \times \frac{m_1 \times m_2}{r^2}$$

where F is the force between the bodies, G is the gravitational constant, m_1 and m_2 are the masses of the two bodies and r is the distance between them.

Why don't planes fall out of the sky?

This is explained by Bernoulli's principle, which states that as the speed of a fluid increases, its pressure decreases. This is what causes the pressure differential between the top and bottom of an aircraft wing, as shown in the diagram on the left.

In its simplest form, the equation can be written as:

$$p + q = p_0$$

where p = static pressure, q = dynamic pressure and p_0 is the total pressure.

How can a couple of kilograms of plutonium have enough energy to wipe out a city?

This is explained by Einstein's theory of special relativity, which states that the speed of light is the same for all observers, even if one of them is moving at half the speed of light. It also connects mass and energy in the equation:

$$E = mc^2$$

where E is the energy, m is the mass and c is the speed of light. As the speed of light is nearly 300 000 kilometres per second, the amount of energy in a small mass is huge. If this can be released, it can be used for good (as in nuclear power stations) or harm (as in nuclear bombs).

Algebra: Equations and inequalities

1 Solving simple linear equations

2 Solving equations with brackets

3 Equations with the variable on both sides

4 Rearranging formulae

5 Solving linear inequalities

This chapter will show you ...

to **F** **E** how to solve linear equations with the variable on one side only

C how to solve linear equations with the variable on both sides

C how to rearrange simple formulae

C how to solve simple linear inequalities

Visual overview

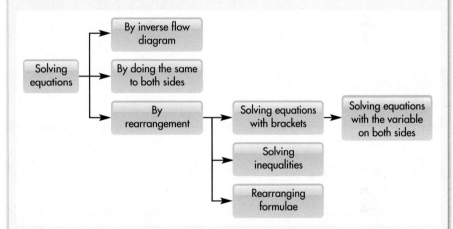

What you should already know

- The basic language of algebra (KS3 level 5, GCSE grade C)
- How to expand brackets and collect like terms (KS3 level 6, GCSE grade D)
- That addition and subtraction are opposite (inverse) operations (KS3 level 4, GCSE grade F)
- That multiplication and division are opposite (inverse) operations (KS3 level 4, GCSE grade F)

Quick check

1 **a** Simplify $5x + 3x - 2x$.   **b** Expand $4(3x - 1)$.

   c Expand and simplify $2(3x - 1) + 3(4x + 3)$.

2 What number can go in the box to make the calculation true?

   a $13 + \square = 9$ **b** $4 \times \square = 10$

This section will show you how to:

- solve a variety of simple linear equations, such as $3x - 1 = 11$, where the variable only appears on one side
- use inverse operations and inverse flow charts
- solve equations by doing the same on both sides
- deal with negative numbers
- solve equations by rearrangement

Key words

do the same to both sides
equation
inverse flow diagram
inverse operations
rearrangement
solution
variable

A teacher gave these instructions to her class.

What algebraic expression represents the teacher's statement? (See Chapter 10.)

- Think of a number.
- Double it.
- Add 3.

This is what two of her students said.

Can you work out Kim's answer and the number that Freda started with?

Kim's answer will be $2 \times 5 + 3 = 13$.

My final answer was 10.

Freda's answer can be set up as an **equation**.

An equation is formed when an expression is put equal to a number or another expression. You are expected to deal with equations that have only one **variable** or letter.

I chose the number 5.

The **solution** to an equation is the value of the variable that makes the equation true.
For example, the equation for Freda's answer is

$$2x + 3 = 10$$

where x represents Freda's number.

The value of x that makes this true is $x = 3\frac{1}{2}$.

Freda

Kim

FM Functional Maths **AU** (AO2) Assessing Understanding **PS** (AO3) Problem Solving

To solve an equation, you have to 'undo' it. That is, you have to reverse the processes that set up the equation in the first place.

Freda did two things. First she multiplied by 2 and then she added 3. The reverse process is *first* to subtract 3 and *then* to divide by 2. So, to solve:

$$2x + 3 = 10$$

Subtract 3 $\qquad 2x + 3 - 3 = 10 - 3$

$$2x = 7$$

Divide by 2 $\qquad \dfrac{2x}{2} = \dfrac{7}{2}$

$$x = 3\dfrac{1}{2}$$

The problem is knowing how an equation is set up in the first place, so that you can undo it in the right order.

There are four ways to solve equations:

- **inverse operations**
- **inverse flow diagrams**
- **'doing the same to both sides'**
- **rearrangement**.

They are all essentially the same. You will have to decide which method you prefer, although you should know how to use all four.

There is one rule about equations that you should *always* follow.

Check that your answer works in the original equation.

For example, to check the answer to Freda's equation, put $x = 3\frac{1}{2}$ into Freda's equation. This gives:

$2 \times 3\frac{1}{2} + 3 = 7 + 3 = 10$

which is correct.

Inverse operations

One way to solve equations is to use **inverse operations**. The opposite or inverse operation to addition is subtraction (and vice versa) and the opposite or inverse operation to multiplication is division (and vice versa).

That means you can 'undo' the four basic operations by using the inverse operation.

EXAMPLE 1

Solve these equations.

a $w + 7 = 9$ **b** $x - 8 = 10$ **c** $2y = 8$ **d** $\dfrac{z}{5} = 3$

    a The opposite operation to $+7$ is -7, so the solution is $9 - 7 = 2$.

        Check: $2 + 7 = 9$

    b The opposite operation to -8 is $+8$, so the solution is $10 + 8 = 18$.

        Check: $18 - 8 = 10$

    c $2y$ means $2 \times y$. The opposite operation to $\times 2$ is $\div 2$, so the solution is $8 \div 2 = 4$.

        Check: $2 \times 4 = 8$

    d $\dfrac{z}{5}$ means $z \div 5$. The opposite operation to $\div 5$ is $\times 5$, so the solution is $3 \times 5 = 15$.

        Check: $15 \div 5 = 3$

EXERCISE 8A

1 Solve the following equations by applying the inverse of the operation on the left-hand side to the right-hand side.

    a $x + 6 = 10$ **b** $w - 5 = 9$

    c $y + 3 = 8$ **d** $p - 9 = 1$

    e $2x = 10$ **f** $3x = 18$

    g $\dfrac{z}{3} = 8$ **h** $4x = 10$

    i $\dfrac{q}{4} = 1$ **j** $x + 9 = 10$

    k $r - 7 = 21$ **l** $\dfrac{s}{6} = 2$

> **HINTS AND TIPS**
>
> **Remember** to perform the inverse operation on the number on the right-hand side.

PS 2 The solution to the equation $5x = 20$ is $x = 4$.

Write down two different equations for which the solution is 4.

AU 3 Here are three equations.

    a $x + 6 = 12$

    b $x - 1 = 5$

    c $x + 5 = 9$

Give a reason why each one could be the odd one out.

> **HINTS AND TIPS**
>
> Solve the equations to give you a clue and look for similarities.

4 Set up an equation to represent the following. Use x for the variable.

My sister is 3 years older than I am. She is 17 years old. How old am I?

5 Set up an equation to represent the following. Use y for the variable.

Six apples cost £1.80. How much is one apple?

Inverse flow diagrams

Another way to solve simple linear equations is to use inverse flow diagrams.

This flow diagram represents the instructions that their teacher gave to Kim and Freda.

$$\longrightarrow \boxed{\times 2} \longrightarrow \boxed{+ 3} \longrightarrow$$

The **inverse flow diagram** looks like this.

$$\longleftarrow \boxed{\div 2} \longleftarrow \boxed{- 3} \longleftarrow$$

Running Freda's answer through this gives:

$$\overset{3\frac{1}{2}}{\longleftarrow} \boxed{\div 2} \overset{7}{\longleftarrow} \boxed{- 3} \overset{10}{\longleftarrow}$$

So, Freda started with $3\frac{1}{2}$ to get an answer of 10.

EXAMPLE 2

Use an inverse flow diagram to solve the following equation.

$3x - 4 = 11$

Flow diagram:

$$\longrightarrow \boxed{\times 3} \longrightarrow \boxed{- 4} \longrightarrow$$

Inverse flow diagram:

$$\longleftarrow \boxed{\div 3} \longleftarrow \boxed{+ 4} \longleftarrow$$

Put through the value on the right-hand side of the equals sign.

$$\overset{5}{\longleftarrow} \boxed{\div 3} \overset{15}{\longleftarrow} \boxed{+ 4} \overset{11}{\longleftarrow}$$

So, the answer is $x = 5$.

Checking the answer gives:

$3 \times 5 - 4 = 11$

which is correct.

E

EXERCISE 8B

1 Use inverse flow diagrams to solve each of the following equations. Remember to check that each answer works for its original equation.

a $3x + 5 = 11$

b $3x - 13 = 26$

c $3x - 7 = 32$

d $4y - 19 = 5$

e $3a + 8 = 11$

f $2x + 8 = 14$

g $2y + 6 = 18$

h $8x + 4 = 12$

i $2x - 10 = 8$

j $\dfrac{x}{5} + 2 = 3$

k $\dfrac{t}{3} - 4 = 2$

l $\dfrac{y}{4} + 1 = 7$

m $\dfrac{k}{2} - 6 = 3$

n $\dfrac{h}{8} - 4 = 1$

o $\dfrac{w}{6} + 1 = 4$

p $\dfrac{x}{4} + 5 = 7$

q $\dfrac{y}{2} - 3 = 5$

r $\dfrac{f}{5} + 2 = 8$

> **HINTS AND TIPS**
>
> **Remember** the rules of BIDMAS/BODMAS.
> So $3x + 5$ means do $x \times 3$ first then $+5$ in the flow diagram. Then do the opposite (inverse) operations in the inverse flow diagram.

PS **2** The diagram shows a two-step number machine.

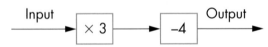

Find a value for the input that gives the same value for the output.

AU **3** A man buys two apples and gets 46p change from £1.

He wants to know the cost of each apple.

By setting up a flow diagram (or otherwise) work out the cost of one apple.

Doing the same to both sides

You need to know how to solve equations by performing the same operation on both sides of the equals sign.

Mary had two bags of marbles, each of which contained the same number of marbles, and five spare marbles.

She put them on scales and balanced them with 17 single marbles.

How many marbles were there in each bag?

If x is the number of marbles in each bag, then the equation representing Mary's balanced scales is:

$2x + 5 = 17$

Take five marbles from each pan:

$2x + 5 - 5 = 17 - 5$
$2x = 12$

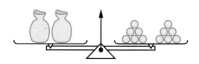

Now halve the number of marbles on each pan.

That is, divide both sides by 2:

$\dfrac{2x}{2} = \dfrac{12}{2}$

$x = 6$

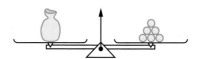

Checking the answer gives $2 \times 6 + 5 = 17$, which is correct.

EXAMPLE 3

Solve each of these equations by 'doing the same to both sides'.

a $3x - 5 = 16$ **b** $\dfrac{x}{2} + 2 = 10$

Add 5 to both sides.

$3x - 5 + 5 = 16 + 5$

$3x = 21$

Divide both sides by 3.

$\dfrac{3x}{3} = \dfrac{21}{3}$

$x = 7$

Checking the answer gives:

$3 \times 7 - 5 = 16$

which is correct.

Subtract 2 from both sides.

$\dfrac{x}{2} + 2 - 2 = 10 - 2$

$\dfrac{x}{2} = 8$

Multiply both sides by 2.

$\dfrac{x}{2} \times 2 = 8 \times 2$

$x = 16$

Checking the answer gives:

$16 \div 2 + 2 = 10$

which is correct.

Dealing with negative numbers

The solution to an equation may be a negative number. You need to know that when a negative number is multiplied or divided by a positive number, then the answer is also a negative number. For example:

$-3 \times 4 = -12$ and $-10 \div 5 = -2$

Check these on your calculator.

E

EXERCISE 8C

1 Solve each of the following equations by 'doing the same to both sides'. Remember to check that each answer works for its original equation.

a $x + 4 = 60$

b $3y - 2 = 4$

c $3x - 7 = 11$

d $5y + 3 = 18$

e $7 + 3t = 19$

f $5 + 4f = 15$

g $3 + 6k = 24$

h $4x + 7 = 17$

i $5m - 3 = 17$

j $\dfrac{w}{3} - 5 = 2$

k $\dfrac{x}{8} + 3 = 12$

l $\dfrac{m}{7} - 3 = 5$

m $\dfrac{x}{5} + 3 = 3$

n $\dfrac{h}{7} + 2 = 1$

o $\dfrac{w}{3} + 10 = 4$

p $\dfrac{x}{3} - 5 = 7$

q $\dfrac{y}{2} - 13 = 5$

r $\dfrac{f}{6} - 2 = 8$

> **HINTS AND TIPS**
>
> Be careful with negative numbers.

AU 2
PS

Think of a number. Divide it by 3 and subtract 6.

My answer is −1.

My starting number is 6.

Teacher

Mandy

Andy

a What answer did Andy get?

b What number did Mandy start with?

AU 3 The solution of the equation $4x + 17 = 9$ is $x = -2$.

Make up two more different equations of the form $ax + b = c$ where a, b and c are positive whole numbers, for which the answer is also −2.

AU 4 A teacher asked her class to solve the equation $2x - 1 = 7$.

Amanda wrote:

$2x - 1 = 7$

$2x - 1 - 1 = 7 - 1$

$2x = 6$

$2x - 2 = 6 - 2$

$x = 4$

Betsy wrote:

$2x - 1 = 7$

$2x - 1 + 1 = 7 + 1$

$2x = 8$

$2x \div 2 = 8 \div 2$

$x = 4$

When the teacher read out the correct answer of 4, both students ticked their work as correct.

a Which student used the correct method?

b Explain the mistakes the other student made.

ACTIVITY

Balancing with unknowns

Suppose you want to solve an equation such as:

$2x + 3 = x + 4$

You can imagine it as a balancing problem with marbles.

2 bags + 3 marbles = 1 bag + 4 marbles

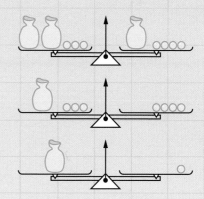

Take one bag from each side.

Take three marbles from each side.

There must be one marble in the bag.

This means that $x = 1$.

Checking the answer gives $2 \times \mathbf{1} + 3 = \mathbf{1} + 4$, which is correct.

Set up each of the following problems as a 'balancing picture' and solve it by 'doing the same to both sides'. Remember to check that each answer works. The first two problems include the pictures to start you off.

1 $2x + 6 = 3x + 1$

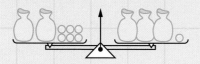

2 $4x + 2 = x + 8$

3 $5x + 1 = 3x + 11$

4 $x + 9 = 2x + 7$

5 $3x + 8 = 2x + 10$

6 $5x + 7 = 3x + 21$

7 $2x + 12 = 5x + 6$

8 $3x + 6 = x + 9$

(Some of the marbles could be broken in half!)

9 Explain why there is no answer to this problem:

$x + 3 = x + 4$

10 One of the bags of marbles on the left-hand pan has had three marbles taken out.

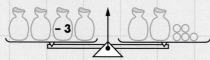

Try to draw the pictures to solve this problem:

$4x - 3 = 2x + 5$

Rearrangement

Solving equations by rearrangement is the most efficient method and the one used throughout the rest of this chapter. The terms of the equation are rearranged until the variable is on its own – usually on the left-hand side of the equals sign.

EXAMPLE 4

Solve $4x + 3 = 23$.

Move the 3 to give: $\qquad\qquad\qquad\qquad\qquad 4x = 23 - 3 = 20$

Now divide both sides by 4 to give: $\qquad\qquad x = \dfrac{20}{4} = 5$

So, the solution is $x = 5$.

EXAMPLE 5

Solve $\dfrac{y - 4}{5} = 3$.

Move the 5 to give: $\qquad\qquad\qquad\qquad\qquad y - 4 = 3 \times 5 = 15$

Now move the 4 to give: $\qquad\qquad\qquad\qquad y = 15 + 4 = 19$

So, the solution is $y = 19$.

EXERCISE 8D

1 Solve each of the following equations. Remember to check that each answer works for its original equation.

a $2x + 4 = 6$ $\qquad\qquad$ **b** $2t + 7 = 13$

c $3x + 10 = 16$ $\qquad\quad$ **d** $4y + 15 = 23$

e $2x - 8 = 10$ $\qquad\quad$ **f** $4t - 3 = 17$

g $5x - 6 = 24$ $\qquad\quad$ **h** $7 - x = 3$

i $12 - 3y = 6$ $\qquad\quad$ **j** $2k + 8 = 4$

k $\dfrac{x}{3} + 7 = 15$ $\qquad\quad$ **l** $\dfrac{t}{5} + 3 = 5$

m $\dfrac{w}{3} - 5 = 2$ $\qquad\quad$ **n** $\dfrac{x}{8} + 3 = 12$

o $\dfrac{m}{7} - 3 = 5$

> **HINTS AND TIPS**
>
> When a variable changes sides of the equals sign, it also changes signs, that is, plus becomes minus and vice versa, multiply becomes divide and vice versa. This is sometimes called 'Change sides, change signs'.

AU 2
PS
A teacher read out the following to her class:

"I am thinking of a number. I subtract 5 from it and then divide the result by 4. The answer is 7. What number did I think of to start with?"

a What was the number the teacher thought of?

b Bryn misunderstood the instructions and got the operations the wrong way round. What number did Bryn think the teacher started with?

3 Solve each of the following equations. Remember to check that each answer works for its original equation.

a $\dfrac{k+1}{2} = 3$

b $\dfrac{h-4}{8} = 3$

c $\dfrac{w+1}{6} = 1$

d $\dfrac{x+5}{4} = 10$

e $\dfrac{y-3}{6} = 5$

f $\dfrac{f+2}{5} = 5$

FM 4 At a conference, doughnuts are delivered for the mid-morning break. The waiter takes five off the tray before taking them to the delegates. There are 10 delegates and they each get two doughnuts.

How many doughnuts were delivered?

> **HINTS AND TIPS**
>
> Begin by associating a letter with 'doughnuts'. Then use algebra to work out your answer.

8.2 Solving equations with brackets

This section will show you how to:
- solve equations that include brackets

Key words
expand
inverse operations

When an equation contains brackets, you must first **expand** or multiply out the brackets and then solve the equation by using one of the previous methods. This example uses **inverse operations**.

EXAMPLE 6

Solve $5(x + 3) = 25$.

First multiply out the brackets: $5x + 15 = 25$

Rearrange. $5x + 15 - 15 = 25 - 15 = 10$

Divide by 5. $\dfrac{5x}{5} = \dfrac{10}{5}$

                                    $x = 2$

EXAMPLE 7

Solve $3(2x - 7) = 15$.

Multiply out the brackets: $6x - 21 = 15$

Add 21 to both sides. $6x = 36$

Divide both sides by 6. $x = 6$

EXERCISE 8E

1   Solve each of the following equations. Some of the answers may be decimals or negative numbers. Remember to check that each answer works for its original equation. Use your calculator if necessary.

a $2(x + 5) = 16$ b $5(x - 3) = 20$

c $3(t + 1) = 18$ d $4(2x + 5) = 44$

e $2(3y - 5) = 14$ f $5(4x + 3) = 135$

g $4(3t - 2) = 88$ h $6(2t + 5) = 42$

i $2(3x + 1) = 11$ j $4(5y - 2) = 42$

k $6(3k + 5) = 39$ l $5(2x + 3) = 27$

m $5(2x - 1) = -45$ n $7(3y + 5) = -7$

> **HINTS AND TIPS**
>
> Once the brackets have been expanded the equations are the same sort as those you have already been dealing with. Remember to multiply everything inside the brackets by what is outside.

AU 2   Fill in values for a, b and c so that the answer to this equation is $x = 4$.

$a(bx + 3) = c$

PS 3   My son is x years old. In five years' time I will be twice his age and both our ages will be multiples of 10. The sum of our ages will be between 50 and 100.

How old am I now?

> **HINTS AND TIPS**
>
> Set up an equation and put it equal to 60, 70, 80, etc. Solve the equation and see if the answer fits the conditions.

PS 4   A rectangle with sides 4 cm and $3x + 1$ cm has a smaller rectangle with sides 2 cm and $2x - 1$ cm cut from it. The remaining area is 32 cm^2.

What is the value of x?

3x + 1 cm

2x – 1 cm

4 cm   2 cm

Equations with the variable on both sides

This section will show you how to:

- solve equations where the variable appears on both sides of the equation

Key word

variable

When a letter or **variable** appears on both sides of an equation, it is best to use the 'do the same to both sides' method of solution and collect all the terms containing the variable on the left-hand side of the equation. If there are more of the variable on the right-hand side, it is easier to turn the equation round. When an equation contains brackets, they must be multiplied out first.

EXAMPLE 8

Solve $5x + 4 = 3x + 10$.

There are more xs on the left-hand side, so you do not need to turn the equation round.

Subtract $3x$ from both sides.	$2x + 4 = 10$
Subtract 4 from both sides.	$2x = 6$
Divide both sides by 2.	$x = 3$

EXAMPLE 9

Solve $2x + 3 = 6x - 5$.

There are more xs on the right-hand side, so turn the equation round.

$$6x - 5 = 2x + 3$$

Subtract $2x$ from both sides.	$4x - 5 = 3$
Add 5 to both sides.	$4x = 8$
Divide both sides by 4.	$x = 2$

EXAMPLE 10

Solve $3(2x + 5) + x = 2(2 - x) + 2$.

Multiply out both brackets.	$6x + 15 + x = 4 - 2x + 2$
Simplify both sides.	$7x + 15 = 6 - 2x$

There are more xs on the left-hand side, so you do not need to turn the equation round.

Add $2x$ to both sides.	$9x + 15 = 6$
Subtract 15 from both sides.	$9x = -9$
Divide both sides by 9.	$x = -1$

EXERCISE 8F

1 Solve each of the following equations.

a $2x + 3 = x + 5$ **b** $5y + 4 = 3y + 6$

c $4a - 3 = 3a + 4$ **d** $5t + 3 = 2t + 15$

e $7p - 5 = 3p + 3$ **f** $6k + 5 = 2k + 1$

g $2t - 7 = 4t - 3$ **h** $2p - 1 = 9 - 3p$

> **HINTS AND TIPS**
>
> **Remember** the rule 'Change sides, change signs'. Show all your working on this type of question. Rearrange before you simplify. If you try to rearrange and simplify at the same time, you will probably get it wrong.

PS 2

 I am thinking of a number. I multiply it by 3 and subtract 2.

 I am thinking of a number. I multiply it by 2 and add 5.

> **HINTS AND TIPS**
>
> Set up an expression for each of them. Put them equal and solve the equation.

Terry and June find that they both thought of the same number and both got the same final answer.

What number did they think of?

3 Solve each of the following equations.

a $2(d + 3) = d + 12$ **b** $5(x - 2) = 3(x + 4)$ **c** $3(2y + 3) = 5(2y + 1)$

d $3(h - 6) = 2(5 - 2h)$ **e** $4(3b - 1) + 6 = 5(2b + 4)$

f $2(5c + 2) - 2c = 3(2c + 3) + 7$

PS 4 Wilson has eight coins, all of the same value, as well as seven pennies.

Chloe has 11 coins of the same value as those that Wilson has and she also has five pennies.

Wilson says, "If you give me one of your coins and four pennies, we will each have the same amount of money."

What is the value of the coins that Wilson and Chloe have?

> **HINTS AND TIPS**
>
> Call the coin x and set up the expressions, e.g. Wilson has $8x + 7$. Then take one x and 4 from Chloe's expression and add one x and 4 to Wilson's. Then put the expressions equal and solve.

AU 5 Explain why the equation $3(2x + 1) = 2(3x + 5)$ cannot be solved.

AU 6 Explain why there are an infinite number of solutions to the equation

$2(6x + 9) = 3(4x + 6)$.

> **HINTS AND TIPS**
>
> Expand the brackets and collect terms on one side as usual. Explain what happens.

Rearranging formulae

This section will show you how to:

- rearrange formulae, using the same methods as for solving equations

Key words

expression
rearrange
subject
transpose
variable

The **subject** of a formula is the **variable** (letter) in the formula that stands on its own, usually on the left-hand side of the 'equals' sign. For example, x is the subject of each of the following.

$$x = 5t + 4 \qquad x = 4(2y - 7) \qquad x = \frac{1}{t}$$

If you need to change the existing subject to a different variable, you have to **rearrange** (**transpose**) the formula to get that variable on the left-hand side.

You do this by using the same rule as for solving equations, that is, move the terms concerned from one side of the 'equals' sign to the other.

The main difference is that when you solve an equation each step gives a numerical value. When you rearrange a formula each step gives an algebraic **expression**.

EXAMPLE 11

Make m the subject of $T = m - 3$.

Move the 3 so that the m is on its own. $T + 3 = m$

Reverse the formula. $m = T + 3$

EXAMPLE 12

From the formula $P = 4t$, express t in terms of P.

(This is another common way of asking you to make t the subject.)

Divide both sides by 4. $\dfrac{P}{4} = \dfrac{4t}{4}$

Reverse the formula. $t = \dfrac{P}{4}$

EXAMPLE 13

From the formula $C = 2m + 3$, make m the subject.

Move the 3 so that the $2m$ is on its own. $\qquad C - 3 = 2m$

Divide both sides by 2. $\qquad \dfrac{C - 3}{2} = \dfrac{2m}{2}$

Reverse the formula. $\qquad m = \dfrac{C - 3}{2}$

EXERCISE 8G

1 $T = 3k$ Make k the subject.

2 $P = m + 7$ Make m the subject.

3 $X = y - 1$ Express y in terms of X.

4 $Q = \dfrac{p}{3}$ Express p in terms of Q.

> **HINTS AND TIPS**
>
> **Remember** about inverse operations and the rule 'Change sides, change signs'.

5 $p = m + t$
   a Make m the subject.
   b Make t the subject.

6 $t = 2k + 7$ Express k in terms of t.

7 $g = \dfrac{m}{v}$ Make m the subject.

8 $t = m^2$ Make m the subject.

9 $C = 2\pi r$ Make r the subject.

10 $A = bh$ Make b the subject.

11 $P = 2l + 2w$ Make l the subject.

12 $m = p^2 + 2$ Make p the subject.

FM 13 Kieran notices that the price of five cream buns is 75p more than the price of nine mince pies.

Let the price of a cream bun be x pence and the price of a mince pie be y pence.

a Express the cost of one mince pie, y, in terms of the price of a cream bun, x.

b If the price of a cream bun is 60p, how much is a mince pie?

> **HINTS AND TIPS**
>
> Set up a formula using the first sentence of information, then rearrange it.

PS 14 Distance, speed and time are connected by the formula:

distance = speed × time

A delivery driver drove 126 km in 1 hour and 45 minutes.

On the return journey he was held up at road works, so his average speed decreased by 9 km per hour.

How long was he held up at the road works?

> **HINTS AND TIPS**
>
> Work out the average speed for the first journey, then work out the average speed for the return journey.

FM 15 The formula for converting degrees Celsius to degrees Fahrenheit is $C = \frac{5}{9}(F - 32)$.

a Show that when $F = -40$, C is also equal to -40.

b Find the value of C when $F = 68$.

c Use this flow diagram to establish the formula for converting degrees Fahrenheit to degrees Celsius.

$$°F \longrightarrow \boxed{-32} \longrightarrow \boxed{\times 5} \longrightarrow \boxed{\div 9} \longrightarrow °C$$

8.5 Solving linear inequalities

This section will show you how to:
● solve a simple linear inequality

Key words
inclusive inequality
integer
linear inequality
number line
strict inequality

Inequalities behave similarly to equations, which you have already met. In the case of **linear inequalities**, you can use the same rules to solve them as you use for linear equations. There are four inequality signs, < which means 'less than', > which means 'greater than', ⩽ which means 'less than or equal to' and ⩾ which means 'greater than or equal to'. Be careful. Never replace the inequality sign with an equals sign or you will get the wrong answer and could end up getting no marks in an exam.

EXAMPLE 14

Solve $2x + 3 < 14$.

This is rewritten as:

$$2x < 14 - 3$$

that is $2x < 11$.

Divide both sides by 2.

$$\frac{2x}{2} < \frac{11}{2}$$

$$\Rightarrow \quad x < 5.5$$

This means that x can take any value below 5.5 but it *cannot* take the value 5.5.

< and > are called **strict inequalities**.

Note: The inequality sign used in the problem is the sign to give in the answer.

EXAMPLE 15

Solve $\dfrac{x}{2} + 4 \geqslant 13$.

Solve just like an equation but leave the inequality sign in place of the equals sign.

Subtract 4 from both sides. $\dfrac{x}{2} \geqslant 9$

Multiply both sides by 2. $x \geqslant 18$

This means that x can take any value above 18 and including 18.

$\leqslant$ and $\geqslant$ are called **inclusive inequalities**.

EXERCISE 8H

1 Solve the following linear inequalities.

- **a** $x + 4 < 7$
- **b** $t - 3 > 5$
- **c** $p + 2 \geqslant 12$
- **d** $2x - 3 < 7$
- **e** $4y + 5 \leqslant 17$
- **f** $3t - 4 > 11$
- **g** $\dfrac{x}{2} + 4 < 7$
- **h** $\dfrac{y}{5} + 3 \leqslant 6$
- **i** $\dfrac{t}{3} - 2 \geqslant 4$
- **j** $3(x - 2) < 15$
- **k** $5(2x + 1) \leqslant 35$
- **l** $2(4t - 3) \geqslant 34$

2 Write down the largest value of x that satisfies each of the following.

- **a** $x - 3 \leqslant 5$, where x is a positive integer.
- **b** $x + 2 < 9$, where x is a positive even integer.
- **c** $3x - 11 < 40$, where x is a square number.
- **d** $5x - 8 \leqslant 15$, where x is a positive odd number.
- **e** $2x + 1 < 19$, where x is a positive prime number.

3 Write down the smallest value of x that satisfies each of the following.

- **a** $x - 2 \geqslant 9$, where x is a positive integer.
- **b** $x - 2 > 13$, where x is a positive even integer.
- **c** $2x - 11 \geqslant 19$, where x is a square number.
- **d** $3x + 7 \geqslant 15$, where x is a positive odd number.
- **e** $4x - 1 > 23$, where x is a positive prime number.

FM 4 Ahmed went to town with £20 to buy two CDs. His return bus fare was £3. The CDs were the same price as each other. When he got home he still had some money in his pocket. What was the most each CD could cost?

HINTS AND TIPS

Set up an inequality and solve it.

AU **5** **a** Explain why you cannot make a triangle with three sticks of length 3 cm, 4 cm and 8 cm.

b Three side of a triangle are x, $x + 2$ and 10 cm.
x is a whole number.
What is the smallest value x can take?

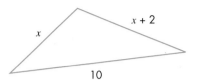

PS **6** These five cards have inequalities and equations marked on them.

$x > 0$	$x < 3$	$x \geqslant 4$	$x = 2$	$x = 6$

The cards are shuffled and then turned over, one at a time.

If two consecutive cards have solutions with any numbers in common, then a point is scored.

If their solutions do not have any numbers in common, then a point is deducted.

a When the cards are laid out in the order below, the first two score –1 because $x = 6$ and $x < 3$ have no numbers in common.

$x = 6$	$x < 3$	$x > 0$	$x = 2$	$x \geqslant 4$

Explain why the total for this combination scores 0.

b Now the cards are laid out in this order. What does this combination score?

$x > 0$	$x = 6$	$x \geqslant 4$	$x = 2$	$x < 3$

c Arrange the five cards to give a maximum score of 4.

The number line

The solution to a linear inequality can be shown on the **number line** by using the following conventions.

 $x \leqslant$  $x \geqslant$  $x <$  $x >$

A strict inequality *does not* include the boundary point but an inclusive inequality *does* include the boundary point.

Below are five examples.

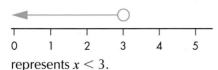

represents $x < 3$.

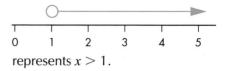

represents $x > 1$.

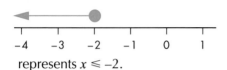

represents $x \leqslant -2$.

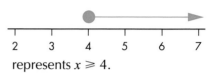

represents $x \geqslant 4$.

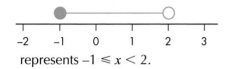

represents $-1 \leqslant x < 2$.

This is a 'between' inequality. It can be written as $x \geqslant -1$ and $x < 2$ but the notation $-1 \leqslant x < 2$ is much neater.

EXAMPLE 16

a Write down the inequality shown by this diagram.

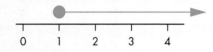

b i Solve the inequality $2x + 3 < 11$. **ii** Mark the solution to **b** on a number line.

c Write down the integers that satisfy both the inequality in **a** and the inequality in **b**.

a The inequality shown is $x \geqslant 1$.

b i $2x + 3 < 11 \Rightarrow 2x < 8 \Rightarrow x < 4$

ii

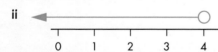

c The integers that satisfy both inequalities are 1, 2 and 3.

EXERCISE 8I

1 Write down the inequality that is represented by each diagram below.

a

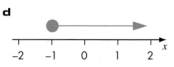

b

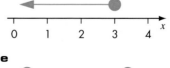

c

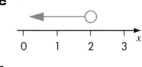

d

e

f

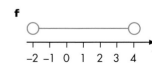

2 Draw diagrams to illustrate the following.

a $x \leqslant 3$ **b** $x > -2$ **c** $x \geqslant 0$ **d** $x < 5$

e $x \geqslant -1$ **f** $2 < x \leqslant 5$ **g** $-1 \leqslant x \leqslant 3$ **h** $-3 < x < 4$

3 Solve the following inequalities and illustrate their solutions on number lines.

a $x + 4 \geqslant 8$ **b** $x + 5 < 3$ **c** $x - 1 \leqslant 2$ **d** $x - 4 > -1$

e $2x > 8$ **f** $3x \leqslant 15$ **g** $4x + 3 < 9$ **h** $\dfrac{x}{2} + 3 \leqslant 2$

i $\dfrac{x}{5} - 2 > 8$ **j** $2x - 1 \leqslant 4$ **k** $\dfrac{x + 2}{3} > 4$ **l** $\dfrac{x - 1}{4} < 3$

AU 4 Copy the number line below twice. On your copies draw two inequalities so that only the integers {−1, 0, 1, 2} are common to both inequalities.

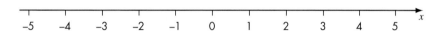

FM 5 Max went to the supermarket with £1.20. He wanted to buy three apples costing x pence each and a chocolate bar costing 54p. When he got to the till, he found he didn't have enough money.

Max took one of the apples back and paid for two apples and the chocolate bar. He counted his change and found he had enough money to buy a 16p chew.

a Explain why $3x + 54 > 120$ and solve the inequality.

b Explain why $2x + 54 \leqslant 104$ and solve the inequality.

c Show the solution to both of these inequalities on a number line.

d What is the possible price of an apple?

PS 6 Here are four inequalities and two sets of integers.

Group them into two sets of three.

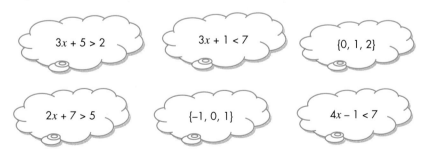

GRADE BOOSTER

F You can solve equations such as $4x = 12$ and $x - 8 = 3$

E You can solve equations such as $3x + 2 = 7$ or $\frac{x}{3} - 7 = 1$

D You can solve equations such as $\frac{x - 2}{3} = 6$ or $3x + 7 = x - 6$

C You can solve equations such as $3(x - 4) = 5x + 8$

C You can solve inequalities such as $3x + 2 < 5$

C You can rearrange simple formulae

What you should know now

- How to solve a variety of linear equations using rearrangement or 'doing the same thing to both sides'
- How to rearrange simple formulae
- How to solve simple inequalities

1 Solve these equations.

  a $4x = 20$

  b $5x - 1 = 9$

  c $3 + x = 9$

  d $4x + 3 = 2x + 13$

2

Jason   My answer is 3.   Teacher   Think of a number double it and subtract 7.   The number I thought of was 15.   Zara

  a What answer did Zara get?

  b What was the number Jason thought of?

3 An orange costs z pence. A lemon costs 4 pence more than an orange.

  a Write down an expression, in terms of z, for the cost of one lemon.

  b Write down an expression, in terms of z, for the total cost of three oranges and one lemon.

  c The total cost of three oranges and one lemon is 60 pence.

  Form an equation in terms of z and solve it to find the cost of one orange.

4 **a** Solve these equations.

  i $2x = 9$

  ii $3x - 8 = 13$

  iii $6x + 9 = x + 24$

  b Simplify these expressions.

  i $5q + 6q + 2q$

  ii $5n + 4p + 2n - p$

5 The length of a rectangle is 10.5 cm.
The perimeter of the rectangle is 28 cm.

10.5 cm

x

  a Using x to represent the length, set up an equation for the information above.

  b Solve the equation to find the width of the rectangle.

6 Solve the following equations.

  a $x + 3 = 10$ (1)

  b $2x = 10$ (1)

  c $3x - 8 = 10$ (2)

  d $5(x + 4) = 10$ (3)

  e $11 + \dfrac{x}{3} = 15$ (2)

(Total 9 marks)

AQA, November 2008, Paper 1 Foundation, Question 15

7 Solve the following equations.

  a $x + 3 = 8$ (1)

  b $3y + 4 = 16$ (2)

  c $2(3z - 1) = 13$ (3)

(Total 6 marks)

AQA, June 2008, Paper 1 Foundation, Question 18

8 **a** Factorise $6x - 10$ (1)

  b Solve the equation $4y + 3 = 8$ (2)

  c Solve the equation $2(t + 5) = 8$ (3)

  d List all the integer solutions of the inequality

  $-6 \leqslant 3n < 5$ (3)

(Total 9 marks)

AQA, November 2008, Paper 2 Foundation, Question 12

9 **a** Solve the equation $7x - 3 = 60$ (2)

b y is an odd integer.

Is each statement below true or false? (3)

$7y - 3$ is never odd

$7y - 3$ is never prime

$7y - 3$ is never a multiple of 7

c Write down one integer which satisfies the inequality $7w > 63$ (1)

(Total 6 marks)

AQA, June 2009, Paper 1 Foundation, Question 10

10 Solve the equations

a $7x - 9 = 3x + 5$ (3)

b $7(y - 9) = 3y + 5$ (3)

(Total 6 marks)

AQA, June 2009, Paper 2 Foundation, Question 22

11 **a** Factorise $x^2 + 4x$ (1)

b Solve the inequality $7y < 3y + 6$ (2)

c Make r the subject of the formula $p = 3 + 2r$ (2)

(Total 5 marks)

AQA, June 2005, Paper 1 Intermediate, Question 16

12 Make t the subject of the formula:

$$u = \frac{t}{3}$$

13 ABC is a triangle with sides, given in centimetres, of x, $2x + 1$ and $3x - 3$.

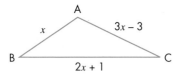

The perimeter of the triangle is 22 cm. Find the value of x.

14 **a** Rearrange the formula:

$C = 6r$

to make r the subject.

b Rearrange the formula:

$A = 3r^2$

to make r the subject.

15 **a** Solve the inequality $3x + 2 \leqslant 8$ (2)

b Write down all the integer values of x satisfying this inequality $-4 \leqslant 2x < 4$ (2)

(Total 4 marks)

AQA, June 2009, Paper 1 Foundation, Question 21

 16 **a** Solve the inequality $3(x - 2) \leqslant 9$ (3)

b The inequality $x \leqslant 3$ is shown on the number line below.

Draw another inequality on the number line so that only the following integers satisfy both inequalities

$\{-2, -1, 0, 1, 2, 3\}$ (1)

(Total 4 marks)

AQA, November 2006, Paper 2 Higher, Question 4

Worked Examination Questions

PS **1** In the table below, the letters a, b, c and d represent different numbers. The total of each row is given at the side of the table.

a	a	a	a	32
a	a	b	b	36
a	a	b	c	33
a	b	c	d	31

Find the values of a, b, c and d.

$a = 8$ —————————————— Look for clues in the question. The first row is all a so $4a = 32$, so $a = 8$

$b = 10$ ——————

$c = 7$ ———————— Use $a = 8$ in the second row to work out b and so on.

$d = 6$

Each correct answer is worth 1 mark.

Total: 4 marks

FM **2** Mark set off for town with a £20 note and £1.50 in change.
His bus fare was 50p.
He bought three CDs for the same price each and a bottle of pop for 98p.
He then came home on the bus, paying 50p for his fare.
Set up an inequality using this information and taking x as the price of a CD.
Work out the most Mark could have paid for each CD.

$3x + 198 < 2150$ ——————— You get 1 method mark and 1 accuracy mark for setting up an equation to show that three CDs plus what Mark spent must be less than what he set off with.

$3x < 1952$ ———————

$x < 650.667$

The most Mark could have paid was £6.50. ——— You get 1 mark for accurately solving the inequality.

Total: 4 marks

You get 1 accuracy mark for taking the next integer value below the inequality as the answer.

AU **3** The solution of the equation $5x + 3 = 3x + 7$ is $x = 2$.

Write down a different equation of the form $ax + b = cx + d$ for which the solution is also $x = 2$.

$6 \times 2 + b = 3 \times 2 + d$ ——————— Start with a value for a and c, say 6 and 3 and work out 6×2 on the left and 3×2 on the right. This is worth 2 marks.

$12 + 7 = 6 + 13$ ——————

So $6x + 7 = 3x + 13$ has a solution of $x = 2$. ———

However, there are many other possible answers.

Add values to both sides to get the same total. This is worth 1 mark.

Total: 4 marks

You get 1 mark for writing down the equation.

Buildings and bridges have to have incredible strength and be able to withstand different weather conditions. Many buildings use the triangle as the basis of their design. Triangles have interesting properties that can be represented using algebra.

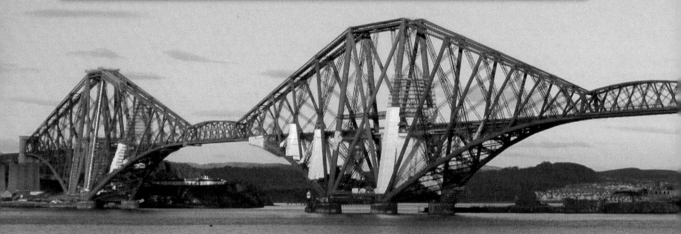

Look at the buildings in the photographs.

How many triangles can you see? What kind of triangles can you spot? Why do you think so many buildings use triangles in their design?

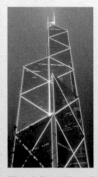

Getting started
You have the following straight pieces of wood.

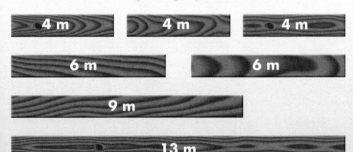

4 m 4 m 4 m
6 m 6 m
9 m
13 m

- What combination can you use to make an equilateral triangular frame?
- What combination can you use to make an isosceles triangular frame?
- What combinations will not make a triangle?

Handy hints
You might find that tables are a useful way to record your results. For example:
Right-angled triangles

a	b	c	a²	b²	c²
6	8	10	36	64	100

Your task

You are going to investigate the relationship between angles and side length in different types of triangle.

Start by constructing several triangles, as described below:

- Draw three triangles, each with one side of 8 cm and one of 6 cm. Use compasses to mark off the lengths.
- Draw one right-angled triangle, one obtuse-angled triangle and one acute-angled triangle. In each case measure the third side.
- Repeat another four times with two other starting lengths.
- Label the two sides you started with a and b and the side you measured c.

Now investigate the relationship between a^2, b^2 and c^2 and whether the angle is a right-angle, obtuse or acute.

Explain your findings by writing down mathematical rules such as these:

- When a triangle with sides a, b and c is right-angled then ...
- When a triangle with sides a, b and c is obtuse-angled then ...
- When a triangle with sides a, b and c is acute-angled then ...

Resources required

- Compasses
- Ruler

Why this chapter matters

For thousands of years, people have tried to solve equations and express them in a symbolic or algebraic way.

1850 BC — The Rhind Papyrus was written. This gave some insight into the mathematics used by the Ancient Egyptians. From this it was clear that they could solve linear equations in one unknown but they had no symbolism. Problems were solved verbally.

1600 BC — In Babylonia mathematicians developed arithmetic methods for solving equations involving squares but they still did not use symbols to represent unknowns.

300 BC — The Cairo Papyrus was written. By this time Egyptian mathematics had moved on to far more complex equations that are met in A level mathematics. They still did not have any symbolism and despite their advanced algebra they only used fractions with a numerator of 1 which made some of their working really difficult.

250 AD — The Greek mathematician, Diophantus, was the first to use a type of symbolism to solve equations although it was many centuries before the verbal method of solution disappeared.

600 AD — Hindu mathematicians were influenced by what the Greek mathematicians had done and did a large amount of work on astronomy. They developed the base 10 (decimal) system and were the first to use the number zero and solve problems using this number. They also introduced negative numbers at this time.

1150 AD — The Hindu mathematician, Bhaskara, was the first to realise that a number such as 9 has two square roots (3 and –3). Hindu mathematicians were the first to use symbols.

1400 AD — Arab mathematicians working in Spain improved the Hindu number and symbolic system into the algebraic notation we use today.

1600 AD — Once symbolism was established mathematics made great advances, with methods for solving complicated equations being found.

1850 AD — With an established method of recording mathematics in a concise form, European mathematicians moved into the world of abstract algebra which deals with concepts such as the square root of –1. (This is beyond GCSE mathematics.)

Today — You are starting to learn a branch of mathematics that has its origins 4000 years ago.

Algebra: Review of algebra

This chapter will show you ...

- **F** how to use letters to represent numbers
- **to** **F** **E** how to solve linear equations with the variable on one side only
- **E** how to form simple algebraic expressions
- **E** how to simplify such expressions by collecting like terms
- **E** how to substitute numbers into expressions and formulae
- **D** how to factorise expressions
- **D** how to express simple rules in algebraic form
- **D** how to solve linear equations with the variable on both sides
- **C** how to solve equations using trial and improvement

Visual overview

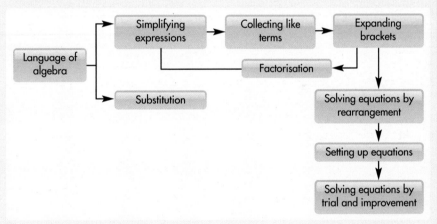

What you should already know

- The BODMAS/BIDMAS rule, which gives the order in which you must do the operations of arithmetic when they occur together (KS3 level 5, GCSE grade F)

Quick check

1 Write down the value of each expression.
   a $(10 - 2) \times 3$ **b** $10 - (2 \times 3)$
2 Work out $(6 - 3) \times (6 + 2 - 1)$
3 **a** Put brackets in the calculation to make the answer 21.
      $2 + 5 \times 3$
   b Put brackets in the calculation to make the answer 47.
      $2 + 5 \times 3 + 6$

Basic algebra

This section will show you how to:	Key words
• recall the rules of algebra	brackets
• simplify algebraic expressions by multiplying terms	equation
	expand
• simplify algebraic expressions by collecting like terms	expand and simplify
	expression
• expand and simplify brackets	factor
• factorise expressions	factorisation
	formula
• substitute numbers into expressions and formulae	like terms
	simplify
	substitution

Work through these examples to remind yourself of some of the algebra you will need for Unit 3.

EXAMPLE 1

State whether each of the following is an expression (E), an equation (Q) or a formula (F).

A: $x + 4 = 9$ **B:** $S = \dfrac{x}{9}$ **C:** $4x - 5y$

A is an equation (Q) as it can be solved to give $x = 5$.

B is a formula (F). This is the formula for the side of a square with a perimeter of x.

C is an expression (E) with two terms.

EXAMPLE 2

What is the perimeter and area of each of these squares?

a side of 8 cm   **b** side of x m

a Perimeter = $4 \times 8 = 32$ cm

Area = $8 \times 8 = 64$ cm^2

b Perimeter = $4 \times x = 4x$ m

Area = $x \times x = x^2$ m^2

EXAMPLE 3

Simplify:

a $2t \times 5t$ **b** $4t^2 \times t$ **c** $2t^2 \times 7t^3$

a $2t \times 5t = 10t^2$

b $4t^2 \times t = 4t^3$

c $2t^2 \times 7t^3 = 14t^5$

EXAMPLE 4

Simplify $6a + 7b + 5c^2 + 2a - 3b + 2c^2$

Write out the expressions with like terms next to each other.

$6a + 2a + 7b - 3b + 5c^2 + 2c^2$

$= 8a + 4b + 7c^2$

EXAMPLE 5

Expand   **a** $3(2x + 7)$ **b** $2x(3x - 4y)$

a $3 \times 2x + 3 \times 7 = 6x + 21$

b $2x \times 3x - 2x \times 4y = 6x^2 - 8xy$

EXAMPLE 6

Expand and simplify $6(2x + 5) - 3(3x - 1)$

$6(2x + 5) - 3(3x - 1) = 12x + 30 - 9x + 3$

$= 12x - 9x + 30 + 3$

$= 3x + 33$

EXAMPLE 7

Factorise the following expressions.

a $3x + 6$ **b** $4my + 12mx$ **c** $5kp - 10k^2p + 15kp^2$

a The common factor is 3, so $3x + 6 = 3(x + 2)$

b The common factor is $4m$, so $4my + 12mx = 4m(y + 3x)$

c The common factor is $5kp$, so $5kp - 10k^2p + 15kp^2 = 5kp(1 - 2k + 3p)$

EXAMPLE 8

The formula for the perimeter of a rectangle is $P = 2l + 2w$

Work out the perimeter when $l = 4.5$ cm and $w = 1.75$ cm

$P = 2(4.5) + 2(1.75) = 9 + 3.5 = 12.5$ cm

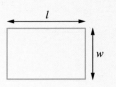

EXERCISE 9A

1 Rex has n marbles.

Stu has three times as many marbles as Rex.

Tamara has five more marbles than Rex.

Ursula has four fewer marbles than Rex.

Vic has six more marbles than Stu.

a How many marbles does each person have?

b How many marbles do they have altogether?

2 My wife is 35 years old and I am Y years old. How old will we be in z years' time?

3 Simplify each of the following expressions.

    a $5 \times 4x$ **b** $3w \times w$

    c $6h \times 3h$ **d** $5x + 7x$

    e $6z - z$ **f** $4y^2 + 5y^2 - 2y^2$

    g $3a + 7b + 8a - 2b$ **h** $8x + 7 - 5x - 9$

PS 4 My son is 24 years old. In 2 years' time he will be half as old as I am.

What age am I now?

AU 5 The answer to $5 \times 6w$ is $30w$.

Write down two **different** expressions for which the answer is $30w$.

PS 6 Alison has £1.25 and Helga has 90p more than Alison.

How much should Helga give to Alison so they both have the same amount?

FM 7 Syd measures the length and width of a football field using a long stick L and a short stick S.

He finds the length is $98L + 2S$ and the width is $52L + S$

a Work out the perimeter in terms of L and S.

b Later he finds that the long stick is 90 cm and the short stick is 20 cm.

What is the actual perimeter of the football field?

Give your answer in metres.

8 Find the value of the following expressions when a = 5, b = 3 and c = –4.

a $4a + 1$ b $6b - a$ c $a^2 + c^2$

d ab e $3ac$ f $4ab - 3bc$

9 State whether each of the following is an expression (E), an equation (Q) or a formula (F).

a $5x - 2$ b $A = s^2$ c $5x - 2 = 13$

AU 10 $5x + 2y + 3x - 5y = 8x - 3y$

Write down two other similar but **different** expressions for which the answer is $8x - 3y$.

FM 11 A taxi company uses the following rule to calculate their fares.

Fare = £3.50 plus £1.20 per kilometre

a How much is a journey of 3 kilometres?

b Frankie pays £8.30 for a taxi ride. How far was the journey?

c Mark knows that his house is 8 kilometres from town. He has £10.50.

Has he got enough for a taxi ride home?

AU 12 a Which of the following expressions are equivalent?

$3m \times 8n$ $2m \times 12n$ $4n \times 6m$ $m \times 24n$

b The expressions $\dfrac{x}{2}$ and x^2 are the same for only one positive value of x.

What is the value?

13 Expand these expressions.

a $5(3 - m)$ b $3(2x + 7)$ c $x(x + 2)$

d $2m(5 - m)$ e $5s(s + 3)$ f $3n(m - p)$

14 Factorise the following expressions.

a $18 - 3m$ b $6x + 12$ c $x^2 + 5x$

d $10m - m^2$ e $15s^2 + 3$ f $3n - pn$

D

AU **15** Find the missing terms to make these equations true.

a $8x + 12y - \boxed{} - \boxed{} = 5x + 4y$

b $3a - 5b - \boxed{} + \boxed{} = a - b$

PS **16** ABCDEF is an L-shape.

AB = DE = x

AF = $4x - 1$ and EF = $3x + 1$

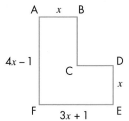

a Explain why the length BC = $3x - 1$.

b Find the perimeter of the shape in terms of x.

c If $x = 6$ cm what is the perimeter of the shape?

PS **17** A square and a rectangle have the same perimeter.

The rectangle has one side that is three times as long as the other.

The square has a side of 8 cm.

What are the dimensions of the rectangle?

AU **18** The formula for the area, A, of a square with side x is $A = x^2$.

The formula for the area, T, of a triangle with base b and height h is $T = \frac{1}{2}bh$

Find **different** values of x, b and h so that $A = T$

AU **19** Kaz knows that x, y and z have the values 3, 6 and 9 but he does not know which variable has which value.

a What is the maximum that the expression $x + 3y - 5z$ could be?

b What is the minimum value that the expression $2x - y + 3z$ could have?

FM **20** A car costs £90 per day to rent.

Some friends decide to rent the car for five days.

a Which of the following formulae would represent the cost per person if there are n people in the car and they share the cost equally?

$\dfrac{450}{5n}$ $\dfrac{450}{n + 5}$ $\dfrac{450}{n}$

b Three friends rent the car.

When they get the bill they find that there is a special price for a five-day rental.

They each find it cost them £20 less than they expected.

How much does a five-day rental cost?

21 Expand each of these expressions.

a $4p^2(3p - q)$ **b** $5t^2(2t^2 + 7)$ **c** $5x(2x + 7y)$

d $2m^2(5 - m^3)$ **e** $8s^3(s + 3t)$ **f** $6nm^2(m - n)$

FM 22 The local supermarket is offering £2 off a large box of chocolates. Madge wants four boxes.

a If the normal price of one box is £t, write down the expressions below that represent how much it will cost Madge to buy four boxes.

$4(t - 2)$ $4t - 1$ $4t - 4$ $4t - 8$

b The original price of a box of chocolates was £8.50. How much will Madge actually pay?

23 Expand and simplify the following expressions.

a $5(4x + 1) + 3(x + 2)$ **b** $4(y - 2) + 5(y + 3)$

c $2(3x - 2) - 4(x + 1)$ **d** $5(2x + 3) + 6(2x - 1)$

e $6x(2x - 3) + 2x(x + 4)$ **f** $3(4x^2 - 3) + x^2(5 + 2x)$

24 Factorise the following expressions.

a $9p^2 + 6pt$ **b** $12mp - 8m^2$

c $16a^2b + 4ab$ **d** $4a^2 - 6a + 2$

e $20xy^2 + 10x^2y + 5xy$ **f** $8mt^2 - 4m^2t$

AU 25 Darren wrote the following: $2(3x - 5) = 5x - 3$

Darren has made two mistakes.

Explain the mistakes that Darren has made.

PS 26 The expansion $3(x + 4) = 3x + 12$ can be shown by the diagram.

	x	4
3	$3x$	12

a What expansion is shown in this diagram?

4	$8y$	16

b Write down an expansion that is shown on this diagram.

$6z$	3

FM 27 In Highville school there are 2000 students. One day a student returns from abroad with an infectious disease. The following day the student is off school. The following day three students are off school with the disease and then each day three times more students than the day before are off. How many days will it be before there are no students left in school?

FM 28 A three-carriage train has $2f$ first-class seats and $2s$ standard-class seats.

A four-carriage train has $3f$ first-class seats and $3s$ standard-class seats.

On a weekday six three-carriage trains and three four-carriage trains travel from Bristol to Bournemouth.

a Write down an expression for the total number of first- and standard-class seats available during the day.

b On average in any day one-third of the first-class seats are used at a cost of £45. On average in any day four-fifths of the standard-class seats are used at a cost of £30.

How much money does the rail company earn in an average day on this route?

Give your answer in terms of f and s.

c $f = 16$ and $s = 50$. It costs the rail company £30 000 per day to operate this route. How much profit or loss do they make on an average day?

PS 29 A rectangle with sides 6 and $3x + 5$ has a smaller rectangle with sides 2 and $x - 2$ cut from it.

Work out the area remaining.

FM 30 Five friends have a meal together. They each have a main course costing £7.25 and a desert costing £2.75.

Colin says that the bill will be $5 \times 7.25 + 5 \times 2.75$.

Kim says that she has an easier way to work out the bill as $5 \times (7.25 + 2.75)$.

a Explain why Colin and Kim's methods both give the correct answer.

b Explain why Kim's method is better.

c What is the total bill?

AU 31 Three students are asked to factorise the expression $16m - 4$.

These are their answers

A	B	C
$2(8m - 2)$	$4(4m - 1)$	$8m\left(2 - \dfrac{1}{2m}\right)$

All the answers are accurately factorised but only one is the normally accepted answer.

a Which student gave the correct answer?

b Explain why the other two students' answers are not acceptable as correct answers.

PS 32 Explain why $7x - 9y$ cannot be factorised.

9.2 Substitution using a calculator

This section will show you how to:
- substitute numbers for letters in formulae and use a calculator to evaluate the resulting expression

Key words
brackets
calculator
formula
substitute

Although you have already **substituted** numbers into **formulae** in the previous exercise, you will meet harder examples where you will need to use a **calculator**.

Sometimes you need to work out the denominator of a fraction. You will need to use **brackets** to do this.

To find w if $w = \dfrac{x}{y + z}$

and $x = 8$, $y = 2.3$ and $z = 4.1$

key into your calculator:

8 ÷ (2 . 3 + 4 . 1) =

You should get the answer 1.25

Although the expression does not include brackets, you need to use them on your calculator.

Now try to work out the value of

$$\frac{4.7^2 - 5.2}{\sqrt{9.6 + 3.8}}$$

You will need to key into your calculator:

(4 · 7 x^2 − 5 · 2) ÷ ((9 · 6 + 3 · 8)) =

You should get the answer 4.614 to 3 dp.

EXERCISE 9B

1 Where $A = 4t + h$, find A when:

    a $t = 2.6$ and $h = 3.9$ **b** $t = 8.4$ and $h = 5.2$

    c $t = 0.8$ and $h = 2.2$

> **HINTS AND TIPS**
>
> With modern calculators you can type in the calculation as it reads. For example, $(1.7)^2 + 3.8$
>
> 1 · 7 x^2 +
> 3 · 8 =

2 Where $P = 5w - 4y$, find P when:

    a $w = 3.6$ and $y = 2.7$ **b** $w = 6.2$ and $y = 4.9$

    c $w = 2.5$ and $y = 0.7$

3 Where $A = b^2 + c$, find A when:

    a $b = 2.4$ and $c = 3.6$ **b** $b = 1.7$ and $c = 3.8$ **c** $b = 0.5$ and $c = 2.75$

4 Where $L = f^2 - g^2$, find L when:

    a $f = 6.4$ and $g = 3.6$ **b** $f = 3.9$ and $g = 2.1$ **c** $f = 5.5$ and $g = 3.5$

5 Where $T = P - n^2$, find T when:

    a $P = 10$ and $n = 1.6$ **b** $P = 5.9$ and $n = 2.3$ **c** $P = 11.4$ and $n = 3.2$

6 Where $A = 180(n - 2)$, find A when:

    a $n = 5.8$ **b** $n = 3.9$ **c** $n = 6.4$

7 Where $t = 10 - \sqrt{P}$, find t when:

    a $P = 1.96$ **b** $P = 27.04$ **c** $P = 2.56$

8 Where $W = v + \dfrac{m}{5}$, find W when:

    a $v = 3.2$ and $m = 7.4$ **b** $v = 2.9$ and $m = 10.6$ **c** $v = -4.6$ and $m = 7.5$

FM 9 The formula for the electricity bill each quarter in a household is £7.50 + £0.07 per unit.

A family uses 6720 units in a quarter.

a How much is their total bill?

b The family pay a direct debit of £120 per month towards their electricity costs.

By how much will they be in credit or debit after the quarter?

AU 10 x and y are different positive numbers.

Choose values for x and y so that the formula

$5x + 3y$

a evaluates to an odd number

b evaluates to a prime number.

> **HINTS AND TIPS**
>
> You will need to remember the prime numbers, 2, 3, 5, 7, 11, 13, 17, 19...

11 Find the value of the following expressions when $x = 1.4$, $y = 2.5$ and $z = 0.8$.

a $\dfrac{3x + 4}{2}$ **b** $\dfrac{x + 2y}{z}$ **c** $\dfrac{y}{z} + x$

FM 12 The formula for the gas bill each quarter in a household is £17.50 + £0.12 per unit.

A family uses 6250 units in a quarter.

a How much is their total bill?

b The family pay a direct debit of £220 per month towards their gas costs.

By how much will they be in credit or debit after the quarter?

AU 13 x and y are different prime numbers.

Choose values for x and y so that the formula

$5x + 2y$

a evaluates to an even number **b** evaluates to an odd number.

PS 14 Marvin hires a car for the day for £40.

He wants to know how much it costs him for each mile he drives.

Petrol is 98p per litre and the car does 10 miles per litre.

Marvin works out the following formula for the cost per mile, C, in pounds for M miles driven.

$$C = 0.098 + \frac{40}{M}$$

a Explain each term of the formula.

b How much will it cost per mile if Marvin drives 200 miles that day?

Solving linear equations

This section will show you how to:

- solve linear equations by rearrangement

Key words

do the same to both sides

inverse operations

rearrangement

solution

variable

Work through these examples to remind yourself how to solve equations.

EXAMPLE 9

Solve these equations.

a $x + 7 = 10$ **b** $5y = 30$ **c** $\dfrac{z}{3} = 5$

a Subtract 7 from both sides

$x + 7 - 7 = 10 - 7$

$x = 3$

b Divide both sides by 5

$\dfrac{5y}{5} = \dfrac{30}{5}$

$y = 6$

c Multiply both sides by 3

$3 \times \dfrac{z}{3} = 3 \times 5$

$z = 15$

Note: Remember to check your answers in the original equations

a $3 + 7 = 10$ ✓   **b** $5 \times 6 = 30$ ✓   **c** $15 \div 3 = 5$ ✓

EXAMPLE 10

Solve the following equations.

a $4x - 5 = 7$ **b** $\dfrac{y}{5} + 7 = 4$ **c** $\dfrac{z+4}{3} = 5$

a Add 5 to both sides $4x = 12$

  Divide both sides by 4 $x = 3$

  Check: $4 \times 3 - 5 = 12 - 5 = 7\ ✓$

b Subtract 7 from both sides $\dfrac{y}{5} = -3$

  Multiply both sides by 5 $y = -15$

  Check: $-15 \div 5 + 7 = -3 + 7 = 4\ ✓$

c Multiply both sides by 3 $z + 4 = 15$

  Subtract 4 from both sides $z = 11$

  Check: $(11 + 4) \div 3 = 15 \div 3 = 5\ ✓$

EXAMPLE 11

Solve the following equations.

a $2(x + 7) = 15$ **b** $4(y - 9) = 12$

a Expand the bracket $2x + 14 = 15$

  Subtract 14 $2x = 1$

  Divide by 2 $x = \dfrac{1}{2}$

  Check: $2 \times \left(\dfrac{1}{2} + 7\right) = 2 \times 7\dfrac{1}{2} = 15\ ✓$

b Expand the bracket $4y - 36 = 12$

  Add 36 $4y = 48$

  Divide by 4 $y = 12$

  Check: $4 \times (12 - 9) = 4 \times 3 = 12\ ✓$

EXAMPLE 12

Solve the following equations.

a $5x + 2 = 3x + 11$ **b** $6(x - 1) = 2x + 14$

a Rearrange the equations to get x terms on one side and
number terms on the other $5x - 3x = 11 - 2$

Collect terms $2x = 9$

Divide by 2 $x = 4\frac{1}{2}$

Check: left-hand side $5 \times 4\frac{1}{2} + 2 = 24\frac{1}{2}$

right-hand side $3 \times 4\frac{1}{2} + 11 = 24\frac{1}{2} =$ left-hand side ✓

b Expand the brackets $6x - 6 = 2x + 14$

Rearrange $6x - 2x = 14 + 6$

Collect terms $4x = 20$

Divide by 4 $x = 5$

Check: $6(5 - 1) = 24, \; 2 \times 5 + 14 = 10 + 14 = 24$ ✓

EXERCISE 9C

Solve the following equations.

1   **a** $x + 5 = 10$ **b** $y - 8 = 9$ **c** $4z = 30$ **d** $\frac{w}{5} = 10$

2   **a** $4x + 5 = 15$ **b** $3x - 7 = 23$ **c** $2x + 9 = 5$ **d** $6x - 5 = 10$

    e $\frac{x}{4} + 7 = 12$ **f** $\frac{x}{3} - 9 = 11$

3   **a** $\frac{x + 2}{5} = 3$ **b** $\frac{x - 7}{6} = 2$ **c** $\frac{x + 3}{2} = 1$ **d** $\frac{x - 1}{8} = 5$

4   **a** $3(x + 7) = 12$ **b** $4(x - 1) = 6$

    c $5(3x + 9) = 45$ **d** $3(2x - 7) = 12$

5   **a** $7x + 9 = 2x + 19$ **b** $4x - 8 = 3x + 7$

    c $3x + 5 = 5x - 9$ **d** $3x - 1 = 6 - 4x$

6   **a** $3(x + 9) = x + 3$ **b** $4(2x - 1) = 3(x + 7)$

    c $2(x + 8) + 3(x - 2) = x + 6$ **d** $5(x - 1) - 2(x - 7) = 4(x + 2)$

Setting up equations

This section will show you how to:

- set up equations from given information and then use the methods already seen to solve them

Key words

equation

variable

Equations are used to represent situations, so that you can solve real-life problems.

EXAMPLE 13

A milkman sets off from the dairy with eight crates of milk each containing b bottles. He delivers 92 bottles to a large factory and finds that he has exactly 100 bottles left on his milk float. How many bottles were in each crate?

The equation is:

$8b - 92 = 100$

$8b = 192$ (Add 92 to both sides.)

$b = 24$ (Divide both sides by 8.)

EXAMPLE 14

The rectangle shown has a perimeter of 40 cm.

Find the value of x.

The perimeter of the rectangle is:

$3x + 1 + x + 3 + 3x + 1 + x + 3 = 40$

This simplifies to $8x + 8 = 40$.

Subtract 8. $8x = 32$

Divide by 8. $x = 4$

$3x + 1$

$x + 3$

EXERCISE 9D

Set up an equation to represent each situation described below. Then solve the equation. Remember to check each answer.

PS 1 A teacher asks her class to think of a number and add 3 to it.

I thought of 6 to start.

My final answer was 11.

   a What was Adrianne's final answer?

   b What was Benjy's original number?

FM 2 Four friends shared the cost, £M, of a meal.

Ken collected each person's share and took the money to the till.

At the till he produced a coupon for £5 without telling his friends.

Ken only paid £7 from his own pocket.

   a Which of the following equations represents this situation?

$$4M - 5 = 7 \qquad \frac{(M - 5)}{4} = 7 \qquad \frac{M}{4} - 5 = 7$$

   b What was the cost, £M, of the meal?

FM 3 A carpet costs £12.75 a square metre. The shop charges £35 for fitting. The final bill was £137.

How many square metres of carpet were fitted?

FM 4 Moshin bought 8 garden chairs. When he got to the till he used a £10 voucher as part payment. His final bill was £56.

   a Set this problem up as an equation using c as the cost in pounds of one chair.

   b Solve the equation to work out the cost of one chair.

D

FM **5** This diagram shows the traffic flow through a one-way system in a town centre.

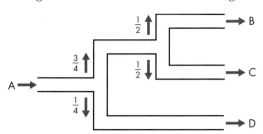

Cars enter at A and at each junction the fractions show the proportion of cars that take each route.

a 1200 cars enter at A. How many come out of each of the exits, B, C and D?

b If 300 cars exit at B, how many cars entered at A?

c If 500 cars exit at D, how many exit at B?

FM **6** A rectangular room is 3 m longer than it is wide. The perimeter is 16 m.

Carpet costs £9.00 per square metre.

How much will it cost to carpet the room?

> **HINTS AND TIPS**
>
> Set up an equation to work out the length and width, then calculate the area.

FM **7** A man buys a daily paper from Monday to Saturday for d pence. On Sunday he buys a Sunday paper for £1.80. His weekly paper bill is £7.20. What is the price of his daily paper?

8 The diagram shows a rectangle.

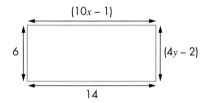

> **HINTS AND TIPS**
>
> Use the letter x for the variable unless you are given a letter to use. Once the equation is set up solve it by the methods above.

a What is the value of x?

b What is the value of y?

PS **9** In this rectangle, the length is 3 cm more than the width. The perimeter is 12 cm.

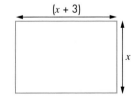

a What is the value of x?

b What is the area of the rectangle?

10 Mary has two bags, each of which contains the same number of sweets. She eats four sweets. She then finds that she has 30 sweets left. How many sweets were there in each bag to start with?

PS 11 A boy is *Y* years old. His father is 25 years older than he is. The sum of their ages is 31. How old is the boy?

PS 12 Another boy is *X* years old. His sister is twice as old as he is. The sum of their ages is 27. How old is the boy?

13 The diagram shows a square. Find *x* if the perimeter is 44 cm.

$(4x - 1)$

PS 14 Max thought of a number. He then multiplied his number by 3. He added 4 to the answer. He then doubled that answer to get a final value of 38. What number did he start with?

15 The angles of a triangle are $2x$, $x + 5°$ and $x + 35°$.

a Write down an equation to show this.

b Solve your equation to find the value of *x*.

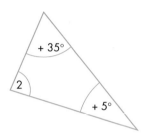

AU 16 The diagram shows two number machines that perform the same operations.

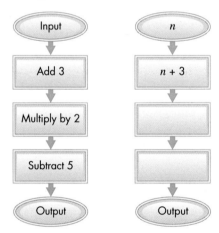

a Starting with an input value of 7 work through the left-hand machine to get the output.

b Find an input value that gives the same value for the output.

c Fill in the algebraic expressions in the right-hand machine for an input of *n*. (The first operation has been filled in for you.)

d Set up an equation for the same input and output and show each step in solving the equation to get the answer in part **b**.

PS 17 A teacher asked her class to find three angles of a triangle that were consecutive even numbers.

Tammy wrote:

$$x + x + 2 + x + 4 = 180$$
$$3x + 6 = 180$$
$$3x = 174$$
$$x = 58$$

So the angles are 58°, 60° and 62°.

The teacher then asked the class to find four angles of a quadrilateral that are consecutive even numbers.

Can this be done? Explain your answer.

FM 18 Mary has a large and a small bottle of pop. The large bottle holds 50 cl more than the small bottle.

From the large bottle she fills 4 cups and has 18 cl left over.

From the small bottle she fills 3 cups and has 1 cl left over.

How much pop does each bottle hold?

> **HINTS AND TIPS**
>
> Do the same type of working as Tammy for a triangle. Work out the value of x. What happens?

> **HINTS AND TIPS**
>
> Set up equations for both using x as the amount of pop in a cup. Put them equal but remember to add 50 to the small bottle equation to allow for the difference. Solve for x, then work out how much is in each bottle.

9.5 Trial and improvement

This section will show you how to:

- use the method of trial and improvement to estimate the answer to equations that do not have exact solutions

Key words

comment
decimal place
guess
trial and
    improvement

Certain equations cannot be solved exactly. However, a close enough solution to such an equation can be found by the **trial and improvement** method. (Sometimes wrongly called the trial and error method.)

The idea is to keep trying different values in the equation to take it closer and closer to the 'true' solution. This step-by-step process is continued until a value is found that gives a solution that is close enough to the accuracy required.

The trial and improvement method is the way in which computers are programmed to solve equations.

EXAMPLE 15

Solve the equation $x^3 + x = 105$, giving the solution correct to one **decimal place**.

Step 1 You must find the two consecutive whole numbers between which x lies. You do this by intelligent guessing.

Try $x = 5$: $125 + 5 = 130$	Too high – next trial needs to be much smaller.
Try $x = 4$: $64 + 4 = 68$	Too low.

So you now know that the solution lies between $x = 4$ and $x = 5$.

Step 2 You must find the two consecutive one-decimal-place numbers between which x lies. Try 4.5, which is halfway between 4 and 5.

This gives $91.125 + 4.5 = 95.625$	Too small.

Now attempt to improve this by trying 4.6.

This gives $97.336 + 4.6 = 101.936$	Still too small.
Try 4.7 which gives 108.523.	This is too high, so the solution is between 4.6 and 4.7.

It looks as though 4.7 is closer but there is a very important final step.

Step 3 Now try the value that is halfway between the two one-decimal-place values. In this case 4.65.

This gives 105.194 625.

This means that 4.6 is nearer the actual solution than 4.7 is, so never assume that the one-decimal-place number that gives the closest value to the solution is the answer.

The diagram on the right shows why this is.

The approximate answer is $x = 4.6$ to 1 decimal place.

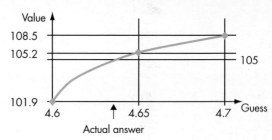

The best way to answer this type of question is to set up a table to show working. There will be three columns: **guess** (the trial); the equation to be solved; and a **comment** whether the value of the equation is too high or too low.

Guess	$x^3 + x$	Comment
4	68	Too low
5	130	Too high
4.5	95.625	Too low
4.6	101.936	Too low
4.7	108.523	Too high
4.65	105.194 625	Too high

EXERCISE 9E

1 Find the two consecutive *whole numbers* between which the solution to each of the following equations lies.

a $x^2 + x = 24$ **b** $x^3 + 2x = 80$ **c** $x^3 - x = 20$

2 Copy and complete the table by using trial and improvement to find an approximate solution to:

$$x^3 + 2x = 50$$

Give your answer correct to 1 decimal place.

Guess	$x^3 + 2x$	Comment
3	33	Too low
4	72	Too high

3 Copy and complete the table by using trial and improvement to find an approximate solution to:

$$x^3 - 3x = 40$$

Give your answer correct to 1 decimal place.

Guess	$x^3 - 3x$	Comment
4	52	Too high

4 Use trial and improvement to find an approximate solution to:

$$2x^3 + x = 35$$

Give your answer correct to 1 decimal place.

You are given that the solution lies between 2 and 3.

> **HINTS AND TIPS**
>
> Set up a table to show your working. This makes it easier for you to show method and the examiner to mark.

5 Use trial and improvement to find an exact solution to:

$$4x^2 + 2x = 12$$

Do not use a calculator.

6 Find a solution to each of the following equations, correct to 1 decimal place.

a $2x^3 + 3x = 35$ **b** $3x^3 - 4x = 52$ **c** $2x^3 + 5x = 79$

PS 7 A rectangle has an area of 100 cm^2. Its length is 5 cm longer than its width.

a Show that, if x is the width, then $x^2 + 5x = 100$.

b Find, correct to 1 decimal place, the dimensions of the rectangle.

8 Use trial and improvement to find a solution to the equation $x^2 + x = 40$.

PS 9 A cuboidal juice carton holds $\frac{1}{2}$ litre (500 cm^3).

The sides of the base are in the ratio 1 : 2.

The height is 8 cm more than the shorter side of the base.

Use trial and improvement to find the dimensions of the carton.

> **HINTS AND TIPS**
>
> Call the length of the side with 'ratio 1' x, write down the other two sides in terms of x and then write down an equation for the volume = 500.

AU **10** A cube of side x cm has a square hole of side $\frac{x}{2}$ and depth 8 cm cut from it.

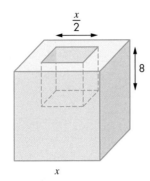

The volume of the remaining solid is 1500 cm³.

a Explain why $x^3 - 2x^2 = 1500$.

b Use trial and improvement to find the value of x to 1 decimal place.

PS **11** Two numbers a and b are such that $ab = 20$ and $a + b = 10$.

Use trial and improvement to find the two numbers to two decimal places.

GRADE BOOSTER

F You can solve equations such as $5x = 15$ and $x + 9 = 12$

E You can solve equations such as $4x + 7 = 3$ or $\dfrac{x}{2} - 5 = 6$

D You can solve equations such as $\dfrac{x - 2}{3} = 8$ or $3x - 8 = x + 7$

D You can set up equations from given information

C You can solve equations such as $4(x - 5) = 2x + 3$

C You can solve equations by trial and improvement

What you should know now

- How to set up equations
- How to simplify and factorise expressions, and substitute numbers into expressions
- How to solve a variety of linear equations using rearrangement or 'doing the same thing to both sides'
- How to solve equations by trial and improvement

1 Are the following statements true (T) or false (F)?

a c multiplied by 3 is written as $3c$ (1)

b d divided by 2 is written as $\frac{d}{2}$ (1)

c a subtracted from b is written as $a - b$ (1)

(Total 3 marks)

AQA, June 2008, Paper 1 Foundation, Question 6

2 a Suki is playing a 'Think of a Number' game.

I think of a number.

I multiply it by 5 and then subtract 3.

The answer is 27.

What number does Suki think of? (2)

b Tim is also playing a 'Think of a Number' game.

I think of a number.

I call it x. I add 2 to my number and then multiply by 5.

The answer is ...

Write down an expression in terms of x for Tim's answer. (2)

(Total 4 marks)

AQA, June 2006, Paper 1 Intermediate, Question 5

3 Alan has some unknown weights labelled a and b and some 5 kg and 10 kg weights.

He finds that the following combinations of weights balance.

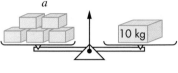

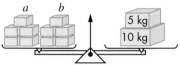

a i Find the value of a. (1)

ii Find the value of b. (2)

b Alan also has some unknown weights labelled c.

He finds that $5c + 2b = c + 6a$

Find the value of c. (4)

(Total 7 marks)

AQA, November 2007, Paper 1 Intermediate, Question 10

4 a Simplify $4c - c + 2c$ (1)

b Solve the equations

i $2x = 24$ (1)

ii $y - 9 = 11$ (1)

iii $\frac{z}{4} = 8$ (1)

iv $4w + 3 = 13$ (2)

(Total 6 marks)

AQA, June 2006, Paper 1 Foundation, Question 17

5 a Simplify $2x + 8 + 4x - 3$ (2)

b Solve the equation $\frac{x}{3} = 5$ (1)

c Tom is investigating the two expressions

$ab + c$ and $a(b + c)$

i He finds that both expressions have the same value when $a = 1$, $b = 3$ and $c = 4$

Show that this is true. (3)

ii Tom says that this means that

$a(b + c) = ab + c$

Explain why Tom is wrong. (2)

(Total 8 marks)

AQA, June 2006, Paper 1 Intermediate, Question 1

C   D   E   F

6 Ali is x cm tall.

a Suki is 5 cm taller than Ali.

Write down an expression in x for Suki's height. (1)

b Ali's sister is 2 cm shorter than Ali.

Write down an expression in x for the height of Ali's sister. (1)

c Ali's father is twice as tall as Ali.

Write down an expression in x for the height of Ali's father. (1)

d Darius has a height, in cm, given by the expression $2x - 65$

He is 115 cm tall.

Solve the equation

$2x - 65 = 115$

to find Darius's height. (2)

(Total 5 marks)

AQA, June 2005, Paper 2 Foundation, Question 19

7 All areas in this question are in square centimetres.

Here is a rectangle of area R, a square of area S and a trapezium of area T.

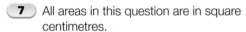

a The area of the shape below is given by $A = R + 2T$

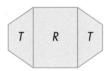

Find the value of A when $R = 7.5$ and $T = 6.3$ (2)

b Here is a different shape.

i Write down an expression for the area of this shape. (1)

ii Which of the following is correct?

$3R = S$ $2R = S$ $R = 2S$ $R = 3S$ (1)

(Total 4 marks)

AQA, November 2007, Paper 1 Intermediate, Question 5

8 Use the formula

$v = u + at$

to find the value of v when $u = -10$, $a = 1.8$ and $t = 3.7$ (2)

(Total 2 marks)

AQA, June 2005, Paper 2 Foundation, Question 24

9 Dean picks three numbers.

His first number is y.

His second number is five more than his first number.

a Write down his second number in terms of y. (1)

b His third number is double his first number.

Write down his third number in terms of y. (1)

c Write down an expression for the sum of the three numbers. (1)

d The sum of the three numbers is 77.

Form an equation and solve it to find the value of y. (3)

(Total 6 marks)

AQA, June 2007, Paper 2 Intermediate, Question 6

10 Complete the following table. (3)

$x = 8$	$3x = 24$
$y = $	$4y = 20$
$3z = 12$	$5z = $ ......

(Total 3 marks)

AQA, June 2007, Paper 1 Foundation, Question 19

11 Two car hire firms use different ways of charging for the hire of a car.

a Cheap Days uses this formula

$$H = 50d + 120$$

H is the hire charge in pounds.

d is the number of days the car is hired.

Work out H when $d = 2$ (2)

b Cheap Miles uses this formula.

$$H = \frac{m + 750}{5}$$

H is the hire charge in pounds.

m is the number of miles the car travels.

Work out m when $H = 200$ (2)

(Total 4 marks)

AQA, June 2007, Paper 1 Intermediate, Question 10

 12 Each expression in this wall is formed by adding the two supporting expressions from the row below.

For example

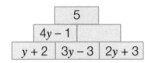

$$x + 3 + 2x + 5 = 3x + 8$$

Use the wall below to find the value of y. (3)

	5	
	$4y - 1$	
$y + 2$	$3y - 3$	$2y + 3$

(Total 3 marks)

AQA, June 2008, Paper 2 Foundation, Question 24

13 Kerry is using trial and improvement to find a solution to the equation

$$8x - x^3 = 5$$

Her first two trials are shown in the table.

x	$8x - x^3$	Comment
2	8	too high
3	−3	too low

Continue the table to find a solution to the equation.

Give your answer to one decimal place. (3)

(Total 3 marks)

AQA, June 2007, Paper 2 Intermediate, Question 11

 14 Use trial and improvement to find a solution to the equation

$$x^3 + 2x = 60$$

Give your answer to 1 decimal place.

You must show your working. (4)

(Total 4 marks)

AQA, November 2007, Paper 2 Intermediate, Question 15

Worked Examination Questions

PS **1** The angles of a quadrilateral are 73°, $2x°$, $3x°$ and 102°.

Find the value of the largest angle in the quadrilateral.

> The key word here is quadrilateral. You should know that the angles in a quadrilateral add up to 360°.

$3x + 2x + 102 + 73 = 360$

> Showing that you know that the sum of the angles equals 360 would get 1 method mark.

$5x + 175 = 360$

> Collecting terms together correctly would get 1 accuracy mark.

Largest angle = 111°

> Solving the equation gives:
> subtracting 175 $5x = 360 - 175 = 185$
> dividing by 5 $x = 37$
> Do not stop at this point as the question asks for the largest angle.
> This is the angle $3x°$, which is $3 \times 37 = 111°$.
> This gets 1 accuracy mark.

Total: 3 marks

2 Mark is x years old.

Nell is eight years older than Mark.

Oliver is twice as old as Nell.

Given that their total age is 44, find Mark's age.

Nell: $x + 8$

Oliver: $2(x + 8)$

> Write down their ages in terms of Mark's age x.
> '8 years older' means an addition, so Nell's age is $x + 8$.
> 'Twice as' means multiply. Use brackets around Nell's age, as $2x + 8$ is wrong.
> This gets 1 method mark.

$x + x + 8 + 2(x + 8)$

$= 2x + 8 + 2x + 16$

$= 4x + 24$

> Write down the total of all the ages.
> Expand the brackets and collect terms.
> This gets 1 accuracy mark.

$4x + 24 = 44$

$4x = 20$

$x = 5$

Mark is five years old.

> Set up the equation and expand the brackets.
> Subtract 24, then divide by 4.
> Setting up the equation gets 1 method mark and solving it correctly gets 1 accuracy mark.

Total: 4 marks

AU **3** The solution of the equation $3x + 7 = 4$ is $x = -1$.

Write down a **different** equation of the form $ax + b = 4$ for which the solution is also $x = -1$.

$4 \times -1 = -4$

> Start with a value for a, say 4, and work out 4×-1.

$-4 + 6 = 2$

So $4x + 6 = 2$ has a solution of $x = -1$

> Add a value, say 6.

(There are many other possible answers.)

> Write down the equation.
> This gets 2 marks for independent working.

Total: 2 marks

In Chapter 4 you may have done a Functional Maths task called 'Walking using Naismith's rule'. As you are reviewing your basic algebra skills, now is a good time to revisit this task.

In 'Revisiting Naismith's rule' you will complete questions similar to those in Chapter 4. You will also investigate Naismith's rule in a new way, developing and extending your algebra skills in a familiar context.

Naismith's rule

Naismith's rule is a rule of thumb that you can use when planning a walk to help you calculate how long it will take. The rule was devised by William Naismith, a Scottish mountaineer, in 1892.

The basic rule is:

Allow 1 hour for every 3 miles (5 km) forward, plus $\frac{1}{2}$ hour for every 1000 feet (300 m) ascent (height).

Getting started

Use algebra to write Naismith's rule. Now, use this rule to copy and complete the table below.

The table that you have completed shows the actual times taken by a school group as they did five different walks in five days. Use this information to work out the following.

● If the group had started at the same times and had the same breaks how long would the group have taken each day, according to Naismith's rule?

● Do you think Naismith's rule is still valid today? Explain your reasons.

Day	Distance (km)	Time (minutes)	Time (hours/ minutes)	Height (m)	Start	Breaks
1	16			250	10.00	2 h
2	18			0	10.00	1 h 30 m
3	11			340	09.30	2 h 30 m
4	13			100	10.30	2 h 30 m
5	14			120	10.30	2 h 30 m

Your task

Use the internet to research the walking times of some of Britain's most famous walks. Produce a report that compares and contrasts Britain's most famous walks, such as Ben Nevis, Snowdon, Helvellyn and the Pennine Way. Your report should contain realistic guidance on how best to approach these walks, including:

- suggested day-by-day plans supported by mathematical evidence
- starting times
- places to rest
- how the walks will vary for walkers of different fitness levels
- how weather conditions could affect the walk and precautions that should be taken.

Using this information, evaluate how similar the walks are, and which walk would be the toughest for an average walker to complete.

One day you will probably want to drive a car.

You might have a shock when you find out how much car insurance costs.

When an insurance company want to decide what to charge for car insurance, they employ mathematicians to help them. People who do this job are called actuaries.

They are very skilled at calculating probabilities, the branch of mathematics that describes the chance of outcomes.

Here is a simple example.

Suppose the chance of an 18-year-old driver having an accident while driving a car for a year is 0.1 or $\frac{1}{10}$.

Suppose also that the average repair cost is £2000. Then a fair charge for year's insurance is $\frac{1}{10}$ of £2000 which is £200. Of course real life is much more complicated than that!

What actuaries do is compile statistics about road accidents. They look at all the factors they can think of. These include age and gender of driver and where they live, the type and age of car, repair costs, and so on. They then use probabilities to calculate the likely amount the insurance company will have to pay out and use that to set the cost of the insurance. The cost is called the premium.

This is a very complicated and difficult task. Different companies might come up with different answers because they make judgements according to their beliefs and experience, which is one reason why premiums can vary from one company to another.

Actuaries work out premiums for all types of insurance, not just car insurance. Whenever you buy an insurance policy you can be sure that actuaries have used mathematical probabilities to decide on the premium you should pay.

UNITED KINGDOM, FEMALES
Period expectations of life (years)
Based on historical mortality rates from 1981 to 2008
Produced by the Office for National Statistics

Attained age (years)	2009	2010	2011	2012	2013	2014	2015	2016	2017	2018
11	71.6	71.9	72.2	72.4	72.7	72.9	73.1	73.4	73.6	73.8
12	70.6	70.9	71.2	71.4	71.7	71.9	72.1	72.4	72.6	72.8
13	69.6	69.9	70.2	70.4	70.7	70.9	71.1	71.4	71.6	71.8
14	68.6	68.9	69.2	69.4	69.7	69.9	70.1	70.4	70.6	70.8
15	67.6	67.9	68.2	68.4	68.7	68.9	69.2	69.4	69.6	69.8
16	66.6	66.9	67.2	67.4	67.7	67.9	68.2	68.4	68.6	68.8
17	65.6	65.9	66.2	66.4	66.7	66.9	67.2	67.4	67.6	67.8
18	64.7	65	65.2	65.5	65.7	65.9	66.2	66.4	66.6	66.9
19	63.7	64	64.2	64.5	64.7	65	65.2	65.4	65.6	65.9
20	62.7	63	63.2	63.5	63.7	64	64.2	64.4	64.7	64.9

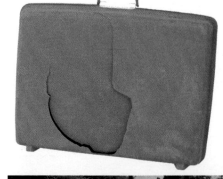

Actuaries do not just work for insurance companies, they are employed by the government too. They compile and use expected life tables like the one illustrated, from the Office of National Statistics. There are many reasons for this. For example, the government uses the information to estimate how much the cost of providing state pensions will be in the future.

Actuaries use mathematical probabilities for many different situations.

Probability: Probability and events 2

This chapter will show you ...

- **E** how to use two-way tables
- **E** how to calculate probability when two events happen at the same time
- **D** how to work out the probability of mutually exclusive events, using the addition rule
- **C** how to calculate experimental probabilities
- **C** how to predict outcomes, using theoretical models, and compare experimental and theoretical data

Visual overview

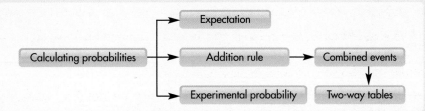

What you should already know

- How to add, subtract and cancel fractions **(KS3 level 4, GCSE grade F)**
- That outcomes of events cannot always be predicted and that the laws of chance apply to everyday events **(KS3 level 4, GCSE grade F)**
- How to list all the outcomes of an event in a systematic manner **(KS3 level 5, GCSE grade E)**

Quick check

1 Cancel the following fractions.

   a $\frac{6}{8}$ **b** $\frac{6}{36}$ **c** $\frac{3}{12}$ **d** $\frac{8}{10}$ **e** $\frac{6}{9}$ **f** $\frac{5}{20}$

2 Calculate the following.

   a $\frac{1}{8}+\frac{3}{8}$ **b** $\frac{5}{12}+\frac{3}{12}$ **c** $\frac{5}{36}+\frac{3}{36}$ **d** $\frac{2}{9}+\frac{1}{6}$ **e** $\frac{3}{5}+\frac{3}{20}$

3 Frank likes to wear brightly coloured hats and socks.

   He has two hats, one is green and the other is yellow.
   He has three pairs of socks, which are red, purple and pink.

   Write down all the six possible combinations of hats and socks Frank could wear.

   For example, he could wear a green hat and red socks.

This section will show you how to:
- work out the probability of two outcomes of events such as P(A) or P(B)

Key words

event

mutually exclusive

You have used this rule already but it has not yet been formally defined.

Mutually exclusive events are ones that cannot happen at the same time, such as throwing an odd number and an even number on a roll of a dice.

When two events are mutually exclusive, you can work out the probability of either of them occurring by adding up the separate probabilities. Mutually exclusive events are events for which, when one occurs, it does not have any effect on the probability of other events.

EXAMPLE 1

A bag contains 12 red balls, 8 green balls, 5 blue balls and 15 black balls. A ball is drawn at random. What is the probability that it is the following:

    a red   **b** black   **c** red or black   **d** not green?

a $P(\text{red}) = \frac{12}{40} = \frac{3}{10}$

b $P(\text{black}) = \frac{15}{40} = \frac{3}{8}$

c $P(\text{red or black}) = P(\text{red}) + P(\text{black}) = \frac{3}{10} + \frac{3}{8} = \frac{27}{40}$

d $P(\text{not green}) = \frac{32}{40} = \frac{4}{5}$

EXERCISE 10A

1. Iqbal throws an ordinary dice. What is the probability that he throws:

   a a 2   **b** a 5   **c** a 2 or a 5?

2. Jennifer draws a card from a pack of cards. What is the probability that she draws:

   a a Heart   **b** a Club   **c** a Heart or a Club?

3. A letter is chosen at random from the letters in the word PROBABILITY. What is the probability that the letter will be:

   a B   **b** a vowel   **c** B or a vowel?

4. A bag contains 10 white balls, 12 black balls and 8 red balls. A ball is drawn at random from the bag. What is the probability that it will be:

   a white   **b** black

   c black or white   **d** not red

   e not red or black?

   > **HINTS AND TIPS**
   >
   > You can only add fractions with the same denominator.

5. At the local School Fayre the tombola stall gives out a prize if you draw from the drum a numbered ticket that ends in 0 or 5. There are 300 tickets in the drum altogether and the probability of getting a winning ticket is 0.4.

   a What is the probability of getting a losing ticket?

   b How many winning tickets are there in the drum?

6. John needs his calculator for his mathematics lesson. It is always in his pocket, bag or locker. The probability it is in his pocket is 0.35 and the probability it is in his bag is 0.45. What is the probability that:

   a he will have the calculator for the lesson

   b his calculator is in his locker?

7. Aneesa has 20 unlabelled CDs, 12 of which are rock, 5 are pop and 3 are classical. She picks a CD at random. What is the probability that it will be:

   a rock or pop

   b pop or classical

   c not pop?

AU 8. The probability that it rains on Monday is 0.5. The probability that it rains on Tuesday is 0.5 and the probability that it rains on Wednesday is 0.5. Kelly argues that it is certain to rain on Monday, Tuesday or Wednesday because 0.5 + 0.5 + 0.5 = 1.5, which is bigger than 1 so it is a certain event. Explain why she is wrong.

FM 9 In a TV game show, contestants throw darts at the dartboard shown.

The angle at the centre of each black sector is 15°.

If a dart lands in a black sector the contestant loses.

Any dart missing the board is rethrown.

What is the probability that a contestant throwing a dart at random does not lose?

PS 10 There are 45 patients sitting in the hospital waiting room.

8 patients are waiting for Dr Speed.
12 patients are waiting for Dr Mayne.
9 patients are waiting for Dr Kildare.
10 patients are waiting for Dr Pattell.
6 patients are waiting for Dr Stone.

A patient suddenly has to go home.

What is the probability that the patient who left was due to see Dr Speed?

AU 11 The probability of it snowing on any one day in February is $\frac{1}{4}$.

One year, there was no snow for the first 14 days.

Ciara said: "The chance of it snowing on any day in the rest of February must now be $\frac{1}{2}$."

Explain why Ciara is wrong.

AU 12
PS At morning break, Pauline has a choice of coffee, tea or hot chocolate.
She also has a choice of a ginger biscuit, a rich tea biscuit or a doughnut.

The probabilities that she chooses each drink and snack are:

Drink	Coffee (C)	Tea (T)	Hot chocolate (H)
Probability	0.2	0.5	0.3

Snack	Ginger biscuit (G)	Rich tea biscuit (R)	Doughnut (D)
Probability	0.3	0.1	0.6

a Leon says that the probability that Pauline has coffee and a ginger biscuit is 0.5. Explain why Leon is wrong.

b There are nine possible combinations of drink and snack. Two of these are coffee and a ginger biscuit (C, G), coffee and rich tea biscuit (C, R).

Write down the other seven combinations.

c Leon now says 'I was wrong before. As there are nine possibilities the probability of Pauline having coffee and a ginger biscuit is $\frac{1}{9}$.'

Explain why Leon is wrong again.

d In fact the probability that Pauline chooses coffee and a ginger biscuit is $0.2 \times 0.3 = 0.06$, and the probability that she chooses coffee and a rich tea biscuit is $0.2 \times 0.1 = 0.02$.

   i Work out the other seven probabilities.

   ii Add up all nine probabilities. Explain the result.

Experimental probability

This section will show you how to:

- calculate experimental probabilities and relative frequencies from experiments
- recognise different methods for estimating probabilities

Key words

bias

equally likely

experimental data

experimental probability

historical data

relative frequency

trials

ACTIVITY

Heads or tails?

Toss a coin 10 times and record the results like this.

Record how many heads you obtained.

Now repeat the above so that altogether you toss the coin 50 times. Record your results and count how many heads you obtained.

Now toss the coin another 50 times and once again record your results and count the heads.

It helps if you work with a partner. First, your partner records while you toss the coin. Then you swap over and record, while your partner tosses the coin. Add the number of heads you obtained to the number your partner obtained.

Now find three more people to do the same activity and add together the number of heads that all five of you obtained.

Now find five more people and add their results to the previous total.

Combine as many results together as possible.

You should now be able to fill in a table like the one on the next page. The first column is the number of times coins were tossed. The second column is the number of heads obtained. The third column is the number in the second column divided by the number in the first column.

The results below are from a group who did the same experiment.

Number of tosses	Number of heads	Number of heads / Number of tosses
10	6	0.6
50	24	0.48
100	47	0.47
200	92	0.46
500	237	0.474
1000	488	0.488
2000	960	0.48
5000	2482	0.4964

If you drew a graph of these results, plotting the first column against the last column, it would look like this.

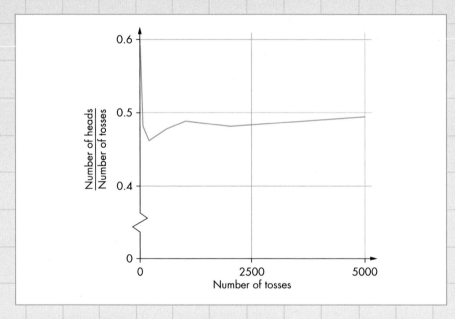

Your results should look very similar.

What happens to the value of $\dfrac{\text{number of heads}}{\text{number of tosses}}$ as the total number of tosses increases?

You should find that it gets closer and closer to 0.5.

The value of 'number of heads ÷ number of tosses' is called an **experimental probability**. As the number of **trials**, or experiments, increases, the value of the experimental probability gets closer to the true or theoretical probability.

Experimental probability is also known as the **relative frequency** of an event.
The relative frequency of an event is an estimate for the theoretical probability. It is given by:

$$\text{relative frequency of an outcome or event} = \frac{\text{frequency of the outcome or event}}{\text{total number of trials}}$$

EXAMPLE 2

The frequency table shows the speeds of 160 vehicles that pass a radar speed check on a dual carriageway.

Speed (mph)	20–29	30–39	40–49	50–59	60–69	70+
Frequency	14	23	28	35	52	8

a   What is the experimental probability that a car is travelling faster than 70 mph?

b   If 500 vehicles pass the speed check, estimate how many will be travelling faster than 70 mph.

a   The experimental probability is the relative frequency, which is $\frac{8}{160} = \frac{1}{20}$.

b   The number of vehicles travelling faster than 70 mph will be $\frac{1}{20}$ of 500.

That is:

   $500 \div 20 = 25$

Finding probabilities

There are three ways in which the probability of an event can be found.

- **First method** If you can work out the theoretical probability of an outcome or event – for example, drawing a King from a pack of cards – this is called using **equally likely** outcomes.

- **Second method** Some probabilities, such as people buying a certain brand of dog food, cannot be calculated using equally likely outcomes. To find the probabilities for such an event, you can perform an experiment such as the one in the Activity on pages 213–214, or conduct a survey. This is called collecting **experimental data**. The more data you collect, the better the estimate is.

- **Third method** The probabilities of some events, such as an earthquake occurring in Japan, cannot be found by either of the above methods. One of the things you can do is to look at data collected over a long period of time and make an estimate (sometimes called a 'best guess') at the chance of the event happening. This is called looking at **historical data**.

EXAMPLE 3

Which method (A, B or C) would you use to estimate the probabilities for the events **a** to **e**?

A: Use equally likely outcomes

B: Conduct a survey or collect data

C: Look at historical data

a Someone in your class will go abroad for a holiday this year.

b You will win the National Lottery.

c Your bus home will be late.

d It will snow on Christmas Day.

e You will pick a red seven from a pack of cards.

a You would have to ask all the members of your class what they intended to do for their holidays this year. You would therefore conduct a survey, Method B.

b The odds on winning are about 14 million to 1, so this is an equally likely outcome, Method A.

c If you catch the bus every day, you can collect data over several weeks. This would be Method C.

d If you check whether it snowed on Christmas Day for the last few years, you would be able to make a good estimate of the probability. This would be Method C.

e There are 2 red sevens out of 52 cards, so the probability of picking one can be calculated:

$$P(\text{red seven}) = \frac{2}{52} = \frac{1}{26}$$

This is Method A.

EXERCISE 10B

D

1 Which of these methods would you use to estimate or state the probabilities for each of the events **a** to **h**?

Method A: Use equally likely outcomes

Method B: Conduct a survey or experiment

Method C: Look at historical data

a How people will vote in the next election.

b A drawing pin dropped on a desk will land point up.

c A Premier League team will win the FA Cup.

d You will win a school raffle.

e The next car to drive down the road will be red.

f You will throw a 'double six' with two dice.

g Someone in your class likes classical music.

h A person picked at random from your school will be a vegetarian.

2 Naseer throws a fair, six-sided dice and records the number of sixes that he gets after various numbers of throws. The table shows his results.

Number of throws	10	50	100	200	500	1000	2000
Number of sixes	2	4	10	21	74	163	329

a Calculate the experimental probability of scoring a 6 at each stage that Naseer recorded his results.

b How many ways can a dice land?

c How many of these ways give a 6?

d What is the theoretical probability of throwing a 6 with a dice?

e If Naseer threw the dice a total of 6000 times, how many sixes would you expect him to get?

3 Marie made a five-sided spinner, like the one shown in the diagram. She used it to play a board game with her friend Sarah.

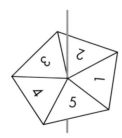

The girls thought that the spinner was not very fair as it seemed to land on some numbers more than others. They threw the spinner 200 times and recorded the results. The results are shown in the table.

Side spinner lands on	1	2	3	4	5
Number of times	19	27	32	53	69

a Work out the experimental probability of each number.

b How many times would you expect each number to occur if the spinner is fair?

c Do you think that the spinner is fair? Give a reason for your answer.

4 A sampling bottle contains 20 balls. The balls are either black or white. (A sampling bottle is a sealed bottle with a clear plastic tube at one end into which one of the balls can be tipped.) Kenny conducts an experiment to see how many black balls are in the bottle. He takes various numbers of samples and records how many of them showed a black ball.

The results are shown in the table.

Number of samples	Number of black balls	Experimental probability
10	2	
100	25	
200	76	
500	210	
1000	385	
5000	1987	

a Copy the table and complete it by calculating the experimental probability of getting a black ball at each stage.

b Using this information, how many black balls do you think there are in the bottle?

5 Use a set of number cards from 1 to 10 (or make your own set) and work with a partner. Take turns to choose a card and keep a record each time of what card you get. Shuffle the cards each time and repeat the experiment 60 times. Put your results in a copy of this table.

Score	1	2	3	4	5	6	7	8	9	10
Total										

a How many times would you expect to get each number?

b Do you think you and your partner conducted this experiment fairly?

c Explain your answer to part **b**.

6 A four-sided dice has faces numbered 1, 2, 3 and 4. The 'score' is the face on which it lands. Five students throw the dice to see if it is biased. They each throw it a different number of times. Their results are shown in the table.

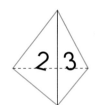

Student	Total number of throws	Score			
		1	2	3	4
Alfred	20	7	6	3	4
Brian	50	19	16	8	7
Caryl	250	102	76	42	30
Deema	80	25	25	12	18
Emma	150	61	46	26	17

a Which student will have the most reliable set of results? Why?

b Add up all the score columns and work out the relative frequency of each score. Give your answers to 2 decimal places.

c Is the dice biased? Explain your answer.

7 If you were about to choose a card from a pack of yellow cards numbered from 1 to 10, what would be the chance of each of the events **a** to **i** occurring? Read each of these statements and describe the probability with a word or phrase chosen from 'impossible', 'not likely', '50–50 chance', 'quite likely', or 'certain'.

a The next card chosen will be a 4.

b The next card chosen will be pink.

c The next card chosen will be a seven.

d The next card chosen will be a number less than 11.

e The next card chosen will be a number bigger than 11.

f The next card chosen will be an even number.

g The next card chosen will be a number more than 5.

h The next card chosen will be a multiple of 1.

i The next card chosen will be a prime number.

PS **8** Andrew made an eight-sided spinner.

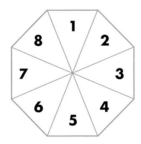

He tested it to see if it was fair.

He spun the spinner and recorded the results.

Unfortunately his little sister spilt something over his results table, so he could not see the middle part.

Number spinner lands on	1	2	3		6	7	8
Frequency	18	19	22		19	20	22

Assuming the spinner was a fair one, try to complete the missing parts of the table for Andrew.

FM **9** At a computer factory, tests were carried out to see how many faulty computer chips were produced in one week.

	Monday	Tuesday	Wednesday	Thursday	Friday
Sample	850	630	1055	896	450
Number faulty	10	7	12	11	4

On which day was it most likely that the highest number of faulty computer chips were produced?

AU **10** Steve tossed a coin 1000 times to see how many heads he got.

He said: "If this is a fair coin, I should get 500 heads."

Explain why he is wrong.

Biased spinner

You need a piece of stiff card, a cocktail stick and some sticky tack.

You may find that it is easier to work in pairs.

Make a copy of this hexagon on the card and push the cocktail stick through its centre to make a six-sided spinner. The size of the hexagon does not really matter, but it does need to be *accurately* drawn.

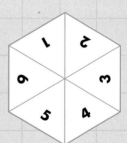

Stick a small piece of tack underneath one of the numbers. You now have a **biased** spinner.

Spin it 100 times and record your results in a frequency table.

Estimate the experimental probability of getting each number.

How can you tell that your spinner is biased?

Put some tack underneath a different number and see whether your partner can predict the number towards which the spinner is biased.

Combined events

This section will show you how to:
- work out the probabilities for two events occurring at the same time

Key words
probability space diagram
sample space diagram

There are many situations where two events occur together. Four examples are given below.

Throwing two dice

Imagine that two dice, one red and one blue, are thrown. The red dice can land with any one of six scores: 1, 2, 3, 4, 5 or 6. The blue dice can also land with any one of six scores. This gives a total of 36 possible combinations. These are shown in the left-hand diagram below, where combinations are given as (2, 3) and so on. The first number is the score on the blue dice and the second number is the score on the red dice.

The combination (2, 3) gives a total of 5. The total scores for all the combinations are shown in the diagram on the right-hand side. Diagrams that show all the outcomes of combined events are called **sample space diagrams** or **probability space diagrams**.

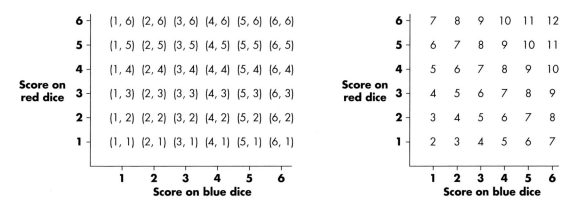

From the diagram on the right, you can see that there are two ways to get a score of 3. This gives a probability of scoring 3 as:

$$P(3) = \frac{2}{36} = \frac{1}{18}$$

From the diagram on the left, you can see that there are six ways to get a 'double'. This gives a probability of scoring a double as:

$$P(\text{double}) = \frac{6}{36} = \frac{1}{6}$$

Throwing coins

Throwing one coin
There are two equally likely outcomes, head or tail:

Throwing two coins together
There are four equally likely outcomes:

Hence:

$$P(2 \text{ heads}) = \frac{1}{4}$$

$$P(\text{head and tail}) = 2 \text{ ways out of } 4 = \frac{2}{4} = \frac{1}{2}$$

Dice and coins

Throwing a dice and a coin

Outcome on coin

	1	2	3	4	5	6
H	(1, H)	(2, H)	(3, H)	(4, H)	(5, H)	(6, H)
T	(1, T)	(2, T)	(3, T)	(4, T)	(5, T)	(6, T)

Score on dice

Hence:

$$P (\text{head and an even number}) = 3 \text{ ways out of } 12 = \frac{3}{12} = \frac{1}{4}$$

EXERCISE 10C

1 To answer these questions, use the diagram on page 259 for all the possible scores when two fair dice are thrown together.

  a What is the most likely score?

  b Which two scores are least likely?

  c Write down the probabilities of throwing all the scores from 2 to 12.

  d What is the probability of a score that is:

  i bigger than 10

  ii between 3 and 7

  iii even

  iv a square number

  v a prime number

  vi a triangular number?

2 Use the diagram on page 259 that shows the outcomes when two fair, six-sided dice are thrown together as coordinates. What is the probability that:

a the score is an even 'double'

b at least one of the dice shows 2

c the score on one dice is twice the score on the other dice

d at least one of the dice shows a multiple of 3?

3 Use the diagram on page 259 that shows the outcomes when two fair, six-sided dice are thrown together as coordinates. What is the probability that:

a both dice show a 6

b at least one of the dice will show a 6

c exactly one dice shows a 6?

4 The diagram shows the scores for the event 'the difference between the scores when two fair, six-sided dice are thrown'. Copy and complete the diagram.

For the event described above, what is the probability of a difference of:

a 1   **b** 0

c 4   **d** 6

e an odd number?

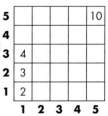

Score on second dice

6	5	4			1	0
5	4	3				1
4	3					
3	2					
2	1					
1	0					
	1	**2**	**3**	**4**	**5**	**6**

Score on first dice

5 When two fair coins are thrown together, what is the probability of:

a two heads   **b** a head and a tail

c at least one tail   **d** no tails?

Use the diagram of the outcomes when two coins are thrown together, on page 222.

6 Two five-sided spinners are spun together and the total score of the faces that they land on is worked out. Copy and complete the probability space diagram shown.

Score on second spinner

5					10
4					
3	4				
2	3				
1	2				
	1	**2**	**3**	**4**	**5**

Score on first spinner

a What is the most likely score?

b When two five-sided spinners are spun together, what is the probability that:

i the total score is 5

ii the total score is an even number

iii the score is a 'double'

iv the total score is less than 7?

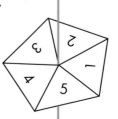

E

PS 7 Two eight-sided dice showing the numbers 1 to 8 were thrown at the same time.

What is the probability that the product of the two dice is an even square number?

AU 8 Isaac rolls two dice and multiplies both numbers to give their product. He wants to know the probability of rolling two dice that will give him a product between 19 and 35.

Explain why a probability space diagram will help him.

FM 9 Nic went to a garden centre to buy some roses.

She found they came in six different colours – white, red, orange, yellow, pink and copper.

She also found they came in five different sizes – dwarf, small, medium, large climbing and rambling.

a Draw a probability space diagram to show all the options.

b She buys a random rose for Auntie Janet. Auntie Janet only likes red and pink roses and she does not like climbing or rambling roses.

What is the probability that Nic has bought for Auntie Janet a rose:

   i that she likes

   ii that she does not like?

C

PS 10 Mrs Roberts asked: "What is the probability of rolling dice and getting a total of 10 or less. You must choose from impossible, very unlikely, unlikely, even chance, likely, very likely, or certain."

Evie replied: "It depends on how many dice I use."

For how many dice is each of the possible choices the answer?

10.4 Expectation

This section will show you how to:
- predict the likely number of successful events, given the number of trials and the probability of any one outcome

Key word
expect

When you know the probability of an outcome of an event, you can predict how many times you would expect that outcome to happen in a certain number of trials.

Note that this is what you **expect**. It is not necessarily what is going to happen. If what you expected always happened, life would be very dull and boring, and the National Lottery would be a waste of time!

EXAMPLE 4

A bag contains 20 balls, 9 of which are black, 6 white and 5 yellow. A ball is drawn at random from the bag, its colour is noted and then it is put back in the bag. This is repeated 500 times.

    a How many times would you expect a black ball to be drawn?

    b How many times would you expect a yellow ball to be drawn?

    c How many times would you expect a black or a yellow ball to be drawn?

a P(black ball) = $\frac{9}{20}$

Expected number of black balls = $\frac{9}{20} \times 500 = 225$

b P(yellow ball) = $\frac{5}{20} = \frac{1}{4}$

Expected number of yellow balls = $\frac{1}{4} \times 500 = 125$

c Expected number of black or yellow balls = 225 + 125 = 350

EXERCISE 10D

1 **a** What is the probability of throwing a 6 with an ordinary dice?

    b I throw an ordinary dice 150 times. How many times can I expect to get a score of 6?

2 **a** What is the probability of tossing a head with a coin?

    b I toss a coin 2000 times. How many times can I expect to get a head?

3 **a** A card is taken at random from a pack of cards. What is the probability that it is:

    i a black card   **ii** a King   **iii** a Heart   **iv** the King of Hearts?

    b I draw a card from a pack of cards and replace it. I do this 520 times. How many times would I expect to get:

    i a black card   **ii** a King   **iii** a Heart   **iv** the King of Hearts?

4 The ball in a roulette wheel can land in one of 37 spaces that are marked with numbers from 0 to 36 inclusive. I always bet on the same number, 13.

    a What is the probability of the ball landing in 13?

    b If I play all evening and there are exactly 185 spins of the wheel in that time, how many times could I expect to win?

5 In a bag there are 30 balls, 15 of which are red, 5 yellow, 5 green, and 5 blue. A ball is taken out at random and then replaced. This is done 300 times. How many times would I expect to get:

a a red ball

b a yellow or blue ball

c a ball that is not blue

d a pink ball?

6 The experiment described in question 5 is carried out 1000 times. Approximately how many times would you expect to get:

a a green ball

b a ball that is not blue?

7 A sampling bottle (as described in question 4 of Exercise 7E) contains red and white balls. It is known that the probability of getting a red ball is 0.3. If 1500 samples are taken, how many of them would you expect to give a white ball?

8 Josie said, "When I throw a dice, I expect to get a score of 3.5."

"Impossible," said Paul, "You can't score 3.5 with a dice."
"Do this and I'll prove it," said Josie.

a An ordinary dice is thrown 60 times. Fill in the table for the expected number of times each score will occur.

Score						
Expected occurrences						

b Now work out the average score that is expected over 60 throws.

c There is an easy way to get an answer of 3.5 for the expected average score. Can you see what it is?

FM 9 The probabilities of some cloud types being seen on any day are given below.

Cumulus	0.3
Stratocumulus	0.25
Stratus	0.15
Altocumulus	0.11
Cirrus	0.05
Cirrcocumulus	0.02
Nimbostratus	0.005
Cumulonimbus	0.004

a What is the probability of *not* seeing one of the above clouds in the sky?

b On how many days of the year would you expect to see altocumulus clouds in the sky?

PS 10 Every evening Anne and Chris cut a pack of cards to see who washes up.

If they cut a King or a Jack, Chris washes up.

If they cut a Queen, Anne washes up.

Otherwise, they wash up together.

In a year of 365 days, how many days would you expect them to wash up together?

AU 11 A market gardener is supplied with tomato plant seedlings and knows that the probability that any plant will develop a disease is 0.003.

How will she find out how many of the tomato plants she should expect to develop a disease?

10.5 Two-way tables

This section will show you how to:	Key word
● read two-way tables and use them to do probability and other mathematics	two-way table

A **two-way table** is a table that links together two variables. For example, the following table shows how many boys and girls there are in a form and whether they are left- or right-handed.

	Boys	Girls
Left-handed	2	4
Right-handed	10	13

This table shows the colour and make of cars in the school car park.

	Red	Blue	White
Ford	2	4	1
Vauxhall	0	1	2
Toyota	3	3	4
Peugeot	2	0	3

One variable is shown in the rows of the table and the other variable is shown in the columns of the table.

EXAMPLE 5

Use the first two-way table on the previous page to answer the following.

    a How many left-handed boys are there in the form?

    b How many girls are there in the form, in total?

    c How many students are there in the form altogether?

    d How many students altogether are right-handed?

    e What is the probability that a student selected at random from the form is:

        i a left-handed boy   **ii** right-handed?

a	2 boys	Read this value from where the 'Boys' column and the 'Left-handed' row meet.
b	17 girls	Add up the 'Girls' column.
c	29 students	Add up all the numbers in the table.
d	23	Add up the 'Right-handed' row.
e **i**	P (left-handed boy) $= \frac{2}{29}$	Use the answers to parts **a** and **c**.
ii	P (right-handed) $= \frac{23}{29}$	Use the answers to parts **c** and **d**.

EXAMPLE 6

Use the second two-way table on the previous page to answer the following.

    a How many cars were in the car park altogether?

    b How many red cars were in the car park?

    c What percentage of the cars in the car park were red?

    d How many cars in the car park were white?

    e What percentage of the white cars were Vauxhalls?

a	25	Add up all the numbers in the table.
b	7	Add up the 'Red' column.
c	28%	7 out of 25 is the same as 28 out of 100.
d	10	Add up the 'White' column.
e	20%	2 out of 10 is 20%.

EXERCISE 10E

1 The following table shows the top five clubs in the top division of the English Football League at the end of the season for the years 1965, 1975, 1985, 1995 and 2005.

		Year				
		1965	**1975**	**1985**	**1995**	**2005**
Position	**1st**	Man Utd	Derby	Everton	Blackburn	Chelsea
	2nd	Leeds	Liverpool	Liverpool	Man Utd	Arsenal
	3rd	Chelsea	Ipswich	Tottenham	Notts Forest	Man Utd
	4th	Everton	Everton	Man Utd	Liverpool	Everton
	5th	Notts Forest	Stoke	Southampton	Leeds	Liverpool

a Which team was in fourth place in 1975?

b Which three teams are in the top five for four of the five years?

c Which team finished three places lower in 1995 than in 1965?

2 Here is a display of 10 cards.

a Complete the two-way table.

		Shaded	Unshaded
Shape	**Circles**		
	Triangles		

b One of the cards is picked at random. What is the probability it shows either a shaded triangle or an unshaded circle?

3 The two-way table shows the number of doors and the number of windows in each room in a primary school.

		Number of doors		
		1	**2**	**3**
Number of windows	**1**	5	4	2
	2	4	5	4
	3	0	4	6
	4	1	3	2

a How many rooms are there in the school altogether?

b How many rooms have two doors?

c What percentage of the rooms in the school have two doors?

d What percentage of the rooms that have one door also have two windows?

e How many rooms have the same number of windows as doors?

E

4 Three cards are lettered A, B and C. Three discs are numbered 4, 5 and 6.

| A | B | C | 4 | 5 | 6 |

One card and one disc are chosen at random.

If the card shows A, 1 is deducted from the score on the disc.
If the card shows B, the score on the disc stays the same.
If the card shows C, 1 is added to the score on the disc.

a Copy and complete the table to show all the possible scores.

		Number on disc		
		4	**5**	**6**
	A	3		
Letter on card	**B**	4		
	C	5		

b What is the probability of getting a score that is an even number?

c In a different game the probability of getting a total that is even is $\frac{2}{3}$.
What is the probability of getting a total that is an odd number?

D

5 The two-way table shows the ages and sexes of a sample of 50 students in a school.

		Age (years)					
		11	**12**	**13**	**14**	**15**	**16**
Sex	**Boys**	4	3	6	2	5	4
	Girls	2	5	3	6	4	6

a How many students are aged 13 years or less?

b What percentage of the students in the table are 16?

c A student from the table is selected at random. What is the probability that the student will be 14 years of age? Give your answer as a fraction in its lowest form.

d There are 1000 students in the school. Use the table to estimate how many boys are in the school altogether.

6 The two-way table shows the numbers of adults and the numbers of cars in 50 houses in one street.

		Number of adults			
		1	**2**	**3**	**4**
	0	2	1	0	0
Number of cars	**1**	3	13	3	1
	2	0	10	6	4
	3	0	1	4	2

a How many houses have exactly two adults and two cars?

b How many houses altogether have three cars?

c What percentage of the houses have three cars?

d What percentage of the houses with just one car have three adults living in the house?

7 Jane has two four-sided spinners.
Spinner A has the numbers 1 to 4 on it and
Spinner B has the numbers 5 to 8 on it.

Both spinners are spun together.

The two-way table shows all the ways the two
spinners can land.

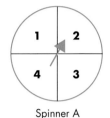

Spinner A

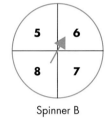

Spinner B

Some of the total scores are filled in.

		Score on Spinner A			
		1	**2**	**3**	**4**
	5	6	7		
Score on Spinner B	**6**	7			
	7				
	8				

a Copy and complete the table to show all the possible total scores.

b How many of the total scores are 9?

c When the two spinners are spun together, what is the probability that the total score
will be:

i 9

ii 8

iii a prime number?

8 The table shows information about the number of items in Zara's wardrobe.

		Type of item		
		Shoes (pairs)	**Trousers**	**T-shirts**
	Blue	6	5	2
Colour	**Black**	12	9	13
	Red	2	2	0

a How many pairs of blue shoes does Zara have?

b How many pairs of trousers does Zara have?

c How many black items does Zara have?

d If a T-shirt is chosen at random from all the T-shirts, what is the probability that it will
be a black T-shirt?

9 Zoe throws a fair coin and rolls a fair dice.

If the coin shows a head, she records the score on the dice.
If the coin shows tails, she doubles the number on the dice.

a Copy and complete the two-way table to show Zoe's possible scores.

		Number on dice					
		1	**2**	**3**	**4**	**5**	**6**
Coin	**Head**	1	2				
	Tail	2	4				

b How many of the scores are square numbers?

c What is the probability of getting a score that is a square number?

PS 10 Here is a two way table for the members of a sports club.

	Men	**Women**	**Children**
Left footed			
Right footed			

Altogether there are 60 members.

4 men are left footed.

The ratio of men to women to children is 5 : 3 : 4.

There are 10 more right footed children then left footed children.

One-fifth of the members are left footed.

Copy and complete the table.

AU 11 A gardener plants some sunflower seeds in a greenhouse and some in the garden.
After they have fully grown, he measures the diameter of the sunflower heads.
The table shows his results.

		Greenhouse	**Garden**
Diameter	**Mean diameter**	16.8 cm	14.5 cm
	Range of diameter	3.2 cm	1.8 cm

a The gardener, who wants to enter competitions, says, "The sunflowers from the greenhouse are better."

Using the data in the table, give a reason to justify this statement.

b The gardener's wife, who does flower arranging, says, "The sunflowers from the garden are better."

Using the data in the table, give a reason to justify this statement.

AU 12 Reyki plants some tomato plants in her greenhouse, while her husband Daniel plants some in the garden.

After the summer they compared their tomatoes.

		Garden	Greenhouse
Diameter	**Mean diameter**	1.8 cm	4.2 cm
	Mean number of tomatoes per plant	24.2	13.3

Use the data in the table to explain who had the better crop of tomatoes.

PS 13 Two hexagonal spinners are spun.

Spinner A is numbered 3, 5, 7, 9, 11 and 13.
Spinner B is numbered 4, 5, 6, 7, 8 and 9.

What is the probability that when the two spinners are spun, the result of multiplying two numbers together will give a product greater than 40?

14 Here are two fair spinners.

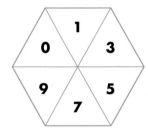 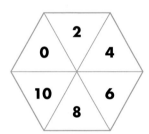

The spinners are spun.

The two numbers obtained are added together.

a Draw a probability sample space diagram.

b What is the most likely score?

c What is the probability of getting a total of 12?

d What is the probability of getting a total of 11 or more?

e What is the probability of getting a total that is an odd number?

GRADE BOOSTER

D You can understand that the total probability of all possible outcomes in a paticular situation is 1

C You can predict the expected number of successes from a given number of trials if the probability of one success is known

C You can calculate relative frequency from experimental evidence and compare this with the theoretical probability

What you should know now

- How to calculate theoretical probabilities from different situations

- How to calculate relative frequency and understand that the reliability of experimental results depends on the number of experiments carried out

1 A survey is to be carried out on how teenagers prefer to buy their music.

They buy their music from CDs (C) or from downloads (D).

A pilot survey of ten teenagers is carried out first.

Boy	C	Girl	C
Girl	D	Boy	D
Girl	D	Girl	D
Boy	C	Boy	C
Boy	D	Girl	D

a Construct a two-way table to show these results. (3)

b In the full survey, the probability of a teenager preferring CDs is 0.3

What is the probability of a teenager **not** preferring CDs? (1)

(Total 4 marks)

AQA, November 2008, Paper 2 Foundation, Question 16

2 A fair six-sided dice and a fair coin are thrown at the same time. This shows the outcome 1H or (1, head).

a Complete the list of all the possible outcomes.

b What is the probability of getting a head and an even number?

c What is the probability of getting a tail *or* an odd number *or* both?

3 Jacob has 3 red counters and 7 blue counters.

Tony has 10 red counters.

Emily has only blue counters.

a Jacob puts his counters into a bag.

What is the probability of choosing a red counter from the bag? (1)

b Tony adds his counters to the bag.

What is the probability of choosing a red counter now? (2)

c Emily adds her counters to the bag.

The probability of choosing a red counter now is $\frac{1}{2}$.

How many blue counters did Emily have? (2)

(Total 5 marks)

AQA, November 2008, Paper 1 Foundation, Question 12

4 A triangular spinner has sections coloured white (W), green (G) and blue (B).

The spinner is spun 20 times and the colour it lands on each time is recorded.

W W B G G W B
G G W G B G B
G W G B G B

a Copy and complete the relative frequency table.

Colour	White (W)	Green (G)	Blue (B)
Relative frequency			

b The table below shows the relative frequencies after this spinner has been spun 100 times.

Colour	White (W)	Green (G)	Blue (B)
Relative frequency	$\frac{21}{100}$	$\frac{52}{100}$	$\frac{27}{100}$

Which of the two relative frequencies for white gives the better estimate of the probability of the spinner landing on white? Give a reason for your answer.

Worked Examination Questions

FM **1** In a raffle 400 tickets have been sold. There is only one prize.

Mr Raza buys 5 tickets for himself and sells another 40.
Mrs Raza buys 10 tickets for herself and sells another 50.
Mrs Hewes buys 8 tickets for herself and sells just 12 others.

  a What is the probability of:

   i   Mr Raza winning the raffle

   ii Mr Raza selling the winning ticket

   iii Mr Raza either winning the raffle or selling the winning ticket?

  b What is the probability of either Mr or Mrs Raza selling the winning ticket?

  c What is the probability of Mrs Hewes not winning the lottery?

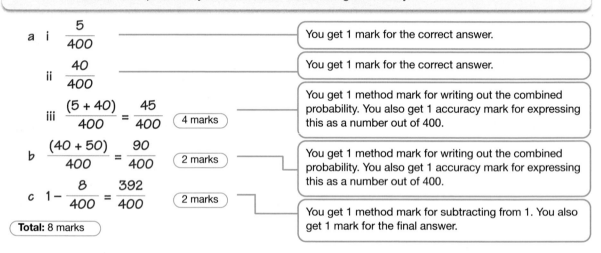

a **i** $\dfrac{5}{400}$ — You get 1 mark for the correct answer.

 ii $\dfrac{40}{400}$ — You get 1 mark for the correct answer.

 iii $\dfrac{(5 + 40)}{400} = \dfrac{45}{400}$ (4 marks) — You get 1 method mark for writing out the combined probability. You also get 1 accuracy mark for expressing this as a number out of 400.

b $\dfrac{(40 + 50)}{400} = \dfrac{90}{400}$ (2 marks) — You get 1 method mark for writing out the combined probability. You also get 1 accuracy mark for expressing this as a number out of 400.

c $1 - \dfrac{8}{400} = \dfrac{392}{400}$ (2 marks) — You get 1 method mark for subtracting from 1. You also get 1 mark for the final answer.

Total: 8 marks

Worked Examination Questions

2 Simon has a bag containing blue, green, red, yellow and white marbles.

   a Complete the table to show the probability of each colour being chosen at random.

   b Which colour of marble is most likely to be chosen at random?

   c Calculate the probability that a marble chosen at random is blue or white.

Colour of marble	Probability
Blue	0.3
Green	0.2
Red	0.15
Yellow	
White	0.1

a   0.25

  2 marks

> The probabilities in the table should add up to 1 so,
> $0.3 + 0.2 + 0.15 + 0.1 + ? = 1$ $0.75 + ? = 1$
> This calculation gets you 1 method mark. Using the calculation, the missing probability is 0.25. This answer gets you 1 mark.

b   Blue marble

  1 mark

> The blue marble is most likely as it has the largest probability of being chosen. The correct answer is worth 1 mark.

c   0.4

  2 marks

> The word 'or' means you add probabilities of separate events.
> The calculation P(blue or white) = P(blue) + P(white) = 0.3 + 0.1 = 0.4 is worth 1 method mark and the correct answer is worth 1 mark.

Total: 5 marks

Joe had a stall at the local fair and wanted to make a reasonable profit from the game below.

In this game, the player rolls two balls down the sloping board and wins a prize if they land in slots that total more than seven.

Joe wants to know how much he should charge for each go and what the price should be. He would also like to know how much profit he is likely to make.

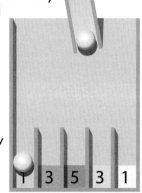

1 3 5 3 1

Getting started

Practise calculating probabilities using the spinner and questions below.

- What is the probability of spinning the spinner and getting:
  - a one
  - a five
  - a two
  - a number other than five?

  Give your answer as a fraction and as a decimal.

- If you spun the spinner 20 times, how many times would you expect to get:
  - a five
  - not five
  - an odd number?

- If you spun the spinner 100 times, how many times would you expect to get:
  - a two
  - not a two
  - a prime number?

Now, think about which probabilities you must calculate, in order to help Joe, and to design your own profitable game.

Your task

For this task you can work individually or in pairs.

1 Write a report for Joe which includes:

- a diagram to show the probability of each outcome
- the probability of at least three different outcomes
- the probability of winning
- at least three different ways of setting up the game, showing the cost of one go, the value of the prize and the expected profit for varying numbers of people playing the game.

  Based on this information, advise Joe how he should set up the game. Justify your decision.

2 Design your own fairground game. Describe the rules for winning, show the probability of winning using a diagram, and give costings and the projected profit.

  Explain why you think your game should be used at the next local fair.

Why this chapter matters

We have already seen that patterns appear in numbers – prime numbers, square numbers and multiples all form patterns. Number patterns are not only of mathematical value – they also make the study of nature and geometric patterns a little more intriguing.

There are many mathematical patterns that appear in nature. The most famous of these is probably the Fibonacci series:

1   1   2   3   5   8   13   21   …

This is formed by adding the two previous terms to get the next term.

The sequence was discovered in 1202 by the Italian Leonardo Fibonacci (c1170–1250), when he was investigating the breeding patterns of rabbits!

Since then, the pattern has been found in many other places in nature. The spirals found in a nautilus shell and in the seed heads of a sunflower plant also follow the Fibonacci series.

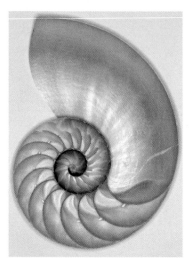
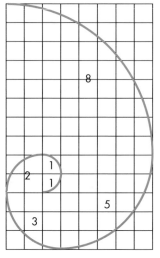

Fractals form another kind of pattern.

Fractals are geometric patterns that are continuously repeated on a smaller scale.

A good example of a fractal is this: start with an equilateral triangle and draw an equilateral triangle, a third the size of the original, on the middle of each side. Keep on repeating this and we get an increasingly complex-looking shape.

The final pattern shown here is called the Koch snowflake. It is named after the Swedish mathematician Helge von Koch (1870–1924).

Fractals are commonly found in nature, a good example being the complex patterns found in plants, such as the fern above right.

Do you know of any other numerical or geometric patterns that you have come across in everyday life?

Algebra: Patterns

This chapter will show you ...

- **E** some of the common sequences of numbers
- **E** how to recognise rules for sequences
- **E** how to express the rule for a sequence in words
- **D** how to work out a sequence, given a formula for the nth term
- **C** how to express the rule for a sequence algebraically

Visual overview

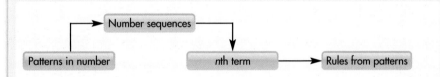

What you should already know

- Basic algebra and how to use letters for numbers (KS3 level 4, GCSE grade F)
- How to substitute numbers into algebraic expressions (KS3 level 5, GCSE grade E)
- How to solve simple linear equations (KS3 level 5, GCSE grade E)

Quick check

1 Angela is x years old. Write down expressions for the ages of the following people, in terms of x.

   a Angela's brother Bill, who is three years older than her.

   b Angela's mother Carol, who is twice as old as Angela.

   c Angela's father Dick, whose age is the sum of Angela's age and Bill's age.

2 Work out the value of the expression $2n + 3$ for:

   a $n = 1$

   b $n = 2$

   c $n = 3$.

Patterns in number

This section will show you how to:

● recognise patterns in number sequences

Key words

pattern

sequence

Look at these number **patterns**.

$$0 \times 9 + 1 = 1$$
$$1 \times 9 + 2 = 11$$
$$12 \times 9 + 3 = 111$$
$$123 \times 9 + 4 = 1111$$
$$1234 \times 9 + 5 = 11111$$

$$1 \times 8 + 1 = 9$$
$$12 \times 8 + 2 = 98$$
$$123 \times 8 + 3 = 987$$
$$1234 \times 8 + 4 = 9876$$
$$12345 \times 8 + 5 = 98765$$

$$1 \times 3 \times 37 = 111$$
$$2 \times 3 \times 37 = 222$$
$$3 \times 3 \times 37 = 333$$
$$4 \times 3 \times 37 = 444$$

$$7 \times 7 = 49$$
$$67 \times 67 = 4489$$
$$667 \times 667 = 444889$$
$$6667 \times 6667 = 44448889$$

Check that the patterns you see there are correct and then try to continue each pattern without using a calculator. The numbers form a **sequence**. Check them with a calculator afterwards.

Spotting patterns is an important part of mathematics. It helps you to see rules for making calculations.

EXERCISE 11A

In questions 1 to 10, look for the pattern and then write the next two lines. Check your answers with a calculator afterwards.

You might find that some of the answers are too big to fit in a calculator display. This is one of the reasons why spotting patterns is important.

AU **1**

$$1 \times 1 = 1$$
$$11 \times 11 = 121$$
$$111 \times 111 = 12321$$
$$1111 \times 1111 = 1234321$$

HINTS AND TIPS

Look for symmetries in the number patterns.

AU **2**

$$9 \times 9 = 81$$
$$99 \times 99 = 9801$$
$$999 \times 999 = 998001$$
$$9999 \times 9999 = 99980001$$

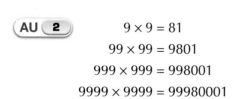

FM Functional Maths   **AU** (AO2) Assessing Understanding   **PS** (AO3) Problem Solving

AU 3 $3 \times 4 = 3^2 + 3$
$4 \times 5 = 4^2 + 4$
$5 \times 6 = 5^2 + 5$
$6 \times 7 = 6^2 + 6$

AU 4 $10 \times 11 = 110$
$20 \times 21 = 420$
$30 \times 31 = 930$
$40 \times 41 = 1640$

HINTS AND TIPS

Think of the numbers as 1 10, 4 20, 9 30, 16 40 …

AU 5
$1 = 1 = 1^2$
$1 + 2 + 1 = 4 = 2^2$
$1 + 2 + 3 + 2 + 1 = 9 = 3^2$
$1 + 2 + 3 + 4 + 3 + 2 + 1 = 16 = 4^2$

AU 6
$1 = 1 = 1^3$
$3 + 5 = 8 = 2^3$
$7 + 9 + 11 = 27 = 3^3$
$13 + 15 + 17 + 19 = 64 = 4^3$

AU 7
$1 = 1$
$1 + 1 = 2$
$1 + 2 + 1 = 4$
$1 + 3 + 3 + 1 = 8$
$1 + 4 + 6 + 4 + 1 = 16$
$1 + 5 + 10 + 10 + 5 + 1 = 32$

AU 8 $12\ 345\ 679 \times 9 = 111\ 111\ 111$
$12\ 345\ 679 \times 18 = 222\ 222\ 222$
$12\ 345\ 679 \times 27 = 333\ 333\ 333$
$12\ 345\ 679 \times 36 = 444\ 444\ 444$

AU 9 $1^3 = 1^2 = 1$
$1^3 + 2^3 = (1 + 2)^2 = 9$
$1^3 + 2^3 + 3^3 = (1 + 2 + 3)^2 = 36$

HINTS AND TIPS

$4 + 5 = 9 = 3^2$
$12 + 13 = 25 = 5^2$
$24 + 25 = 49 = 7^2$

AU 10
$3^2 + 4^2 = 5^2$
$10^2 + 11^2 + 12^2 = 13^2 + 14^2$
$21^2 + 22^2 + 23^2 + 24^2 = 25^2 + 26^2 + 27^2$

From your observations on the number patterns in questions 1 to 10, answer questions 11 to 19 without using a calculator.

HINTS AND TIPS

Look for clues in the patterns from questions 1 to 10, for example, $1111 \times 1111 = 1234321$. This is four 1s times four 1s, so what will it be for nine 1s times nine 1s?

PS 11 $111\ 111\ 111 \times 111\ 111\ 111 =$

PS 12 $999\ 999\ 999 \times 999\ 999\ 999 =$

PS 13 $12 \times 13 =$

PS 14 $90 \times 91 =$

PS 15 $1 + 2 + 3 + 4 + 5 + 6 + 7 + 8 + 9 + 8 + 7 + 6 + 5 + 4 + 3 + 2 + 1 =$

PS 16 $57 + 59 + 61 + 63 + 65 + 67 + 69 + 71 =$

PS 17 $1 + 9 + 36 + 84 + 126 + 126 + 84 + 36 + 9 + 1 =$

PS 18 $12\ 345\ 679 \times 81 =$

PS 19 $1^3 + 2^3 + 3^3 + 4^3 + 5^3 + 6^3 + 7^3 + 8^3 + 9^3 =$

C

PS **20** The letters of the alphabet are written as the pattern:

ABBCCCDDDDEEEEEFFFFFFGGGGGGG …

so that the number of times each letter is written matches its place in the alphabet.

So, for example, as J is the 10th letter in the alphabet, there will be 10 Js in the list.

When the pattern gets to the 26th Z, it repeats.

What letter will be the 1000th in the list?

HINTS AND TIPS

Work out how many letters there are in the sequence from ABB … to … ZZZ, then work out how many of these sequences are needed to get past 1000.

11.2 Number sequences

This section will show you how to:	Key words
● recognise how number sequences are building up	consecutive difference sequence term term-to-term

A number **sequence** is an ordered set of numbers with a rule for finding every number in the sequence. The rule that takes you from one number to the next could be a simple addition or multiplication, but often it is more tricky than that. So you need to look *very* carefully at the pattern of a sequence.

Each number in a sequence is called a **term** and is in a certain position in the sequence.

Look at these sequences and their rules.

3, 6, 12, 24, … doubling the previous term each time … 48, 96, …

2, 5, 8, 11, … adding 3 to the previous term each time … 14, 17, …

1, 10, 100, 1000, … multiplying the previous term by 10 each time … 10 000, 100 000

1, 8, 15, 22, … adding 7 to the previous term each time … 29, 36, …

These are all quite straightforward once you have looked for the link from one term to the next (**consecutive** terms). Sequences like this may be called **term-to-term** sequences.

Differences

For some sequences you need to look at the **differences** between consecutive terms to determine the pattern.

EXAMPLE 1

Find the next two terms of the sequence 1, 3, 6, 10, 15, … .

Looking at the differences between consecutive terms:

```
1    3    6    10    15
  ↑    ↑    ↑    ↑
  2    3    4    5
```

> This is a special sequence of numbers. Do you recognise it? You will meet it again, later in the chapter.

So the sequence continues as follows.

```
1    3    6    10    15    21    28
  ↑    ↑    ↑    ↑
  2    3    4    5    +6    +7
```

So the next two terms are 21 and 28.

The differences usually form a number sequence of their own, so you need to find the *sequence of the differences* before you can expand the original sequence.

EXERCISE 11B

AU 1 Look at the following number sequences. Write down the next three terms in each and explain how each sequence is formed.

a 1, 3, 5, 7, …	**b** 2, 4, 6, 8, …
c 5, 10, 20, 40, …	**d** 1, 3, 9, 27, …
e 4, 10, 16, 22, …	**f** 3, 8, 13, 18, …
g 2, 20, 200, 2000, …	**h** 7, 10, 13, 16, …
i 10, 19, 28, 37, …	**j** 5, 15, 45, 135, …
k 2, 6, 10, 14, …	**l** 1, 5, 25, 125, …

2 By considering the differences in the following sequences, write down the next two terms in each case.

a 1, 2, 4, 7, 11, …	**b** 1, 2, 5, 10, 17, …
c 1, 3, 7, 13, 21, …	**d** 1, 4, 10, 19, 31, …
e 1, 9, 25, 49, 81, …	**f** 1, 2, 7, 32, 157, …
g 1, 3, 23, 223, 2223, …	**h** 1, 2, 4, 5, 7, 8, 10, …
i 2, 3, 5, 9, 17, …	**j** 3, 8, 18, 33, 53, …

E

3 Look at the sequences below. Find the rule for each sequence and write down its next three terms.

a 3, 6, 12, 24, …   b 3, 9, 15, 21, 27, …

c 128, 64, 32, 16, 8, …   d 50, 47, 44, 41, …

e 2, 5, 10, 17, 26, …   f 5, 6, 8, 11, 15, 20, …

g 5, 7, 8, 10, 11, 13, …   h 4, 7, 10, 13, 16, …

i 1, 3, 6, 10, 15, 21, …   j 1, 2, 3, 4, …

k 100, 20, 4, 0.8, …   l 1, 0.5, 0.25, 0.125, …

D

AU **4** Look carefully at each number sequence below. Find the next two numbers in the sequence and try to explain the pattern.

a 1, 1, 2, 3, 5, 8, 13, …

b 1, 4, 9, 16, 25, 36, …

c 3, 4, 7, 11, 18, 29, …

d 1, 8, 27, 64, 125, …

> **HINTS AND TIPS**
>
> These patterns do not go up by the same value each time so you will need to find another connection between the terms.

5 Triangular numbers are found as follows.

1   3   6   10

Find the next four triangular numbers.

6 Hexagonal numbers are found as follows.

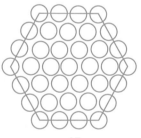

1   7   19   37

Find the next three hexagonal numbers.

AU **7** On the first day of Christmas my true love sent to me:

> a partridge in a pear tree

On the second day of Christmas my true love sent to me:

> two turtle doves
> and a partridge in a pear tree

and so on until …

On the twelfth day of Christmas my true love sent to me:

> twelve drummers drumming
> eleven pipers piping
> ten lords a-leaping
> nine ladies dancing
> eight maids a-milking
> seven swans a-swimming
> six geese a-laying
> five golden rings
> four calling birds
> three French hens
> two turtle doves
> and a partridge in a pear tree.

How many presents were given in total on the twelve days of Christmas?

Work out the pattern for the number of presents each day and find the total number of presents after each day. Describe any patterns that you spot.

PS **8** The first term that these two sequences have in common is 17:

> 8, 11, 14, 17, 20, ...
>
> 1, 5, 9, 13, 17, ...

What are the next two terms that the two sequences have in common?

AU **9** Two sequences are:

> 2, 5, 8, 11, 14, ...
>
> 3, 6, 9, 12, 15, ...

Will the two sequences ever have a term in common? Yes or no?
Justify your answer.

FM **10** The sequence 1, 1, 2, 3, 5, 8, 13, 21, 34, … can be used to give an approximate conversion of miles to kilometres.

For example, 5 miles ≈ 8 kilometres, 8 miles ≈ 13 kilometres, and so on.

a Convert 55 miles to kilometres

b Convert 68 kilometres to miles.

The nth term of a sequence

This section will show you how to:
- recognise how number sequences are built up
- generate sequences, given the nth term
- find an algebraic rule for the nth term of a sequence

Key words
coefficient
consecutive
difference
linear sequence
nth term

Finding the rule

When using a number sequence, you sometimes need to know, say, its 50th term, or even a higher term in the sequence. To do so, you need to find the rule that produces the sequence in its general form.

It may be helpful to look at the problem backwards. That is, take a rule and see how it produces a sequence. The rule is given for the general term, which is called the **nth term**.

EXAMPLE 2

A sequence is formed by the rule $3n + 1$, where $n = 1, 2, 3, 4, 5, 6, \ldots$. Write down the first five terms of the sequence.

Substituting $n = 1, 2, 3, 4, 5$ in turn:

$(3 \times 1 + 1), (3 \times 2 + 1), (3 \times 3 + 1), (3 \times 4 + 1), (3 \times 5 + 1), \ldots$
    4   7   10   13   16

So the sequence is 4, 7, 10, 13, 16, … .

Notice that in Example 2 the **difference** between each term and the next is always 3, which is the **coefficient** of n (the number attached to n). Also, the constant term is the difference between the first term and the coefficient, that is, $4 - 3 = 1$.

EXAMPLE 3

The nth term of a sequence is $4n - 3$. Write down the first five terms of the sequence.

Substituting $n = 1, 2, 3, 4, 5$ in turn:

$(4 \times 1 - 3), (4 \times 2 - 3), (4 \times 3 - 3), (4 \times 4 - 3), (4 \times 5 - 3)$
    1   5   9   13   17

So the sequence is 1, 5, 9, 13, 17, … .

Notice that in Example 3 the difference between each term and the next is always 4, which is the coefficient of n.

Also, the constant term is the difference between the first term and the coefficient, that is, $1 - 4 = -3$.

EXERCISE 11C

1 Use each of the following rules to write down the first five terms of a sequence.

  a $2n + 1$ for $n = 1, 2, 3, 4, 5$ **b** $3n - 2$ for $n = 1, 2, 3, 4, 5$

  c $5n + 2$ for $n = 1, 2, 3, 4, 5$ **d** n^2 for $n = 1, 2, 3, 4, 5$

  e $n^2 + 3$ for $n = 1, 2, 3, 4, 5$

2 Write down the first five terms of the sequence that has as its nth term:

  a $n + 3$ **b** $3n - 1$ **c** $5n - 2$

  d $n^2 - 1$ **e** $4n + 5$.

> **HINTS AND TIPS**
>
> Substitute numbers into the expressions until you can see how the sequence works.

3 The first two terms of the sequence of fractions $\dfrac{n - 1}{n + 1}$ are:

  $n = 1$: $\dfrac{1 - 1}{1 + 1} = \dfrac{0}{2} = 0$ $n = 2$: $\dfrac{2 - 1}{2 + 1} = \dfrac{1}{3}$

  Work out the next five terms of the sequence.

4 A sequence is formed by the rule $\frac{1}{2} \times n \times (n + 1)$ for $n = 1, 2, 3, 4, \ldots$.

  The first term is given by $n = 1$: $\frac{1}{2} \times 1 \times (1 + 1) = 1$

  The second term is given by $n = 2$: $\frac{1}{2} \times 2 \times (2 + 1) = 3$

  a Work out the next five terms of this sequence.

  b This is a well-known sequence you have met before. What is it?

FM 5 A haulage company uses this formula to calculate the cost of transporting n pallets.

  For $n \leqslant 5$, the cost will be £$(40n + 50)$

  For $6 \leqslant n \leqslant 10$, the cost will be £$(40n + 25)$

  For $n \geqslant 11$, the cost will be £$40n$

  a How much will the company charge to transport 7 pallets?

  b How much will the company charge to transport 15 pallets?

  c A company is charged £170 for transporting pallets. How many pallets did they transport?

  d Another haulage company uses the formula £$50n$ to calculate the cost for transporting n pallets.

  At what value of n do the two companies charge the same amount?

PS **6** The formula for working out a series of fractions is $\dfrac{2n + 1}{3n + 1}$.

   a Work out the first three fractions in the series.

   b **i** Work out the value of the fraction as a decimal when $n = 1\,000\,000$.

   ii What fraction is equivalent to this decimal?

   iii How can you tell this from the original formula?

AU **7** The nth term of a sequence is $3n + 7$.
The nth term of another sequence is $4n - 2$.

These two series have several terms in common but only one term that is common and has the same position in the sequence.

Without writing out the sequences, show how you can tell, using the expressions for the nth term, that this is the 9th term.

8 5! means factorial 5, which is $5 \times 4 \times 3 \times 2 \times 1 = 120$.

7! means $7 \times 6 \times 5 \times 4 \times 3 \times 2 \times 1 = 5040$.

   a Calculate 2!, 3!, 4! and 6!.

   b If your calculator has a factorial button, check that it gives the same answers as you get for part **a**. What is the largest factorial you can work out with your calculator before you get an error?

Finding the nth term of a linear sequence

In a **linear sequence** the *difference* between one term and the next is always the same.

For example:

   2, 5, 8, 11, 14, … difference of 3

The nth term of this sequence is given by $3n - 1$.

Here is another linear sequence.

   5, 7, 9, 11, 13, … difference of 2

The nth term of this sequence is given by $2n + 3$.

So, you can see that the nth term of a linear sequence is *always* of the form $An + b$, where:

- A, the coefficient of n, is the difference between each term and the next term (**consecutive** terms)

- b is the difference between the first term and A.

EXAMPLE 4

Find the *n*th term of the sequence 5, 7, 9, 11, 13, … .

The difference between consecutive terms is 2. So the first part of the *n*th term is 2*n*.

Subtract the difference, 2, from the first term, 5, which gives $5 - 2 = 3$.

So the *n*th term is given by $2n + 3$. (You can test it by substituting $n = 1, 2, 3, 4, …$.)

EXAMPLE 5

Find the *n*th term of the sequence 3, 7, 11, 15, 19, … .

The difference between consecutive terms is 4. So the first part of the *n*th term is 4*n*.

Subtract the difference, 4, from the first term, 3, which gives $3 - 4 = -1$.

So the *n*th term is given by $4n - 1$.

EXAMPLE 6

From the sequence 5, 12, 19, 26, 33, … , find:

a the *n*th term   **b** the 50th term.

a The difference between consecutive terms is 7. So the first part of the *n*th term is 7*n*.

Subtract the difference, 7, from the first term, 5, which gives $5 - 7 = -2$.

So the *n*th term is given by $7n - 2$.

b The 50th term is found by substituting $n = 50$ into the rule, $7n - 2$.

$$50\text{th term} = 7 \times 50 - 2 = 350 - 2$$
$$= 348$$

EXERCISE 11D

1 Find the next two terms and the *n*th term in each of these linear sequences.

a 3, 5, 7, 9, 11, …   **b** 5, 9, 13, 17, 21, …

c 8, 13, 18, 23, 28, …   **d** 2, 8, 14, 20, 26, …

e 5, 8, 11, 14, 17, …   **f** 2, 9, 16, 23, 30, …

g 1, 5, 9, 13, 17, …   **h** 3, 7, 11, 15, 19, …   **i** 2, 5, 8, 11, 14, …

j 2, 12, 22, 32, …   **k** 8, 12, 16, 20, …   **l** 4, 9, 14, 19, 24, …

HINTS AND TIPS

Remember to look at the differences and the first term.

2 Find the *n*th term and the 50th term in each of these linear sequences.

   a 4, 7, 10, 13, 16, …   **b** 7, 9, 11, 13, 15, …   **c** 3, 8, 13, 18, 23, …

   d 1, 5, 9, 13, 17, …   **e** 2, 10, 18, 26, …   **f** 5, 6, 7, 8, 9, …

   g 6, 11, 16, 21, 26, …   **h** 3, 11, 19, 27, 35, …   **i** 1, 4, 7, 10, 13, …

   j 21, 24, 27, 30, 33, …   **k** 12, 19, 26, 33, 40, …   **l** 1, 9, 17, 25, 33, …

3 For each sequence **a** to **j**, find:

   i the *n*th term   **ii** the 100th term.

   a 5, 9, 13, 17, 21, …   **b** 3, 5, 7, 9, 11, 13, …   **c** 4, 7, 10, 13, 16, …

   d 8, 10, 12, 14, 16, …   **e** 9, 13, 17, 21, …   **f** 6, 11, 16, 21, …

   g 0, 3, 6, 9, 12, …   **h** 2, 8, 14, 20, 26, …   **i** 7, 15, 23, 31, …

   j 25, 27, 29, 31, …

FM 4 An online CD retail company uses the following price chart. The company charges a standard basic price for a single CD, including postage and packing.

n	1	2	3	4	5	6	7	8	9	10	11	12	13	14	15
Charge (£)	10	18	26	34	42	49	57	65	73	81	88	96	104	112	120

   a Using the charges for 1 to 5 CDs, work out an expression for the *n*th term.

   b Using the charges for 6 to 10 CDs, work out an expression for the *n*th term.

   c Using the charges for 11 to 15 CDs, work out an expression for the *n*th term.

   d What is the basic charge for a CD?

PS 5 Look at this series of fractions.

$$\frac{31}{109}, \frac{33}{110}, \frac{35}{111}, \frac{37}{112}, \frac{39}{113}, \ldots$$

> **HINTS AND TIPS**
>
> Use algebra to set up an equation.

   a Explain why the *n*th term of the numerators is $2n + 29$.

   b Write down the *n*th term of the denominators.

   c Explain why the terms of the series will eventually get very close to 1.

   d Which term of the series has a value equal to 1?

AU 6 The square numbers are 1, 4, 9, 16, 25, … .

   a Continue the sequence for another five terms.

   b The *n*th term of this sequence is n^2. Give the *n*th term of these sequences.

     i 2, 5, 10, 17, 26, …   **ii** 2, 8, 18, 32, 50 …   **iii** 0, 3, 8, 15, 24, …

General rules from given patterns

This section will show you how to:
- find the nth term from practical problems
- recognise some special sequences and how they are built up

Many problem-solving situations that you are likely to meet involve number sequences. So you do need to be able to formulate general rules from given number patterns.

EXAMPLE 7

The diagram shows a pattern of squares building up.

a How many squares will there be in the nth pattern?

b What is the largest pattern number you can make with 200 squares?

a

Pattern number	1	2	3	4	5
Number of squares	1	3	5	7	9

Looking at the difference between consecutive patterns, you should see it is always 2 squares. So, use $2n$. Subtract the difference, 2, from the first number, which gives $1 - 2 = -1$. So the number of squares in the nth pattern is $2n - 1$.

b
$$2n - 1 = 200$$
$$2n = 201$$
$$n = 100.5$$

So, 200 squares will make up to the 100th pattern.

There are some number sequences that occur frequently. It is useful to know these as they are very likely to occur in examinations.

Even numbers: The even numbers are 2, 4, 6, 8, 10, 12, … . The nth term of this sequence is $2n$.

Odd numbers: The odd numbers are 1, 3, 5, 7, 9, 11, … . The nth term of this sequence is $2n - 1$.

Square numbers: The square numbers are 1, 4, 9, 16, 25, 36, … . The nth term of this sequence is n^2.

Triangular numbers: The triangular numbers are 1, 3, 6, 10, 15, 21, … . The nth term of this sequence is $\frac{1}{2}n(n + 1)$.

Powers of 2: The powers of 2 are 2, 4, 8, 16, 32, 64, … . The nth term of this sequence is 2^n.

Powers of 10: The powers of 10 are 10, 100, 1000, 10 000, 100 000, 1 000 000, … . The nth term of this sequence is 10^n.

Prime numbers: The first 20 prime numbers are 2, 3, 5, 7, 11, 13, 17, 19, 23, 29, 31, 37, 41, 43, 47, 53, 59, 61, 67, 71.

A prime number is a number that has only two factors, 1 and itself.

There is no pattern to the prime numbers so they do not have an nth term.

Remember: There is only one even prime number, and that is 2.

EXERCISE 11E

1 A pattern of squares is built up from matchsticks as shown.

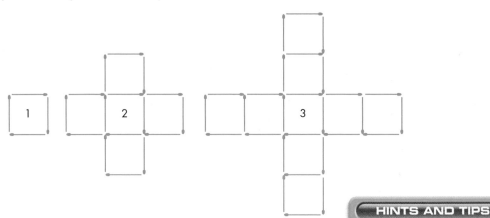

HINTS AND TIPS

Write out the number sequences to help you see the patterns.

a Draw the fourth diagram.

b How many squares are there in the *n*th diagram?

c How many squares are there in the 25th diagram?

d With 200 squares, which is the biggest diagram that could be made?

2 A pattern of triangles is built up from matchsticks.

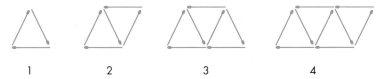

a Draw the fifth set of triangles in this pattern.

b How many matchsticks are needed for the *n*th set of triangles?

c How many matchsticks are needed to make the 60th set of triangles?

d If there are only 100 matchsticks, which is the largest set of triangles that could be made?

AU 3 A conference centre had tables each of which could sit six people. When put together, the tables could seat people as shown.

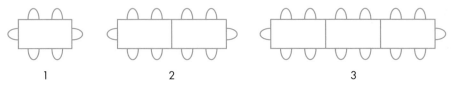

a How many people could be seated at four tables?

b How many people could be seated at *n* tables put together in this way?

c At a conference, 50 people wished to use the tables in this way. How many tables would they need?

4 Regular pentagons of side length 1 cm are joined together to make a pattern as shown.

1 2 3 4

Copy this pattern and write down the perimeter of each shape.

a What is the perimeter of patterns like this made from:

i 6 pentagons **ii** n pentagons **iii** 50 pentagons?

b What is the largest number of pentagons that can be put together like this to have a perimeter less than 1000 cm?

FM 5 Lampposts are put at the end of every 100 m stretch of a motorway, as shown.

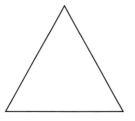

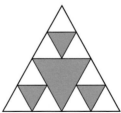

1 2 3

a How many lampposts are needed for:

i 900 m of this motorway **ii** 8 km of this motorway?

b The contractor building the M99 motorway has ordered 1598 lampposts. How long is the M99?

PS 6 Draw an equilateral triangle.

Mark the midpoints of each side and draw and shade in the equilateral triangle formed by joining these points.

Repeat this with the three unshaded triangles remaining.

Keep on doing this with the unshaded triangles that are left.

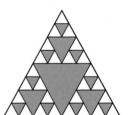

The pattern is called a Sierpinski triangle and is one of the earliest examples of a fractal type pattern.

The shaded areas in each triangle are $\frac{1}{4}, \frac{7}{16}, \frac{37}{64}, \frac{175}{256}$.

It is very difficult to work out an nth term for this series of fractions.

Use your calculator to work out the **unshaded** area, e.g. $\frac{3}{4}, \frac{9}{16}$ ...

You should be able to write down a formula for the nth term of this pattern.

Pick a large value for n.

Will the shaded area ever cover all of the original triangle?

AU 7 Thom is building three different patterns with matches.

He builds the patterns in steps.

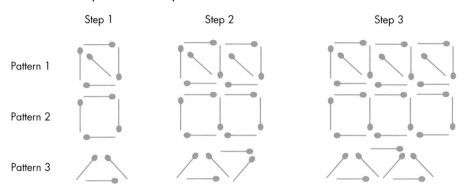

Step 1   Step 2   Step 3

Pattern 1

Pattern 2

Pattern 3

Thom has five boxes of matches which are labelled 'Average contents 42 matches'.

Will Thom have enough matches to get to step 20 with all these patterns together?

Show your working.

8 a The powers of 2 are 2, 4, 8, 16, 32, … .

What is the nth term of this sequence?

b A supermarket sells four different-sized bottles of water.

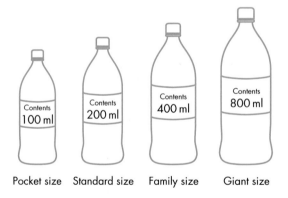

Contents 100 ml   Contents 200 ml   Contents 400 ml   Contents 800 ml

Pocket size   Standard size   Family size   Giant size

i Describe the number pattern that the contents follow.

ii The supermarket introduces a super giant size, which is the next sized bottle in the pattern. How much water is there in this bottle?

9 The powers of 10 are 10^1, 10^2, 10^3, 10^4, 10^5, … .

This gives the sequence 10, 100, 1000, 10 000, 100 000, … .

The nth term is given by 10^n.

a Describe the connection between the number of zeros in each term and the power of the term.

b If $10^n = 1\,000\,000$, what is the value of n?

10 The number p is odd and the number q is even. State if the following are odd or even.

a $p + 1$ **b** $q + 1$ **c** $p + q$

GRADE BOOSTER

F You can give the next term in a sequence and describe how the pattern is building up

E You can find any term in a number sequence and recognise patterns in number calculations

D You can substitute numbers into an nth term rule

D You can understand how odd and even numbers interact in addition, subtraction and multiplication problems

C You can give the nth term of a linear sequence

What you should know now

- How to recognise a number pattern and explain how the pattern is made
- How to recognise a linear sequence and find its nth term

1 a Write down the next two terms in the sequence.

   31, 27, 23, 19   (2)

b What is the rule for continuing the sequence?   (1)

c Jane says that the sequence will eventually include –3.

   Explain why Jane is wrong.   (1)

   (Total 4 marks)

   AQA, June 2009, Paper 2 Foundation, Question 13

2 a Write down the next two numbers in the following sequences.

   i   60   54   48   42   36   ....

   ii   1   2   4   8   16   ....   (4)

b Another sequence begins

   1   3   7   15   31

   Explain the rule for continuing this sequence.   (1)

   (Total 5 marks)

   AQA, June 2008, Paper 1 Foundation, Question 9

3 A pattern is formed from squares.

Pattern 1   Pattern 2   Pattern 3   Pattern 4

   a Draw Pattern 4.   (1)

   b Find the number of squares in Pattern 6.   (1)

   (Total 2 marks)

   AQA, June 2008, Paper 2 Foundation, Question 4

4 The first ten prime numbers are 2, 3, 5, 7, 11, 13, 17, 19, 23, 29.

   P is a prime number.
   Q is an odd number.

State whether each of the following is always odd, always even or could be either odd or even.

   a $P(Q + 1)$

   b $Q - P$

5 The nth term of a sequence is given by the expression

   $n^2 + 5$

Write down the first **three** terms of the sequence.   (2)

   (Total 2 marks)

   AQA, November 2008, Paper 2 Foundation, Question 19

6 a A sequence of numbers starts

   9   7   5   3   ....

   Write down the next two numbers in this sequence.   (2)

b A different sequence uses this rule.

   Add the last two numbers then halve the result

   i The sequence of numbers starts

   12   4   8   6   ....

   Work out the next two numbers in this sequence.   (2)

   ii Jenna says that the rule to find the next number in this sequence could also be:

   "Find the mean of the last two numbers."

   Is Jenna right?

   Explain your answer.   (1)

   (Total 5 marks)

   AQA, June 2007, Paper 1 Foundation, Question 9

7 a A sequence has nth term $4n + 1$.

   i Write down the first three terms of this sequence.   (2)

   ii Toms says that 2006 is a term in this sequence.

   Explain why he is wrong.   (1)

b A different sequence has nth term

   $(n + 3)^2 - 9$

   Show that the first term of this sequence is 7.   (1)

   (Total 4 marks)

   AQA, November 2006, Paper 1 Intermediate, Question 8

C   D   E   F

8 p and q are odd numbers.

a Is $p + q$ an odd number, an even number or could it be either? (1)

b Is pq an odd number, an even number or could it be either? (1)

(Total 2 marks)

AQA, June 2005, Paper 1 Foundation, Question 24

9 Here is a number pattern.

Line 1 $\qquad 1 = \dfrac{1 \times 2}{2}$

Line 2 $\qquad 1 + 2 = \dfrac{2 \times 3}{2}$

Line 3 $\qquad 1 + 2 + 3 = \dfrac{3 \times 4}{2}$

Line 4 $\qquad 1 + 2 + 3 + 4 = \dfrac{4 \times 5}{2}$

Line 5 $\quad 1 + 2 + 3 + 4 + 5 =$

a Complete line 5 of the pattern. (1)

b Write down line 6 of the pattern. (1)

c Use the pattern to find the sum of the whole numbers from 1 to 24.

You **must** show your working. (2)

(Total 4 marks)

AQA, June 2007, Paper 2 Intermediate, Question 2

10 Consecutive patterns are put together.

Pattern 1 and 2   Pattern 2 and 3   Pattern 3 and 4

a The numbers of counters in the combined patterns form the sequence:

4, 9, 16, …

How many counters will be in the next combined pattern in the sequence?

b What type of numbers are 4, 9, 16, … ?

c How many counters will be in the combined pattern formed by patterns 9 and 10?

11 Martin says that the square of any number is always bigger than the number. Give an example to show that Martin is wrong.

12 It is known that n is an integer.

a Explain why $2n + 1$ is always an odd number for all values of n.

b Explain why n^2 could be either odd or even.

13 Here are the nth terms of three sequences.

Sequence 1 nth term $4n + 1$
Sequence 2 nth term $3n + 3$
Sequence 3 nth term $3n - 1$

For each sequence state whether the numbers in the sequence are

A   Always multiples of 3
S   Sometimes multiples of 3
N   Never multiples of 3 (3)

(Total 3 marks)

AQA, November 2006, Paper 2 Intermediate, Question 13

14 a Write down the next term for each of the following sequences.

Give the rule for each sequence.

i 7, 13, 19, 25, …… (2)

ii 11, 8, 5, 2, …… (2)

b Find the nth term of this sequence.

6, 10, 14, 18, …… (2)

(Total 6 marks)

AQA, November 2008, Paper 1 Foundation, Question 13

Worked Examination Questions

PS **1** **a** Matches are used to make patterns.

  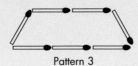 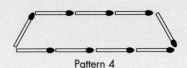

Pattern 1   Pattern 2   Pattern 3   Pattern 4

   i How many matches would be needed for the 10th pattern?

   ii How many matches would be needed for the nth pattern?

b The patterns are used to make a sequence of shapes.

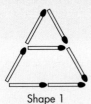 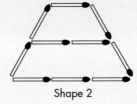 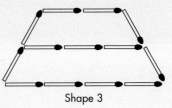

Shape 1   Shape 2   Shape 3

The number of matches needed to make these shapes is

    7, 10, 13, 16, ….

   i How many matches are needed for the nth shape?

   ii What shape number could be made with 64 matches?

---

a **i** The number of matches needed is 21.

   ii $2n + 1$

(2 marks)

b **i** $3n + 4$

   ii 20th shape

(4 marks)

(**Total:** 6 marks)

The number of matches needed is 3, 5, 7, 9, … which is going up by 2 each time.
Continuing the sequence for 10 terms gives 3, 5, 7, 9, 11, 13, 15, 17, 19, 21.
You get 1 mark for the correct answer.

The sequence goes up by 2 each time and the first term is 3, so the nth term is $2n + 1$.
You get you 1 mark for the correct answer.

The sequence goes up by 3 each time and the first term is 7 so the nth term is $3n + 4$. ($3n$ on its own would get 1 method mark. The full answer gets you 1 method mark and 1 mark for accuracy.)

$3n + 4 = 64$, then $3n = 60$, so $n = 20$. (Setting up the equation $3n + 4 = 64$ gets you 1 method mark and the answer of 20 gets 1 mark for accuracy.)

Worked Examination Questions

2 Tom is building fences, using posts and rails.

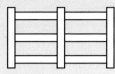

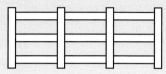

Stage 1
2 posts, 3 rows

Stage 2
3 posts, 6 rows

Stage 3
4 posts, 9 rows

   a How many rails will there be in a fence with 6 posts?

   b How many posts will be needed for a fence with 27 rails?

   c Posts cost £12 and rails costs £5.
      Write down a formula for the cost of a fence with n posts.

a 15 rails

(1 mark)

This answer can be found by 'counting on', i.e.
5 posts, 12 rails
6 posts, 15 rails.
This gets you 1 mark.

b 10 posts

(1 mark)

c £($27n - 15$)

(2 marks)

(**Total:** 4 marks)

This could be found by 'counting on' but it is better to find a rule. The number of rails is in the 3 times table and is the multiple that is 1 less than the number of posts.
$27 = 9 \times 3$, so number of posts is $9 + 1$.
This gets you 1 mark.

It is very important that once a formula is found, it is checked. This is a fundamental part of functional maths. So take one of the given examples at the start of the question:
Stage 2 uses 3 posts and 6 rails, for which the cost is $3 \times 12 + 6 \times 5 = £66$.
The formula gives $27 \times 3 - 15 = 81 - 15 = 66$.

Once the formula is found in part (b) it is a matter of putting the costs and the nth terms together.
Cost = $12n + 3 \times 5 \times (n - 1)$
You get 1 method mark for the equation.
You get 1 accuracy mark for simplifying the answer.

**PS
AU**

3 Here are formulae for the nth terms of three sequences.

    Formula 1: $4n + 1$
    Formula 2: $5n - 2$
    Formula 3: $5n + 10$

    Say if the sequences generated by the nth terms always (A) give multiples of 5, never (N) give multiples of 5 or sometimes (S) give multiples of 5.

Formula 1: Sometimes (S)

Formula 2: Never (N)

Formula 3: Always (A)

(**Total:** 3 marks)

Substitute $n = 1, 2, 3$, etc. until you can be sure of the sequences.
These are:
    5, 9, 13, 17, 21, 25, 29, ...
    3, 8, 13, 18, 23, 28, ...
    15, 20, 25, 30, 35, 40, ...
(You get 1 mark for each part.)

Some sets of square numbers have a special connection. For example:

$$3^2 + 4^2 = 5^2$$

The sets (3, 4, 5) is called a Pythagorean triple. (In Book 2, Chapter 14 you will learn about Pythagoras' theorem, which is connected with right-angled triangles. Pythagorean triples, however, were well known before Pythagoras himself!)

Pythagorean triples were used by ancient people in the construction of great monuments such as Stonehenge and the Egyptian pyramids. They are an excellent example of how we can use patterns.

Getting started

Many patterns, including Pythagorean triples, can be found, using diagrams.

1 Working in pairs, start with the smallest odd number and, going up in consecutive odd numbers, draw diagrams to represent the numbers. Extend your diagrams using a rule.

Have a look at what another group has done. Have they found the same rule as yours? Is there more than one way to represent odd numbers diagrammatically?

Getting started (continued)

2 Represent a square number diagrammatically
– the clue is in the name.

One representation of odd numbers can be
used to show how consecutive odd numbers
add together to make square numbers.
See if you can find this and show it in a
diagram of 3^2.

3 Use the pattern from the diagrams to
complete this sequence:

1 $= 1^2 = 1$
$1 + 3$ $= 2^2 = 4$
$1 + 3 + 5$ $= ..... = ...$
$1 + 3 + 5 + 7 = ... = ...$
               $= ..... = ...$

Look for connections between the last odd
number and the square number.

Use these connections to fill in the missing
numbers in this sequence:

$1 + 3 + 5 + ... + 19 = ... = ...$
$1 + 3 + 5 + ... + ... = 12^2 = 144$
$1 + 3 + 5 + ... + 99 = ... = ...$
$1 + 3 + 5 + ... + ... = 200^2 = 40\,000$

Your task

It is your task to use diagrams to find patterns,
and make and test generalisations.

1 Draw accurate diagrams to show that
(5, 12, 13) is a Pythagorean triple and that it
is true that 32 + 42 = 52.

2 Show, using a diagram, that (5, 12, 13) is a
Pythagorean triple.

3 What is the next Pythagorean triple?

4 Use this method to find at least three more
Pythagorean triples.

(Use a calculator to check that you are right)

Extension

Use the internet to find out about
Pythagoras, his theorem and the
Platonic Formula.

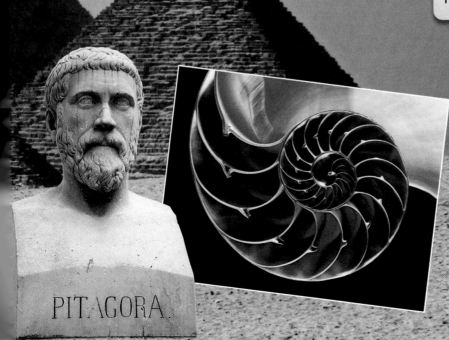

PITAGORA

The biggest moveable volume on Earth is water. There are about 1.4 billion cubic kilometres (km^3) of water on our planet. This water is essential for the natural processes that create and sustain life on Earth. Volume can help us to assess whether the planet has enough of one of the natural resources that is vital to our daily diet as well as to the survival of our whole planet's ecosystem.

The volume of water on Earth is actually increasing as a result of condensation from volcanoes and from comets that enter the Earth's atmosphere. These processes happen all the time, but they add only about a cubic metre ($1 \ m^3$) of water every year.

So, just where can we find the total volume of water on our planet?

The table below shows where it is to be found and how it is used by humans, animals and plants.

About 70% of the Earth's surface area is covered in water!

	Volume (km^3)	Comment
Oceans	1 338 000 000	Salt water in oceans, seas and bays
Ice	24 364 000	Ice caps, glaciers, snow and ground ice
Groundwater	23 400 000	Fresh and salt water underground and in deep wells
Lakes and reservoirs	176 400	Fresh and salt water stored on the Earth's surface and often used for drinking, irrigation and recreation
Soil moisture	16 500	Used by crops, trees and surface vegetation
Water vapour in the atmosphere	12 900	Including clouds, fog and dew
Rivers	2120	Drinking, irrigation and recreation
Swamp water	11 470	Temporary and permanent wetland areas

As you can see most of the total volume of water on Earth is in our oceans. The Atlantic, Pacific, Indian, Arctic and Southern oceans hold 97.3% of the total water on Earth.

Within the waters of the ocean there are over 1 million kilograms of gold, currently worth £61 billion. That is £9 for every person on Earth.

The volumes of water on Earth are very large and difficult to measure. In comparison, the volumes of most shapes we need to measure are much easier to work out.

This chapter leads you through the process of calculating the volumes and surface areas of a variety of shapes.

Chapter

Geometry: Surface area and volume of 3D shapes

1 Units of volume

2 Surface area and volume of a cuboid

3 Surface area and volume of a prism

4 Volume of a cylinder

This chapter will show you ...

G the units used when finding the volume of 3D shapes

E how to calculate the surface area and volume of a cuboid

C how to calculate the surface area and volume of prisms

C how to calculate the volume of a cylinder

Visual overview

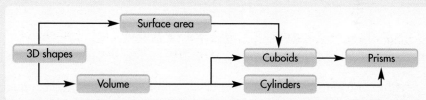

What you should already know

- How to find the area of a rectangle and a triangle (see Chapter 6) (KS3 level 5, GCSE grade E)

- The units used with area (KS3 level 5, GCSE grade E)

- The names of basic 3D shapes (KS3 level 3, GCSE grade G)

- What is meant by the term 'volume' (KS3 level 5, GCSE grade G)

Quick check

What are the mathematical names of these 3D shapes?

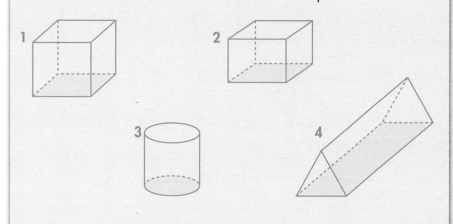

This section will show you how to:
• use the correct units with volume

Key words
cubic centimetre
cubic metre
cubic millimetre
edge
face
vertex
volume

Volume is the amount of space taken up inside a 3D shape.
Volume is measured in **cubic millimetres** (mm^3), **cubic centimetres** (cm^3) or **cubic metres** (m^3).

Length, area and volume

A cube with an **edge** of 1 cm has a volume of 1 cm^3 and each **face** has an area of 1 cm^2.

A cube with an edge of 2 cm has a volume of 8 cm^3 and each face has an area of 4 cm^2.

A cube with an edge of 3 cm has a volume of 27 cm^3 and each face has an area of 9 cm^2.

1 cm

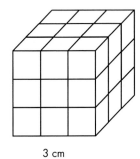

2 cm   3 cm

EXAMPLE 1

How many cubes, each 1 cm by 1 cm by 1 cm, have been used to make these steps? What volume do they occupy?

When you count the cubes, do not forget to include those hidden at the back.

You should count:

6 + 4 + 2 = 12

The volume of each cube is 1 cm^3.

So, the volume of the steps is:

12 × 1 = 12 cm^3

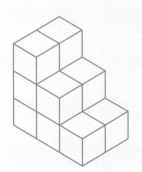

ACTIVITY

Area and volume

For shape A and shape B work out the area of the front face and the volume.
Then copy out the sentences and fill in the missing numbers.

Shape A	Shape B	
  Area of front face = … cm² Volume = … cm³	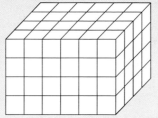 Area of front face = … cm² Volume = … cm³	Each side of shape B is … times as big as each side of shape A. The area of the front face of shape B is … times as big as the front face of shape A. The volume of shape B is …. times as big as the volume of shape A.
  Area of front face = … cm² Volume = … cm³	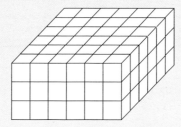 Area of front face = … cm² Volume = … cm³	Each side of shape B is … times as big as each side of shape A. The area of the front face of shape B is … times as big as the front face of shape A. The volume of shape B is …. times as big as the volume of shape A.

EXERCISE 12A

1 Find the volume of each 3D shape, if the edge of each cube is 1 cm.

a

b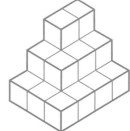

G

c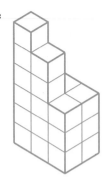

d

F

AU 2 Zoe says, "there are a 100 cubes in the shape."
Explain how she might have calculated this.

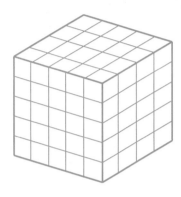

E

PS 3 Each of the blocks in this diagram are centimetre cubes.
What is the total visible surface area of the cubes?

4 David was asked to pack the display of packets,
shown opposite into cartons. All the packets are the
same size, with volume 100 cm³. Each carton
will contain 24 of the packets. How many cartons
will David need?

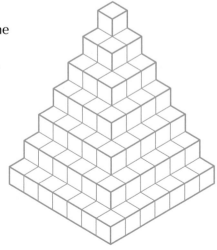

ACTIVITY

Many-faced shapes

All 3D shapes have **faces**, **vertices** and **edges**.
(**Note:** Vertices is the plural of vertex.)

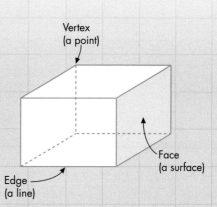

Look at the shapes in the table below and on the next page. Then copy the table and fill it in.

Remember that there are hidden faces, vertices and edges. These are shown with dashed lines.

Look at the numbers in the completed table.

- For each shape, can you find the connection between the following properties?

    - The number of faces, F

    - The number of vertices, V

    - The number of edges, E

- Find some other solid shapes. Does your connection also hold for those?

Shape	Name	Number of faces (F)	Number of vertices (V)	Number of edges (E)
	Cuboid			
	Square-based pyramid			
	Triangular-based pyramid (or tetrahedron)			

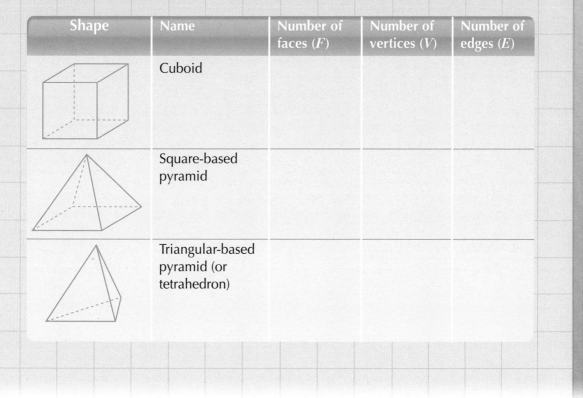

Shape	Name	Number of faces (F)	Number of vertices (V)	Number of edges (E)
	Octahedron			
	Triangular prism			
	Hexagonal prism			
	Hexagon-based pyramid			

This section will show you how to:
- calculate the surface area and volume of a cuboid

Key words

capacity

height

length

litre

surface area

volume

width

A cuboid is a box shape, all six faces of which are rectangles.

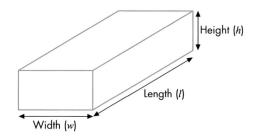

Every day you will come across many examples of cuboids, such as breakfast cereal packets, shoe boxes, DVD players – and even this book.

The **volume** of a cuboid is given by the formula:

volume = **length** × **width** × **height**   *or* $V = l \times w \times h$ *or* $V = lwh$

The **surface area** of a cuboid is calculated by finding the total area of the six faces, which are rectangles. Notice that each pair of opposite rectangles have the same area. So, from the diagram above:

area of top and bottom rectangles = 2 × length × width = $2lw$

area of front and back rectangles = 2 × height × width = $2hw$

area of two side rectangles = 2 × height × length = $2hl$

Hence, the surface area of a cuboid is given by the formula:

surface area = $A = 2lw + 2hw + 2hl$

EXAMPLE 2

Calculate the volume and surface area of this cuboid.

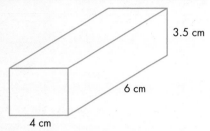

3.5 cm

6 cm

4 cm

Volume = $V = lwh = 6 \times 4 \times 3.5 = 84$ cm^3

Surface area = $A = 2lw + 2hw + 2hl$

$$= (2 \times 6 \times 4) + (2 \times 3.5 \times 4) + (2 \times 3.5 \times 6)$$

$$= 48 + 28 + 42 = 118 \text{ cm}^2$$

Note:

1 cm^3 = 1000 mm^3 and 1 m^3 = 1 000 000 cm^3

The word '**capacity**' is often used for the volumes of liquids or gases.

The unit used for measuring capacity is the **litre**, l, with:

1000 millilitres (ml) = 1 litre

100 centilitres (cl) = 1 litre

1000 cm^3 = 1 litre

1 m^3 = 1000 litres

EXERCISE 12B

1 Find **i** the volume and **ii** the surface area of each of these cuboids.

a

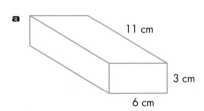

11 cm

3 cm

6 cm

b

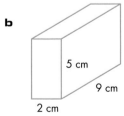

5 cm

9 cm

2 cm

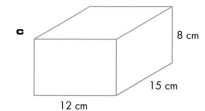

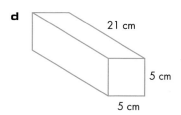

2 Find the capacity of a fish-tank with dimensions: length 40 cm, width 30 cm and height 20 cm. Give your answer in litres.

3 Find the volume of the cuboid in each of the following cases.

a The area of the base is 40 cm² and the height is 4 cm.

b The base has one side 10 cm and the other side 2 cm longer, and the height is 4 cm.

c The area of the top is 25 cm² and the depth is 6 cm.

4 Calculate **i** the volume and **ii** the surface area of each of the cubes with these edge lengths.

a 4 cm   **b** 7 cm   **c** 10 mm

d 5 m   **e** 12 m

FM 5 Safety regulations say that in a room where people sleep there should be at least 12 m³ for each person. A dormitory is 20 m long, 13 m wide and 4 m high. What is the greatest number of people who can safely sleep in the dormitory?

6 Complete the table below for cuboids **a** to **e**.

	Length	Width	Height	Volume
a	8 cm	5 cm	4.5 cm	
b	12 cm	8 cm		480 cm³
c	9 cm		5 cm	270 cm³
d		7 cm	3.5 cm	245 cm³
e	7.5 cm	5.4 cm	2 cm	

7 A tank contains 32 000 litres of water. The base of the tank measures 6.5 m by 3.1 m. Find the depth of water in the tank. Give your answer to one decimal place.

8 A room contains 168 m³ of air. The height of the room is 3.5 m. What is the area of the floor?

9 What are the dimensions of cubes with these volumes?

a 27 cm³   **b** 125 m³   **c** 8 mm³   **d** 1.728 m³

D

10 Calculate the volume of each of these shapes.

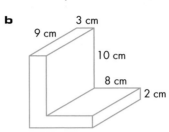

a

2 cm
3 cm
6 cm
2 cm
10 cm
7 cm

b

3 cm
9 cm
10 cm
8 cm
2 cm

AU 11 A cuboid has volume of 125 cm^3 and a total surface area of 160 cm^2.

Is it possible that this cuboid is a cube? Give a reason for your answer.

C

PS 12 The volume of a cuboid is 1000 cm^3. What is the smallest surface area it could have?

12.3 Surface area and volume of a prism

This section will show you how to:
- calculate the surface area and volume of a prism

Key words

cross-section
prism
surface area
volume

A **prism** is a 3D shape that has the same **cross-section** running all the way through it, whenever it is cut perpendicular to its length. Here are some examples.

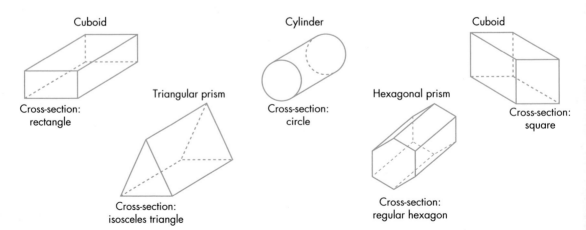

Cuboid

Cross-section: rectangle

Triangular prism

Cross-section: isosceles triangle

Cylinder

Cross-section: circle

Hexagonal prism

Cross-section: regular hexagon

Cuboid

Cross-section: square

The **volume** of a prism is found by multiplying the area of its cross-section by the length of the prism (or height if the prism is stood on end), that is:

volume of prism = area of cross-section × length   *or* $V = Al$

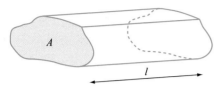

EXAMPLE 3

Calculate the **surface area** and the volume of the triangular prism below.

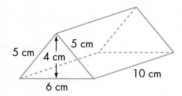

The surface area is made up of three rectangles and two isosceles triangles.

Area of the three rectangles = $10 \times 5 + 10 \times 5 + 10 \times 6 = 50 + 50 + 60 = 160$ cm^2

Area of one triangle = $\dfrac{6 \times 4}{2} = 12$, so area of two triangles = 24 cm^2

Therefore, the total surface area = 184 cm^2

Volume of the prism = Al

Area of the cross-section = area of the triangle = 12 cm^2

So, $V = 12 \times 10 = 120$ cm^3

EXERCISE 12C

1 For each prism shown:

    i   sketch the cross-section

    ii calculate the area of the cross-section

    iii calculate the volume.

> **HINTS AND TIPS**
>
> Look back at page 131 to remind yourself how to calculate the areas of compound shapes.

a

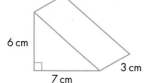

b

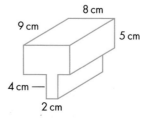

c

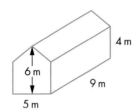

d

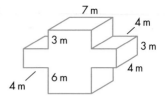

e

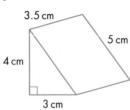

f

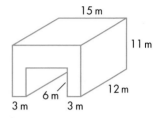

2 Each of these prisms has a constant cross-section in the shape of a right-angled triangle.

a Find the volume of each prism.   **b** Find the total surface area of each prism.

i

ii

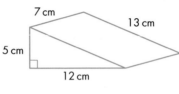

3 The uniform cross-section of a swimming pool is a trapezium with parallel sides of lengths 1 m and 2.5 m, with a perpendicular distance of 30 m between them. The width of the pool is 10 m. How much water is in the pool when it is full? Give your answer in litres.

4 Which of these 3D shapes has the greater volume?

a

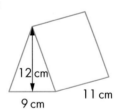

b

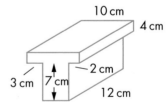

FM **5** Sandra had a swimming pool in her garden. The shallow end is 1 m deep and the deep end 2 m deep, as shown in the diagram.

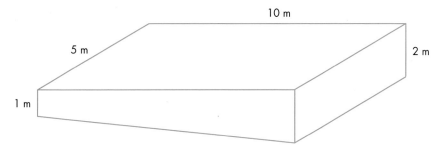

a What is the volume of the pool?

b How many litres of water will it take to fill the pool?

c Sandra has a hosepipe that will deliver water to the pool at the rate of 5 gallons a minute.

   How long will it take to fill the pool? Give your answer in days, hours and minutes.

PS **6** The metal cuboid shown in the diagram is melted down and cast into a cube.

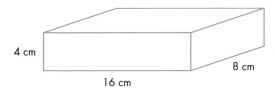

What is the surface area of the cube?

AU **7** Kira needs to find the weight of a metal lintel above a door.

She has calculated the volume of the lintel to be 22 500 cm³.

She knows the weight of 1 cm³ of this metal.

Explain how she can find the approximate weight of the lintel while it is still in place above the door.

This section will show you how to:
- calculate the volume of a cylinder

Key words

π
cylinder
height
length
radius
volume

The **volume** of a **cylinder** is found by multiplying the area of its circular cross-section by its **height**, that is:

volume = area of circle × height   *or* $V = \pi r^2 h$

where r is the **radius** of the cylinder and h is its height or **length**.

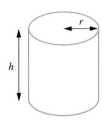

EXAMPLE 4

Calculate the volume of a cylinder with a radius of 5 cm and a height of 12 cm.

Volume = r^2h = × 5^2 × 12 = 942.5 cm³ (to 1 decimal place)

EXERCISE 12D

1 Find the volume of each of these cylinders. Round your answers to a sensible degree of accuracy.

a   base radius 4 cm and height 5 cm

b   base diameter 9 cm and height 7 cm

c   base diameter 13.5 cm and height 15 cm

d   base radius 1.2 m and length 5.5 m.

2 Find the volume of each of these cylinders. Round your answers to a sensible degree of accuracy.

a

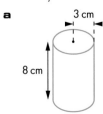

3 cm

8 cm

b

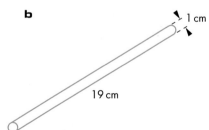

1 cm

19 cm

c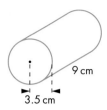

d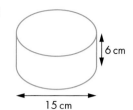

FM 3 The diameter of a cylindrical marble column is 60 cm and its height is 4.2 m. The cost of making this column is quoted as £67.50 per cubic metre. What is the estimated total cost of making the column?

4 A cylindrical container is 65 cm in diameter. Water is poured into the container until it is 1 m deep. How much water is in the container? Give your answer to the nearest litre.

5 A cylindrical can of soup has a diameter of 7 cm and a height of 9.5 cm. How much soup does it hold?

6 A metal bar is 1 m long and has a diameter of 6 cm. What is the volume of the metal bar?

7 What are the volumes of the following cylinders? Give your answers in terms of π.

a with a base radius of 6 cm and a height of 10 cm

b with a base diameter of 10 cm and a height of 12 cm.

PS 8 Copper wire is made by softening and rolling a copper ingot (a piece of cast metal) measuring 15 cm by 15 cm by 60 cm.

What length of wire of radius 0.5 mm can be rolled from the copper ingot? Give your answer in kilometres.

AU 9 Explain how you can tell which of the two cylinders has the larger volume without actually calculating their volumes.

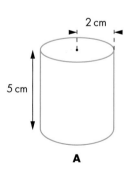

A

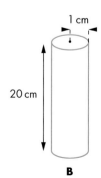

B

GRADE BOOSTER

G You can find the volume of a 3D shape by counting cubes

F You can find the surface area of 3D shapes by counting squares on faces

E You know the formula $V = lbh$ to find the volume of a cuboid

E You can find the surface area of a cuboid

C You can find the surface area and volume of a prism

C You can find the volume of a cylinder

What you should know now

- The units used when finding volume
- How to find the surface area and volume of a cuboid
- How to find the surface area and volume of a prism
- How to find the volume of a cylinder

1 The diagram shows a solid shape made from four 1-centimetre cubes.

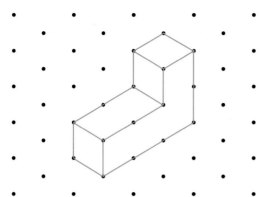

a What is the volume of the solid shape? (2)

b What is the surface area of the solid shape? (3)

(Total 5 marks)

2 A cuboid is made from centimetre cubes.
The area of the base of the cuboid is 5 cm².
The volume of the cuboid is 10 cm³.
Work out the surface area of the cuboid. (3)

(Total 3 marks)

AQA, November 2005, Paper 1 Intermediate, Question 17

3 A cuboid has a volume of 75 cm³.

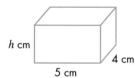

The length is 5 cm.
The width is 4 cm.
Find the height, *h* cm. (2)

(Total 2 marks)

AQA, June 2008, Paper 2 Foundation, Question 15

4 Jasmin has a pond in her garden.
The surface of the pond is a semicircle of radius 1.4 m.

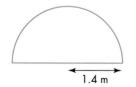

Not to scale

1.4 m

a Calculate the area of a semicircle of radius 1.4 m.
Give your answer to a sensible degree of accuracy.
You **must** show your working. (3)

b The pond is 50 cm deep.
The sides of the pond are vertical.
Calculate the volume of the pond.
Give your answer in cubic metres. (2)

(Total 5 marks)

AQA, November 2005, Paper 2 Intermediate, Question 17

5 The diagram shows a cylinder with a height of 5 cm and a diameter of 16 cm.

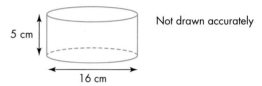

Calculate the volume of the cylinder.
Give your answer in terms of π.
State the units of your answer. (4)

(Total 4 marks)

AQA, May 2008, Paper 1 Foundation, Question 26

6 A triangular prism has dimensions as shown.

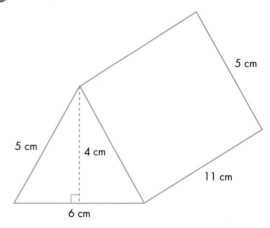

a Calculate the total surface area of the prism.

b Calculate the volume of the prism.

Worked Examination Questions

FM **1** A baker uses square and circular tins to make his cakes.

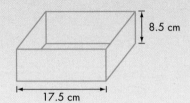 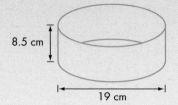

a The edge of the base of the square tin is 17.5 cm. Its height is 8.5 cm.
Calculate the volume of the tin.

b The diameter of the circular tin is 19 cm. Its height is 8.5 cm.
Calculate the volume of the circular tin.

a $V = lbh$

$= 17.5 \times 17.5 \times 8.5$

$= 2600 \text{ cm}^3$ (to the nearest ten)

> You earn 1 method mark for correctly setting up the calculation of $17.5 \times 17.5 \times 8.8$.
> You also earn 1 accuracy mark for an answer that could round to 2600.

(2 marks)

b $V = \pi r^2 h$

The diameter is 19 cm, so $r = 9.5$ cm.

So, $V = \pi \times 9.5^2 \times 8.5$

$= 2410 \text{ cm}^3$ (to the nearest ten)

> You earn 1 method mark for correctly setting up the calculation of $9.5 \times 9.5 \times 8.5$.
> You also earn 1 accuracy mark for an answer that could round to 2410.

(2 marks)

(**Total:** 4 marks)

> There could also be a mark for rounding your numbers to a sensible degree of accuracy. Here this would be to either the nearest ten or the nearest unit.
> In some examinations, there could also be a mark for giving the correct units.

Worked Examination Questions

FM **2** A firm makes small rods that are cylinders of radius 5 mm and length 10 cm.

These are all made from blocks of metal measuring 60 cm by 60 cm by 2 m.

How many rods can be created from one block of metal?

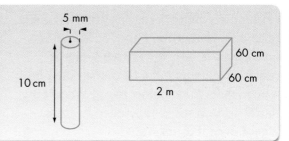

Volume of block = 200 cm × 60 cm × 60 cm

$\qquad$ = 720 000 cm^3

Volume of a rod = $\pi \times 0.5^2 \times 10$

$\qquad$ = 2.5π cm^3

Number of rods = $\dfrac{720\,000}{2.5\pi}$

$\qquad$ = 91 673.247 22

$\qquad$ = 91 673

Total: 9 marks

> You get 1 method mark for converting to common units and 1 method mark for the calculation $200 \times 60 \times 60$.

> You get 1 accuracy mark for finding 720 000.

> You get 1 method mark for converting to common units and 1 method mark for the calculation $\pi \times 0.5^2 \times 10$.

> You get 1 accuracy mark for 2.5π or 7.853 981 634. However, this mark could be lost for rounding too soon in this calculation.

> You get 1 method mark for the calculation $720\,000 \div 2.5\pi$.

> 1 accuracy mark is available for 91673.2 …. (… means the number continues.)

> You get 1 mark for accuracy for a final integer value 91 673. Note that an important part of solving this problem is that you do not round until the end, otherwise you may get an inaccurate result; for example, if you took the volume of the rod to be 7.85 then you would get a result of 91 719.

AU **3** Isaac is asked to find the area of a circle with diameter 2 m. He says it is about 6 m^2. Explain how you can tell he is wrong.

If diameter is 2 m then the radius is 1 m

Area = $\pi \times 1^2$

π is approximately 3, so this works out to be 3 m^2 and not 6 m^2.

Total: 3 marks

> 1 method mark is available for recognising radius = 1 m.

> You also get 1 method mark for realising that area = $\pi \times 1^2$.

> You get 1 mark for accuracy for saying π is approximately 3. Note that if you calculated the area of the circle exactly, you would not get the last mark.

Farmers have to do mathematical calculations almost every day. For example, an arable farmer may need to know how much seed to buy, how much water is required to irrigate the field each day, how much wheat they expect to grow and how much storage space they need to store wheat once it is harvested.

Farming can be filled with uncertainties, including changes in weather, crop disease and changes in consumption. It is therefore important that farmers correctly calculate variables that are within their control, to minimise the impact of changes that are outside their control.

Grain storage
Wheat is stored in large containers called silos. These are usually big cylinders but can also be various other shapes.

Important information about wheat crops (yield data)

- A 1 kg bag of seeds holds 26 500 seeds

- A 1 kg bag of seeds costs 50p

- I want to plant 60 bags of seed

- I need to plant 100 seeds in each square metre (m^2) of field

- I need to irrigate each square metre of the field with 5 litres of water each day

- I expect to harvest 0.7 kg of wheat from each square metre of the field

- Every cubic metre (m^3) of storage will hold 800 kg of wheat

Your task

Rufus, a crop farmer, is going to grow his first field of wheat next summer. Using all the information that he has gathered, help him to plan for his wheat crop. You should consider:

- the size of the field that he will require
- how much seed he will need
- how much water he will need to irrigate the crops, per day, and how it will be stored
- how he will store the seeds and wheat
- how much profit he could make if grain is sold at £92.25 per tonne.

Getting started

Think about these points to help you create your plan.

- What different shapes and sizes of field could the crops be grown in?
- If Rufus needs a reservoir to hold one day's irrigation water, what size cylinder would he need? How would this change if he chose a cuboid? What other shapes and sizes of reservoir could he use?
- What shapes and sizes could the silos be?

Handy hints

Remember: 1000 litres = 1 m^3

1000 kg = 1 tonne

How does a designer go about making a new design?

Where does the inspiration come from – the natural world, or objects made by man?

Two dimensional designs are all around us; on clothes, curtains, furniture fabric, carpets, wallpaper and any flat surface that can be decorated. Often these designs will make use of symmetry to produce a repeated pattern. This makes it easy to manufacture a design on a fabric, for example, where a basic unit can be copied endlessly.

Variation can be achieved by making use of reflections, rotations and translations so that a simple starting point can be used to build lots of different patterns. It is the work of designers in many different fields to produce the patterns we see wherever we look.

There are examples of repeated design on the right from all over the world. Can you find examples of reflections, rotations and translations in them?

You might think that there is an infinite number of repeating 'wallpaper designs' but in fact any repeating wallpaper design can be classified as one of just 17 different possible types, depending on the symmetry it has. Every repeating design you see will be one of these 17 types.

For example, these hexagonal and triangular designs might look different, but both have points where six lines of symmetry meet, points where three lines of symmetry meet and points where two lines of symmetry meet, so they are both examples of the same type of wallpaper design. The different types are studied by scientists who are interested in the structure of crystals.

Czech floor tiles

Dish from Turkey

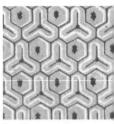
Persian ornament

Egyptian tomb painting

Wall in Spain

Chinese plate design

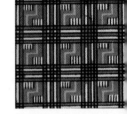
Cloth from Hawaii

13 Geometry: Transformations 2

This chapter will show you ...

- **D** what is meant by a transformation
- **D** how to translate 2D shapes
- **D** how to reflect 2D shapes
- **D** how to rotate 2D shapes
- **D** how to enlarge 2D shapes

Visual overview

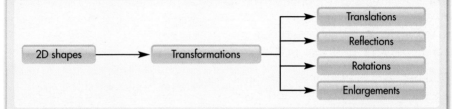

What you should already know

- How to find the lines of symmetry of a 2D shape
  (KS3 level 3, GCSE grade F)
- How to find the order of rotational symmetry of a 2D shape
  (KS3 level 4, GCSE grade F)
- How to find the equation of a line (KS3 level 6, GCSE grade E)

Quick check

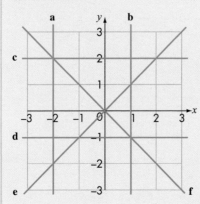

Write down the equations of the lines drawn on the grid.

Translations

This section will show you how to:
- translate a 2D shape

Key words

image
object
transformation
translate
translation
vector

A **transformation** changes the position or the size of a 2D shape in a particular way. You will deal with the four basic ways of using transformations to change a shape: **translation**, reflection, rotation and enlargement.

When a transformation is carried out, the shape in its original position is called the **object** and in its 'new' position it is called the **image**. For translations, reflections and rotations, the object and image are congruent.

A translation is the movement of a shape from one position to another without reflecting it or rotating it. It is sometimes called a 'sliding' transformation, since the shape appears to slide from one position to another.

Every point in the shape moves in the same direction and through the same distance. The object shape **translates** to the image position.

EXAMPLE 1

Describe the following translations.

    a Triangle A to triangle B

    b Triangle A to triangle C

    c Triangle A to triangle D

a Triangle A has been transformed into triangle B by a translation of 5 squares right.

b Triangle A has been transformed into triangle C by a translation of 4 squares up.

c Triangle A has been transformed into triangle D by a translation of 3 squares right and 4 squares up.

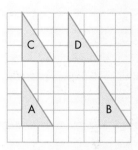

A translation can also be described by using a **vector**. (This is sometimes called a 'column vector'.)

A vector is written in the form $\begin{pmatrix} a \\ b \end{pmatrix}$, where a describes the horizontal movement and b describes the vertical movement.

EXAMPLE 2

Find the vectors for the following translations.

    a A to B   **b** B to C   **c** C to D   **d** D to A

a The vector describing the translation from

    A to B is $\begin{pmatrix} 2 \\ 1 \end{pmatrix}$.

b The vector describing the translation from

    B to C is $\begin{pmatrix} 2 \\ 0 \end{pmatrix}$.

c The vector describing the translation from

    C to D is $\begin{pmatrix} -3 \\ 2 \end{pmatrix}$.

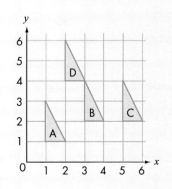

d The vector describing the translation from D to A is $\begin{pmatrix} -1 \\ -3 \end{pmatrix}$.

EXERCISE 13A

1 Copy each of these shapes onto squared paper and draw its image, using the given translation.

a

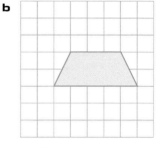

3 squares right

b

3 squares up

c

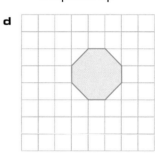

3 squares down

d

3 squares left

2 Copy each of these shapes onto squared paper and draw its image, using the given translation.

a

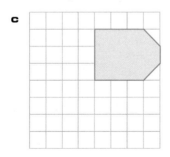

4 squares right and
3 squares down

b

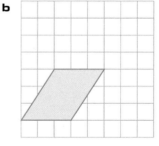

3 squares right and
3 squares up

c

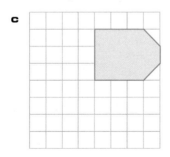

4 squares left and
3 squares down

d

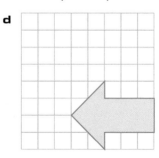

1 square left and
4 squares up

3 Use vectors to describe these translations.

   a **i** A to B   **ii** A to C   **iii** A to D   **iv** A to E   **v** A to F   **vi** A to G

   b **i** B to A   **ii** B to C   **iii** B to D   **iv** B to E   **v** B to F   **vi** B to G

   c **i** C to A   **ii** C to B   **iii** C to D   **iv** C to E   **v** C to F   **vi** C to G

   d **i** D to E   **ii** E to B   **iii** F to C   **iv** G to D   **v** F to G   **vi** G to E

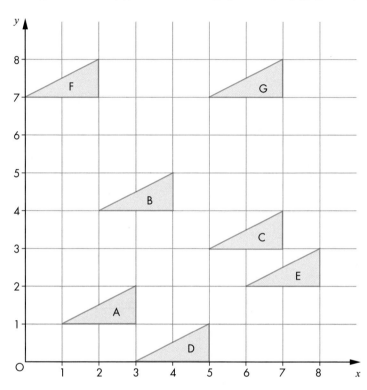

4 Draw a coordinate grid for $-1 \leqslant x \leqslant 6$ and $-4 \leqslant y \leqslant 6$.

   a Draw the triangle with coordinates A(1, 1), B(2, 1) and C(1, 3).

   b Draw the image of triangle ABC after a translation with vector $\begin{pmatrix} 2 \\ 3 \end{pmatrix}$. Label this P.

   c Draw the image of triangle ABC after a translation with vector $\begin{pmatrix} -1 \\ 2 \end{pmatrix}$. Label this Q.

   d Draw the image of triangle ABC after a translation with vector $\begin{pmatrix} 3 \\ -2 \end{pmatrix}$. Label this R.

   e Draw the image of triangle ABC after a translation with vector $\begin{pmatrix} -2 \\ -4 \end{pmatrix}$. Label this S.

5 Using your diagram from question 4, use vectors to describe the following translations.

   a P to Q   **b** Q to R   **c** R to S   **d** S to P

   e R to P   **f** S to Q   **g** R to Q   **h** P to S

PS **6** Use a 10 × 10 grid and draw the triangle with coordinates A(0, 0), B(1, 0) and C(0, 1). How many different translations are there that use integer values only and will move the triangle ABC to somewhere in the grid? (Do not draw them all.)

7 In a game of Snakes and ladders, the snakes and ladders can each be described by a translation.

Use the following vectors.

Ladders $\begin{pmatrix} 1 \\ 2 \end{pmatrix}$, $\begin{pmatrix} 2 \\ 5 \end{pmatrix}$, $\begin{pmatrix} -3 \\ 4 \end{pmatrix}$, $\begin{pmatrix} -2 \\ 3 \end{pmatrix}$, $\begin{pmatrix} 3 \\ 2 \end{pmatrix}$

Snakes $\begin{pmatrix} 1 \\ -3 \end{pmatrix}$, $\begin{pmatrix} 3 \\ -4 \end{pmatrix}$, $\begin{pmatrix} -2 \\ -2 \end{pmatrix}$, $\begin{pmatrix} -1 \\ -3 \end{pmatrix}$, $\begin{pmatrix} 2 \\ -5 \end{pmatrix}$

Put all five ladders and all five snakes onto a 10 × 10 grid to design a Snakes and ladders game board.

AU **8** If a translation is given by:

$$\begin{pmatrix} x \\ y \end{pmatrix}$$

describe the translation that would take the image back to the original.

13.2 Reflections

This section will show you how to:	Key words
• reflect a 2D shape in a mirror line	image
	mirror line
	object
	reflect
	reflection

A **reflection** is a transformation of a 2D shape so that it becomes the mirror **image** of itself.

Notice that each point on the image is the same perpendicular distance from the **mirror line** as the corresponding point on the **object**.

So, if you could fold the whole diagram along the mirror line, every point on the object would coincide with its reflection.

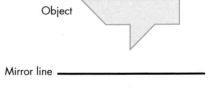

Object

Mirror line —————

Image

EXAMPLE 3

 a Reflect the triangle ABC in the *x*-axis. Label the image P.

 b Reflect the triangle ABC in the *y*-axis. Label the image Q.

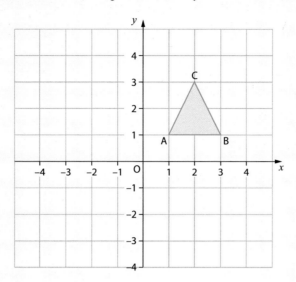

 a The mirror line is the *x*-axis. So, each vertex on triangle P will be the same distance from the *x*-axis as the corresponding vertex on the object.

 b The mirror line is the *y*-axis. So, each vertex on triangle Q will be the same distance from the *y*-axis as the corresponding vertex on the object.

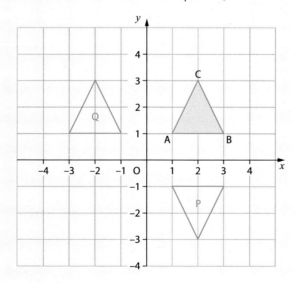

EXERCISE 13B

1 Copy each shape onto squared paper and draw its image after a reflection in the given mirror line.

a
b

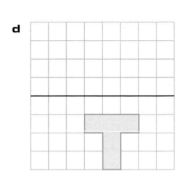

c
d

2 Copy these figures onto squared paper and then draw the reflection of each in the given mirror line.

a 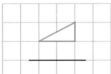 b c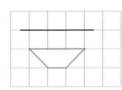

d 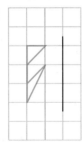 e 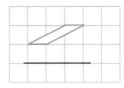 f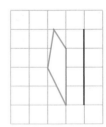

g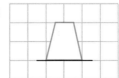

3 Copy these figures onto squared paper and then draw the reflection of each in the given mirror line.

a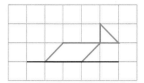

b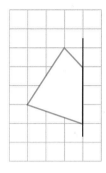

c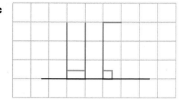

4 Copy this diagram onto squared paper.

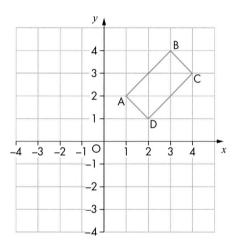

a Reflect the rectangle ABCD in the x-axis. Label the image R.

b Reflect the rectangle ABCD in the y-axis. Label the image S.

c What special name is given to figures that are exactly the same shape and size?

5 **a** Draw a coordinate grid for $-5 \leqslant x \leqslant 5$ and $-5 \leqslant y \leqslant 5$.

b Draw the triangle with coordinates A(1, 1), B(3, 1) and C(4, 5).

c Reflect triangle ABC in the x-axis. Label the image P.

d Reflect triangle P in the y-axis. Label the image Q.

e Reflect triangle Q in the x-axis. Label the image R.

f Describe the reflection that will transform triangle ABC onto triangle R.

6 Copy this diagram onto squared paper.

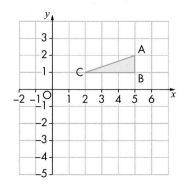

HINTS AND TIPS

Remember that x-lines are parallel to the y-axis and y-lines are parallel to the x-axis.

a Reflect triangle ABC in the line $x = 2$. Label the image X.

b Reflect triangle ABC in the line $y = -1$. Label the image Y.

7 Draw these figures on squared paper and then draw the reflection of each in the given mirror line.

a 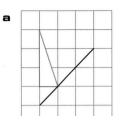   **b**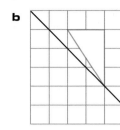

c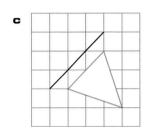

8 Draw these figures on squared paper and then draw the reflection of each in the given mirror line.

a 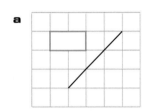   **b**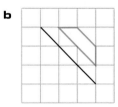

c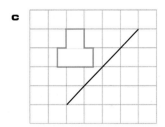

9 A designer used the following instructions to create a design.

- Start with any rectangle ABCD.
- Reflect the rectangle ABCD in the line AC.
- Reflect the rectangle ABCD in the line BD.

Draw a rectangle and use the above to create a design.

PS 10 By using any one of the squares as a starting square ABCD, describe how to keep reflecting the shape to get the final shape in the diagram.

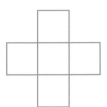

AU 11 If Gill reflects shape T in a line, then reflects the image of T in the same line, explain why the final image is in the same position as the original shape.

12 a Draw a pair of axes for $-5 \leqslant x \leqslant 5$ and $-5 \leqslant y \leqslant 5$. Then draw the lines $y = x$ and $y = -x$, as shown below.

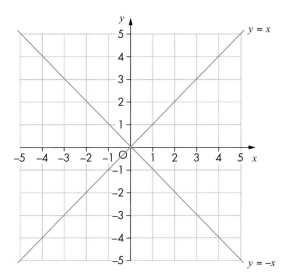

b Draw the triangle with coordinates A(2, 1), B(5, 1) and C(5, 3).

c Draw the reflection of triangle ABC in the x-axis and label the image P.

d Draw the reflection of triangle P in the line $y = -x$ and label the image Q.

e Draw the reflection of triangle Q in the y-axis and label the image R.

f Draw the reflection of triangle R in the line $y = x$ and label the image S.

g Draw the reflection of triangle S in the x-axis and label the image T.

h Draw the reflection of triangle T in the line $y = -x$ and label the image U.

i Draw the reflection of triangle U in the y-axis and label the image W.

j What single reflection will move triangle W to triangle ABC?

13 a Repeat the steps of question 12 but start with any shape you like.

b Is your answer to part **j** the same as before?

c Would your answer to part **j** always be the same, no matter what shape you started with?

This section will show you how to:
- rotate a 2D shape about a point

Key words

angle of rotation
anticlockwise
centre of rotation
clockwise
image
object
rotate
rotation

A **rotation** transforms a 2D shape to a new position by turning it about a fixed point, called the **centre of rotation**.

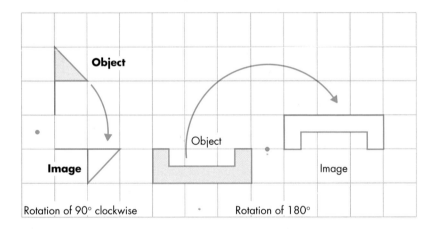

Rotation of 90° clockwise   Rotation of 180°

Note:

- The turn is called the **angle of rotation** and the direction is expressed as **clockwise** or **anticlockwise**.

- The position of the centre of rotation is always specified.

- The angles of rotation that occur in GCSE examinations are a $\frac{1}{4}$-turn or 90°, a $\frac{1}{2}$-turn or 180° and a $\frac{3}{4}$-turn or 270°.

- The rotations 180° clockwise and 180° anticlockwise are the same.

EXAMPLE 4

Draw the **image** of this shape after it has been rotated through 90° clockwise about the point X.

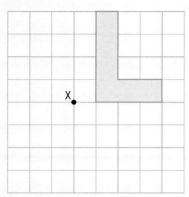

Using tracing paper is always the easiest way to tackle rotations.

First trace the **object** shape and fix the centre of rotation with a pencil point. Then **rotate** the tracing paper through 90° clockwise.

The tracing now shows the position of the image.

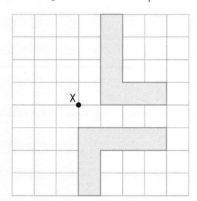

EXERCISE 13C

1 Copy each of these diagrams onto squared paper. Draw each image, using the given rotation about the centre of rotation, X.

a

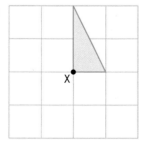

$\frac{1}{2}$-turn

b

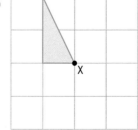

$\frac{1}{4}$-turn clockwise

D

c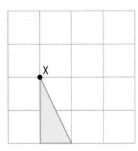

d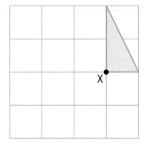

$\frac{1}{4}$-turn anticlockwise

$\frac{3}{4}$-turn clockwise

2 Copy each of these diagrams onto squared paper. Draw each image, using the given rotation about the centre of rotation, X.

a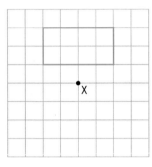

b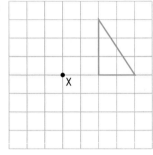

$\frac{1}{2}$-turn

$\frac{1}{4}$-turn clockwise

c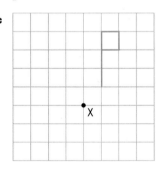

d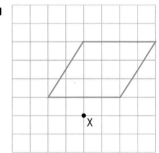

$\frac{1}{4}$-turn anticlockwise

$\frac{3}{4}$-turn clockwise

3 Copy this diagram onto squared paper.

a Rotate the shape through 90° clockwise about the origin O. Label the image P.

b Rotate the shape through 180° clockwise about the origin O. Label the image Q.

c Rotate the shape through 270° clockwise about the origin O. Label the image R.

d What rotation takes R back to the original shape?

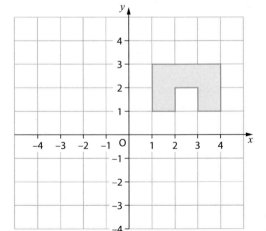

4 Copy this diagram onto squared paper.

a Write down the coordinates of the vertices of the triangle ABC.

b Rotate the triangle ABC through 90° clockwise about the origin O. Label the image S.

Write down the coordinates of the vertices of triangle S.

c Rotate the triangle ABC through 180° clockwise about the origin O. Label the image T.

Write down the coordinates of the vertices of triangle T.

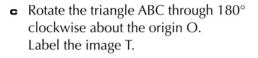

d Rotate the triangle ABC through 270° clockwise about the origin O. Label the image U.

Write down the coordinates of the vertices of triangle U.

e What do you notice about the coordinates of the four triangles?

5 On squared paper, copy these shapes and their centres of rotation.

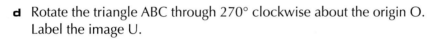

a Rotate each shape about its centre of rotation as follows.

i first by 90° anticlockwise

ii then by a further 180°.

b Describe, in each case, the transformation that would take the original shape to the final image.

6 A graphic designer used the following instructions for creating a design.

• Start with a triangle ABC.

• Reflect the triangle in the line AB.

• Rotate the whole shape about point C clockwise 90°, then a further clockwise 90°, then a further clockwise 90°.

From any triangle of your choice, use the above instructions to create a design.

D

PS **7** By using any one of the squares as a starting square ABCD, describe how to keep rotating the shape to get the final diagram shown.

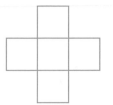

C

AU **8** I rotate a shape S about point P, through 90° clockwise to give an image S′.

I repeat the same rotation on S′ to give S″.

I repeat the rotation on S″ to give S‴.

I repeat the rotation on S‴ to give S‴′.

Explain why S‴′ is in the same position as S.

9 Copy this diagram onto squared paper.

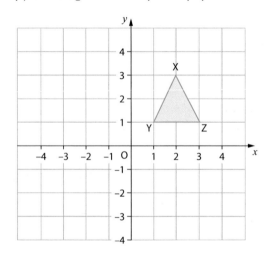

> **HINTS AND TIPS**
>
> Use tracing paper for part **c** and try out different centres until you find the correct one.

a Rotate triangle XYZ through 90° anticlockwise about the point (1, –2). Label the image P.

b Reflect triangle P in the *x*-axis. Label this triangle Q.

c Describe the transformation that maps triangle Q onto triangle XYZ.

10 **a** Draw a coordinate grid, labelling both axes from –5 to 5.

b Draw the triangle with vertices A(2, 1), B(3, 1) and C(3, 5).

c Reflect triangle ABC in the *x*-axis, then reflect the image in the *y*-axis. Label the final position A′B′C′.

d Describe the single transformation that maps triangle ABC onto triangle A′B′C′.

e Will this always happen no matter what shape you start with?

f Will this still happen if you reflect in the *y*-axis first, then reflect in the *x*-axis?

Enlargements

This section will show you how to:

- enlarge a 2D shape by a scale factor

Key words

centre of
    enlargement
enlarge
enlargement
image
object
scale factor

An **enlargement** is a transformation that changes the size of a 2D shape to give a similar **image**. It always has a **centre of enlargement** and a **scale factor**.

The length of each side of the enlarged shape will be:

    length of each side of the **object** × scale factor

The distance of each image point on the enlargement from the centre of enlargement will be:

    distance of original point from centre of enlargement × scale factor

There are two distinct ways to **enlarge** a shape: the ray method and the coordinate method.

EXAMPLE 5

Enlarge the object triangle ABC by a scale factor 3 about O to give the image triangle A'B'C'.

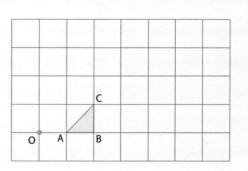

The image triangle A'B'C' is shown below.

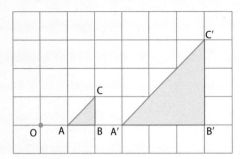

Note

- The length of each side on the enlarged triangle A'B'C' is three times the corresponding length of each side on the original triangle, so that the sides are in the ratio 1 : 3.

- The distance of any point on the enlarged triangle from the centre of enlargement is three times the corresponding distance from the original triangle.

Ray method

This is the *only* way to construct an enlargement when the diagram is not on a grid. The following example shows how to enlarge a triangle ABC by scale factor 3 about a centre of enlargement O by the ray method.

EXAMPLE 6

Enlarge triangle ABC using O as the centre of enlargement and with scale factor 3.

Draw rays from the centre of enlargement, O, to each vertex of the triangle ABC and extend beyond.

Measure the distance from each vertex on triangle ABC to the centre of enlargement and multiply it by 3 to give the distance of each image vertex from the centre of enlargement for triangle A'B'C'.

Once each image vertex has been found, the whole image shape can then be drawn.

Check the measurements and see for yourself how the calculations have been done. Notice again that each line is three times as long in the enlargement.

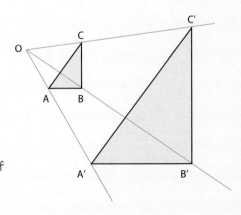

Coordinate method

Triangle A'B'C' is an enlargement of triangle ABC by scale factor 2, with the origin, O, as the centre of enlargement.

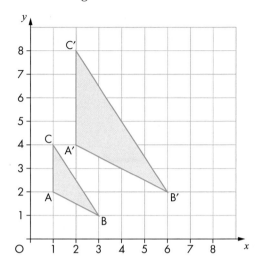

The coordinates of A are (1, 2) and the coordinates of A' are (2, 4). Notice that the coordinates of A' are the coordinates of A multiplied by 2, which is the scale factor of enlargement.

Check that the same happens for the other vertices.

This is a useful method for enlarging shapes on a coordinate grid, when the origin, O, is the centre of enlargement.

EXAMPLE 7

Enlarge the square by scale factor 3, using the origin as the centre of enlargement.

The coordinates of the original square are (1, 1), (2, 1), (2, 2) and (1, 2).

The enlarged square will have these coordinates multiplied by 3.

The coordinates are, therefore, (3, 3), (6, 3), (6, 6) and (3, 6), as shown on the diagram.

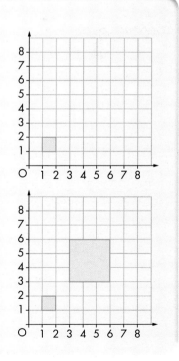

Note: This only works for enlargements centred on the origin. It is not always the case that the origin is the centre of enlargement. Always read the question carefully.

Counting squares

EXAMPLE 8

Enlarge triangle ABC by a scale factor 2, with the point P(0, 1) as the centre of enlargement.

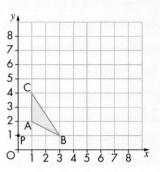

The point A is one square right and one square up from P. As the scale factor is 2, mark A′ two squares right and two squares up from P (2 × 1 = 2).

The point B is three squares right from P, so mark B′ six squares right from P (2 × 3 = 6).

The point C is one square right and three squares up from P, so mark C′ two squares right and six squares up from P (2 × 1 = 2, 2 × 3 = 6).

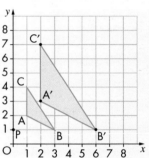

EXERCISE 13D

1 Copy each of these figures, with its centre of enlargement. Then enlarge it by the given scale factor, using the ray method.

a

Scale factor 2

b

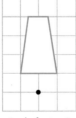

Scale factor 3

c

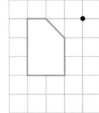

Scale factor 2

d

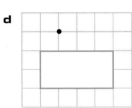

Scale factor 3

2 Copy each of these diagrams onto squared paper and enlarge it by scale factor 2, using the origin as the centre of enlargement.

a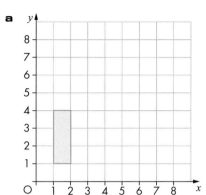

b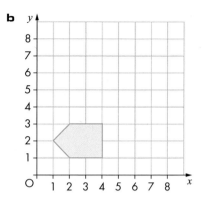

c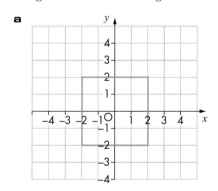

3 Copy each of these diagrams onto squared paper and enlarge it by scale factor 2, using the given centre of enlargement.

a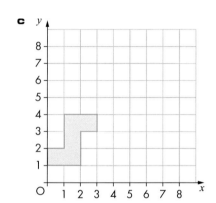

Centre of enlargement (−1, 1)

b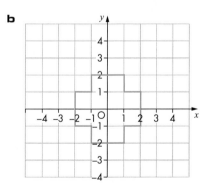

Centre of enlargement (−2, −3)

4 **a** Draw a triangle ABC on squared paper.

b Mark four different centres of enlargement on your diagram:

one above your triangle one below your triangle
one to the left of your triangle one to the right of your triangle.

c From each centre of enlargement, draw an enlargement by scale factor 2.

d What do you notice about each enlarged shape?

D

5 A designer is told to use the following routine.

- Start with a rectangle ABCD.

- Reflect ABCD in the line AC.

- Rotate the whole new shape about C through 180°.

- Enlarge the whole shape by scale factor 2, centre of enlargement point A.

Start with any rectangle of your choice and create the design above.

AU 6 If I enlarge a shape with scale factor 2, the new shape is congruent to the first.

Is this true? Explain your answer.

C

PS 7 If I enlarge a shape by scale factor 3, how many times bigger will the area of the new shape be?

8 'Strange but true'… you can have an enlargement in mathematics that is actually smaller than the original shape! This happens when you 'enlarge' a shape by a fractional scale factor. For example, triangle ABC on the right has been enlarged by scale factor $\frac{1}{2}$ about the centre of enlargement, O, to give the image triangle A'B'C'.

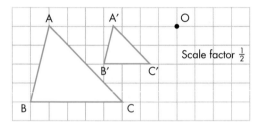

Scale factor $\frac{1}{2}$

Copy the shape below onto squared paper and enlarge it by scale factor $\frac{1}{2}$ about the centre of enlargement, O.

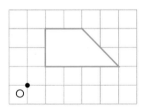

9 Copy this diagram onto squared paper.

a Enlarge the rectangle A by scale factor $\frac{1}{3}$ about the point (–2, 1). Label the image B.

b Write down the ratio of the lengths of the sides of rectangle A to the lengths of the corresponding sides of rectangle B.

c Work out the ratio of the perimeter of rectangle A to the perimeter of rectangle B.

d Work out the ratio of the area of rectangle A to the area of rectangle B.

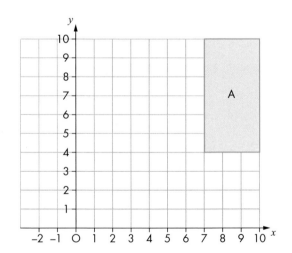

GRADE BOOSTER

D You can translate a 2D shape

D You can reflect a 2D shape in a line $x = a$ or $y = b$

D You can rotate a 2D shape about the origin

D You can enlarge a 2D shape by a whole-number scale factor

C You can translate a 2D shape by a vector

C You can reflect a 2D shape in the line $y = x$ or $y = -x$

C You can rotate a 2D shape about any point

C You can enlarge a 2D shape about any point

What you should know now

- How to translate a 2D shape
- How to reflect a 2D shape
- How to rotate a 2D shape
- How to enlarge a 2D shape

1 Shapes *A*, *B*, *C* and *D* are made from squares of sides 1 cm.

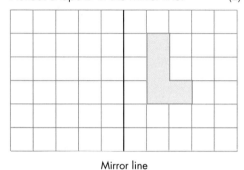

A	*B*	*C*	*D*

a Which **two** shapes are congruent? (1)

b Shape *D* is drawn on the grid.

Reflect shape *D* in the mirror line. (2)

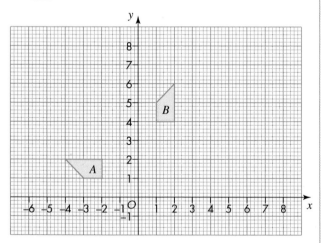

Mirror line

(Total 3 marks)

AQA, November 2008, Paper 2 Foundation, Question 9

2 The diagram shows two shapes, *A* and *B*.

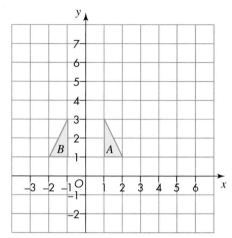

a Describe fully the single transformation that takes shape *A* onto shape *B*. (3)

b Enlarge shape *B* by scale factor 2, with (0, 7) as the centre of enlargement. (3)

(Total 6 marks)

AQA, June 2006, Paper 1 Intermediate, Question 14

 3 This question is about transformations of triangle *A*.

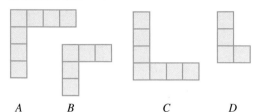

a Describe fully the single transformation that takes triangle *A* onto triangle *B*. (2)

b Translate triangle *A*, 2 units to the left and 3 units down.

Label the new triangle *C*. (1)

c Enlarge triangle *A* by a scale factor of 3, centre (0, 1).

Label the new triangle *D*. (3)

(Total 6 marks)

AQA, June 2007, Paper 2 Intermediate, Question 10

 4

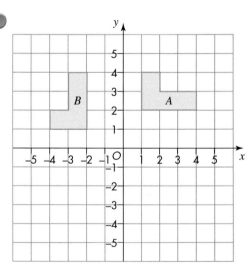

a Describe the **single** transformation that takes shape *A* to shape *B*. (3)

b Reflect shape *B* in the line $y = -1$. (2)

(Total 5 marks)

AQA, June 2008, Paper 2 Foundation, Question 22

5 The diagram shows a shaded flag.

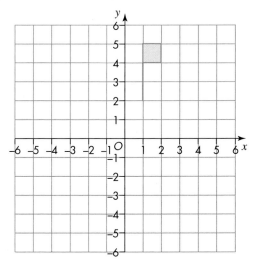

a Rotate the shaded flag 90° anticlockwise about the origin.

Label this new flag with the letter *A*. (3)

b Reflect the original shaded flag in the line $y = 1$.

Label this new flag with the letter *B*. (2)

c Rotate the original shaded flag by a quarter-turn clockwise about (0, 2).

Label this new flag with the letter C. (2)

(Total 7 marks)

AQA, Paper 2 Intermediate, June 2005, Question 12

Worked Examination Questions

1 The grid shows several transformations of the shaded triangle.

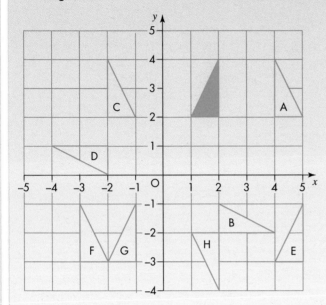

a Write down the letter of the triangle:

   i after the shaded triangle is reflected in the line $x = 3$

   ii after the shaded triangle is translated by the vector $\begin{pmatrix} 3 \\ -5 \end{pmatrix}$

   iii after the shaded triangle is rotated 90° clockwise about O.

b Describe fully the single transformation that takes triangle F onto triangle G.

a i A

    ii E

    iii B

(3 marks)

b A reflection in the line $x = -2$.

(2 marks)

(**Total:** 5 marks)

$x = 3$ is the vertical line passing through $x = 3$ on the x-axis.
The correct answer receives 1 mark.

Move the triangle 3 squares to the right and 5 squares down.
The correct answer receives 1 mark.

Use tracing paper to help you. Trace the shaded triangle, pivot the paper on O holding it in place with your pencil point, and rotate the paper through 90° clockwise.
The correct answer receives 1 mark.

The vertical mirror line passes through $x = -2$ on the x-axis.
You get 1 method mark for identifying the reflection and 1 accuracy mark for the mirror line.

Worked Examination Questions

PS **2** Find the single transformation that is equivalent to a rotation of 90° clockwise about the origin, followed by a reflection in the line $y = x$.

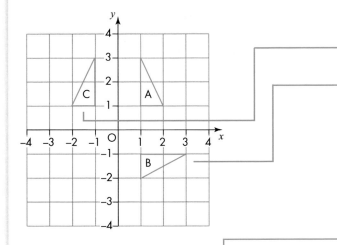

Start with a simple shape on the grid, triangle A.

You get 1 accuracy mark for correctly reflecting triangle B to C.

You get 1 accuracy mark for correctly rotating triangle A to B.

Single transformation is a reflection in the y-axis.

You get 1 method mark for identifying this is a reflection and 1 accuracy mark for identifying the mirror line as the y-axis.

Total: 4 marks

One of the most important numbers in mathematics is the ratio of the circumference of a circle to its diameter – also known as pi (π). Many formulae from mathematics, science and engineering involve this number. It has been found to be a constant, but its exact value has never been calculated.

Mathematicians (or philosophers, as they used to be known) since biblical times have attempted to accurately calculate the value of pi. This was difficult in the times before calculators! It was variously calculated to be 3 (which was almost certainly found by measurement), $4\frac{8}{9}^2$ (by the ancient Egyptians) and $\frac{22}{7}$ (by Archimedes).

However, the European Renaissance brought with it a whole new world of mathematics. By then, mathematicians had discovered formulae for calculating π. The only difficulty in computing π was, and still is, the sheer tedium of continuing the calculation. Mathematicians devoted a vast amount of time and effort to this pursuit. In 1873 a mathematician called Shanks calculated π to 707 places. Soon after this, another mathematician called De Morgan found that Shanks had made an error in the 528th place, after which all his digits were wrong!

In 1949, one of the first computers was used to calculate π to 2000 places.

While the value of π has been computed to more than a trillion digits, elementary applications, such as calculating the circumference of a circle, will rarely require more than a dozen decimal places. For example, the value of π to 11 decimal places is accurate enough to calculate the circumference of a circle the size of the Earth with a precision of a millimetre. The value of π to 39 decimal places is sufficient to calculate the circumference of any circle that fits in the observable universe to a precision comparable to the size of a hydrogen atom.

π can help us calculate the circumference of the 'building blocks' of life, atoms (left) and Earth itself (above).

The value of π to 200 decimal places is:

3.141 592 653 589 793 238 462 643 383 279 502 884 197 169 399 375 105 820 974 944 592 307 816 406 286 208 998 628 034 825 342 117 067 982 148 086 513 282 306 647 093 844 609 550 582 231 725 359 408 128 481 117 450 284 102 701 938 521 105 559 644 622 948 954 930 381 96

So far no one has spotted any patterns in the digits.

The current record for the highest number of decimals places for π is 1 241 100 000 000, set by Yasumasa Kanada of Japan.

Geometry: Circles

1 Drawing circles

2 The circumference of a circle

3 The area of a circle

4 Answers in terms of π

This chapter will show you ...

to **G** **F** how to draw circles

D how to calculate the circumference of a circle

D how to calculate the area of a circle

D how to write answers in terms of π

Visual overview

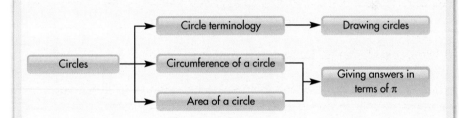

What you should already know

● How to use a pair of compasses to draw a circle (KS3 level 4, GCSE grade G)

● The words 'radius' and 'diameter' (KS3 level 4, GCSE grade G)

● How to use a protractor to draw angles (KS3 level 5, GCSE grade F)

● How to round numbers to a given number of decimal places (KS3 level 4, GCSE grade F)

● How to find the square and square root of a number (KS3 level 5, GCSE grade F)

Quick check

Write down the answer to each of the following, giving your answers to one decimal place.

1 5.21^2 2 8.78^2 3 15.5^2

4 $\sqrt{10}$ 5 $\sqrt{65}$ 6 $\sqrt{230}$

This section will show you how to:
- draw accurate circles
- draw diagrams made from circles

Key words

arc
centre
chord
circumference
diameter
radius
sector
segment
tangent

You need to know the following terms when dealing with circles.

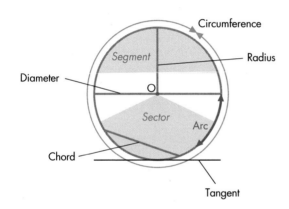

O	The **centre** of a circle.
Diameter	The 'width' of a circle. Any diameter passes through O.
Radius	The distance from O to the edge of a circle. The length of the diameter is twice the length of the radius.
Circumference	The perimeter of a circle.
Chord	A line joining two points on the circumference.
Tangent	A line that touches the circumference at one point only.
Arc	A part of the circumference of a circle.
Sector	A part of the area of a circle, lying between two radii and an arc.
Segment	A part of the area of a circle, lying between a chord and an arc.

When drawing a circle, you first need to set your compasses to a given radius.

EXAMPLE 1

Draw a circle with a radius of 3 cm.

Set your compasses to a radius of 3 cm, as shown in the diagram.

Draw a circle and mark the centre O.

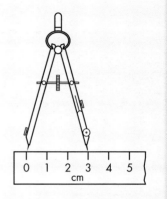

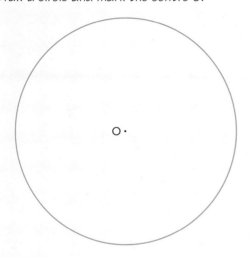

EXERCISE 14A

1 Measure the radius of each of the following circles, giving your answers in centimetres. Write down the diameter of each circle.

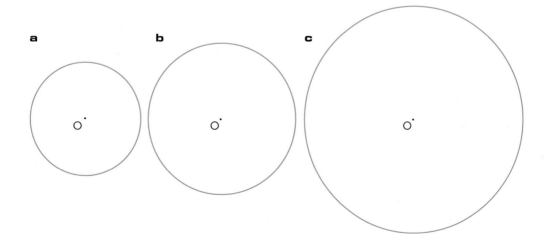

2 Draw circles with the following measurements.

a radius = 2 cm

b radius = 3.5 cm

c diameter = 8 cm

d diameter = 10.6 cm

AU **3** The centre of this circle is O and the four points A, B, C and D are on the circumference.

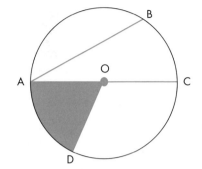

Here are some words that are used with circles.

chord   circumference   diameter
radius   sector   segment
tangent

Use a different one of these words to complete each of these sentences.

a The line AC is a …

b The line AB is a …

c The line OD is a …

d The shaded part is a …

4 Draw the following shapes accurately.

a

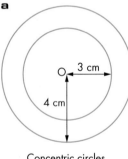

Concentric circles

b

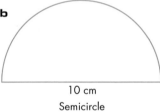

10 cm
Semicircle

c

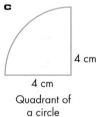

4 cm

4 cm
Quadrant of
a circle

5 Draw accurate copies of these diagrams.

a

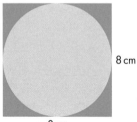

8 cm

8 cm

b

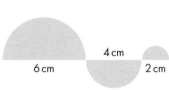

4 cm
6 cm
2 cm

c

5 cm

5 cm

6 **a** Draw a circle of radius 4 cm.

b Keeping your compasses set to a radius of 4 cm, step round the circle making marks on the circumference that are 4 cm apart.

c Join the points with a pencil and ruler to make a polygon.

d What name is given to the polygon you have drawn?

PS **7** **a** Draw a circle of radius 4 cm.

b Draw a tangent at any point on the circumference.

c Draw a radius to meet the tangent.

d Measure the angle between the tangent and the radius.

e Repeat the exercise for circles with different radii.

f Write down what you have found out about a radius touching a tangent at a point.

PS **8** The shape in the diagram is made from three identical semicircles.

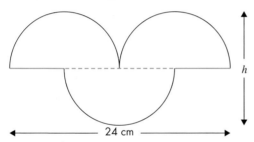

a Work out the radius of one of the semicircles.

b Work out the height, h, marked on the diagram.

14.2 # The circumference of a circle

This section will show you how to:
● calculate the circumference of a circle

Key words
π (pronounced pi)
circumference
diameter
radius

ACTIVITY

Round and round

Find six cylindrical objects – bottles, cans, tubes, or piping will do. You also need about 2 m of string.

Copy this table so that you can fill it in as you do this activity.

Object number	Diameter	Circumference	Circumference ÷ Diameter
1			
2			
3			
4			
5			
6			

Measure, as accurately as you can, the **diameter** of the first object. Write this measurement in your table.

Wrap the string around the object 10 times, as shown in the diagram. Make sure you start and finish along the *same line*. Mark clearly the point on the string where the tenth wrap ends.

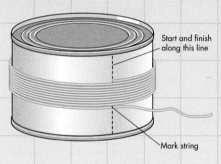

Start and finish along this line

Mark string

Then measure, as accurately as you can, the length of your 10 wraps. This should be the distance from the start end of the string to the mark you made on it.

Next, divide this length of string by 10. You have now found the length of the **circumference** of the first object. Write this in the table.

Repeat this procedure for each of the remaining objects.

Finally, complete the last column in the table by using your calculator to divide the circumference by the diameter. In each case, round your answer to two decimal places.

If your measurements have been accurate, all the numbers you get should be about 3.14.

This is the well-known number that is represented by the Greek letter π. You can obtain a very accurate value for π by pressing the ▢ key on your calculator. Try it and see how close your numbers are to it.

You calculate the circumference, C, of a circle by multiplying its diameter, d, by π, and then rounding your answer to one or two decimal places.

The value of π is found on all scientific calculators, with $\pi = 3.141\,592\,654$, but if it is not on your calculator, then take $\pi = 3.142$.

The circumference of a circle is given by the formula:

circumference $= \pi \times$ diameter   *or* $C = \pi d$

As the diameter is twice the **radius**, r, this formula can also be written as $c = 2\pi r$.

EXAMPLE 2

Calculate the circumference of the circle with a diameter of 4 cm.

Use the formula:

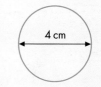
4 cm

$C = \pi d$

$\quad = \pi \times 4$

$\quad = 12.6$ cm (rounded to 1 decimal place)

Remember The length of the radius of a circle is half the length of its diameter. So, when you are given a radius, in order to find a circumference you must first *double* the radius to get the diameter.

EXAMPLE 3

Calculate the diameter of a circle that has a circumference of 40 cm.

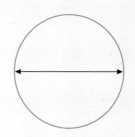

$C = \pi \times d$

$40 = \pi \times d$

$d = \dfrac{40}{\pi} = 12.7$ cm (rounded to 1 decimal place)

EXERCISE 14B

D

1 Calculate the circumference of each circle illustrated below. Give your answers to 1 decimal place.

a
8 cm

b
5 cm

c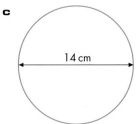
14 cm

d
7 cm

e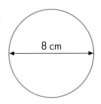
6 cm

f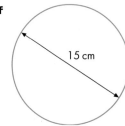
15 cm

g
9.2 cm

h
4.7 cm

2 Find the circumference of each of the following coins. Give your answers to 1 decimal place.

a 1p coin, diameter 2 cm

b 2p coin, diameter 2.6 cm

c 5p coin, diameter 1.7 cm

d 10p coin, diameter 2.4 cm

3 Calculate the circumference of each circle illustrated below. Give your answers to 1 decimal place.

a
5 cm

b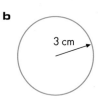
3 cm

c
1.5 cm

d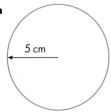
4 cm

e
0.9 cm

f
2.5 cm

g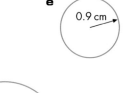
13 cm

h
6.3 cm

4 The radius of the wheels on Tim's bike is 31.5 cm.

   a Calculate the circumference of one of the wheels. Give your answer to the nearest centimetre.

   b Tim rides his bike for 1 km. How many complete revolutions does each wheel make?

FM 5 The diagram represents a race-track on a school playing field. The diameter of each circle is shown.

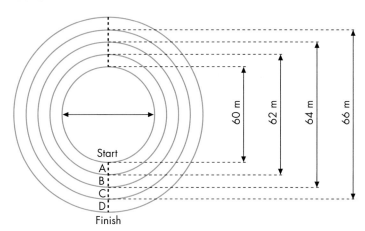

In a race with four runners, each runner starts and finishes on the same inner circle of their lane after completing one circuit.

   a Calculate the distance run by each runner in their lane.

   b How much further than A does D have to run?

6 A rope is wrapped eight times round a capstan (cylindrical post), the diameter of which is 35 cm. How long is the rope?

AU 7 A hamster has a treadmill of diameter 12 cm.

   a What is the circumference of the treadmill?

   b How many centimetres has the hamster run when the wheel has made 100 complete revolutions?

   c Change the answer to part **b** into metres.

   d One night, the hamster runs and runs and runs. He turns the wheel 100 000 times. How many kilometres has he run?

8 A circle has a circumference of 314 cm. Calculate the diameter of the circle.

9 What is the diameter of a circle if its circumference is 76 cm? Give your answer to 1 decimal place.

10 What is the radius of a circle with a circumference of 100 cm? Give your answer to 1 decimal place.

AU 11 A semicircular protractor has a diameter of 10 cm.

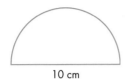

10 cm

Which of the following is the correct length for the perimeter of the protractor?

a 15.7 cm **b** 25.7 cm **c** 31.4 cm **d** 41.4 cm

12 Assume that the human waist is circular.

a What are the distances round the waists of the following people?

Sue: waist radius of 10 cm   Dave: waist radius of 12 cm

Julie: waist radius of 11 cm   Brian: waist radius of 13 cm

b Compare differences between pairs of waist circumferences. What connection do they have to π?

c What would be the difference in length between a rope stretched tightly round the Earth and another rope always held 1 m above it?

13 **a** Calculate the perimeter of each of shapes A and B.

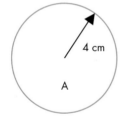

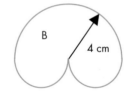

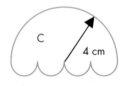

b Write down the perimeter of shape C.

PS 14 A square has sides of length *a* and a circle has radius *r*.

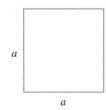

The perimeter of the square is equal to the circumference of the circle.

Show that $r = \dfrac{2a}{\pi}$

FM 15 Ben works in a park and wants to buy enough fencing to go round a semicircular flowerbed that has a diameter of 8 m.

The fencing is sold in 2-m lengths. How many lengths does Ben need?

The area of a circle

This section will show you how to:

- calculate the area of a circle

Key words

π

area

diameter

radius

The **area**, A, of a circle is given by the formula:

$$\text{area} = \pi \times \textbf{radius}^2 \quad or \quad A = \pi \times r \times r \quad or \quad A = \pi r^2$$

Remember This formula uses the radius of a circle. So, when you are given the **diameter** of a circle, you must *halve* it to get the radius.

EXAMPLE 4

Radius given

Calculate the area of a circle with a radius of 7 cm.

$$\begin{aligned} \text{Area} &= \pi r^2 \\ &= \pi \times 7^2 \\ &= \pi \times 49 \\ &= 153.9 \text{ cm}^2 \text{ (rounded to 1 decimal place)} \end{aligned}$$

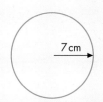

7 cm

EXAMPLE 5

Diameter given

Calculate the area of a circle with a diameter of 12 cm.

First, halve the diameter to get the radius:

$$\text{radius} = 12 \div 2 = 6 \text{ cm}$$

Then, find the area:

$$\begin{aligned} \text{area} &= \pi r^2 \\ &= \pi \times 6^2 \\ &= \pi \times 36 \\ &= 113.1 \text{ cm}^2 \text{ (rounded to 1 decimal place)} \end{aligned}$$

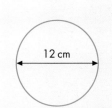

12 cm

EXERCISE 14C

1 Calculate the area of each circle illustrated below. Give your answers to 1 decimal place.

a

5 cm

b

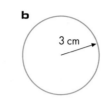
3 cm

c

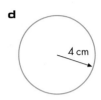
1.5 cm

d

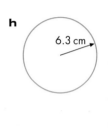
4 cm

e

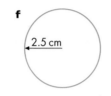
0.9 cm

f

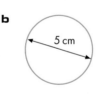
2.5 cm

g

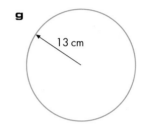
13 cm

h

6.3 cm

2 Find the area of one face of the following coins. Give your answers to 1 decimal place.

 a 1p coin, radius 1 cm

 b 2p coin, radius 1.3 cm

 c 5p coin, radius 0.85 cm

 d 10p coin, radius 1.2 cm

3 Calculate the area of each circle illustrated below.
Give your answers to 1 decimal place.

> **HINTS AND TIPS**
>
> **Remember** to halve the diameter to find the radius. The only formula for the area of a circle is $A = \pi r^2$.

a

8 cm

b

5 cm

c

14 cm

d

7 cm

e

6 cm

f

15 cm

g

9.2 cm

h

4.7 cm

AU **4** Milk-bottle tops are stamped from rectangular strips as shown.

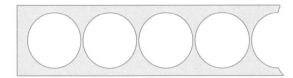

Each milk-bottle top is made from a circle of radius 1.7 cm.
Each rectangular strip measures 4 cm by 500 cm.

a What is the area of one milk-bottle top?

b How many milk-bottle tops can be stamped out of one strip 500 cm long when there is a 0.2 cm gap between adjacent tops?

c What is the area of the rectangular strip?

d What will be the total area of all the milk-bottle tops stamped out of the one strip?

e What waste is produced by one stamping?

FM **5** A young athlete can throw the discus a distance of 35 m but is never too sure of the direction in which he will throw it. What area of the field should be closed while he is throwing the discus?

6 Calculate **i** the circumference and **ii** the area of each of these circles. Give your answers to 1 decimal place.

a
9 cm

b
22 cm

c
6.5 cm

d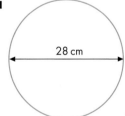
28 cm

7 A circle has a circumference of 60 cm.

a Calculate the diameter of the circle to 1 decimal place.

b What is the radius of the circle to 1 decimal place?

c Calculate the area of the circle to 1 decimal place.

8 Calculate the area of a circle with a circumference of 110 cm.

HINTS AND TIPS

Because π can be taken as 3.14 or 3.142, answers need not be exact. Examiners usually accept a range of answers.

9 Calculate the area of the following shapes. Give your answers to 1 decimal place.

a

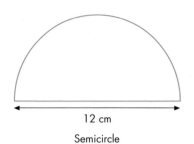

12 cm

Semicircle

b

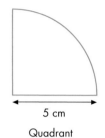

5 cm

Quadrant

10 Calculate the area of the shaded part of each of these diagrams.

a

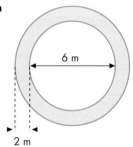

6 m

2 m

b

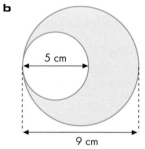

5 cm

9 cm

c

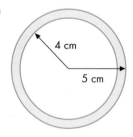

4 cm

5 cm

11 The diagram shows a circular photograph frame.

Work out the area of the frame. Give your answer to 1 decimal place.

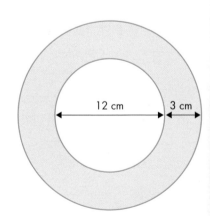

12 cm   3 cm

PS 12 A square has sides of length a and a circle has radius r.

The area of the square is equal to the area of the circle.

Show that $r = \dfrac{a}{\sqrt{\pi}}$

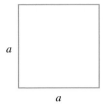

AU 13 A circle fits exactly inside a square of sides 10 cm.

Calculate the area of the shaded region. Give your answer to 1 decimal place.

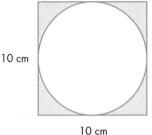

10 cm

10 cm

14.4 Answers in terms of π

This section will show you how to:	Key words
• give answers for circle calculations in terms of π	π
	area
	circumference
	diameter
	radius

There are times when you do not want a numerical answer to a circle problem but need to give the answer in terms of π. (The numerical answer could be evaluated later.)

EXAMPLE 6

What are the **circumference** and **area** of this circle?

Leave your answers in terms of π.

Circumference = $\pi d = \pi \times 14 = 14\pi$ cm

Area = $\pi r^2 = \pi \times 7^2 = \pi \times 49 = 49\pi$ cm^2

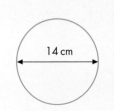

14 cm

If a question asks you to leave an answer in terms of π, it is most likely to be on the non-calculator paper and hence saves you the trouble of using your calculator.

However, if you did, and calculated the numerical answer, you could well lose a mark.

EXERCISE 14D

In this exercise, all answers should be given in terms of π.

1 A circle has diameter 10 cm.

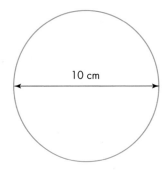

10 cm

State the circumference of the circle.

2 State the circumference of each of the following circles.

   a diameter 4 cm

   b radius 10 cm

   c diameter 15 cm

   d radius 2 cm

3 State the area of each of the following circles.

   a radius 4 cm

   b diameter 10 cm

   c radius 3 cm

   d diameter 18 cm

4 State the radius of the circle with a circumference of 50π cm.

5 State the radius of the circle with an area of 100π cm^2.

6 State the diameter of a circle with a circumference of 200 cm.

7 State the radius of a circle with an area of 25 cm^2.

8 Work out the area for each of the following shapes, giving your answers in terms of π.

a

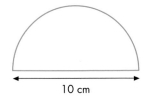

10 cm

b

8 cm

c

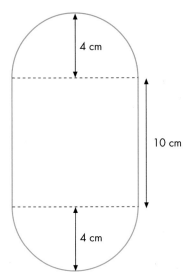

4 cm

10 cm

4 cm

d

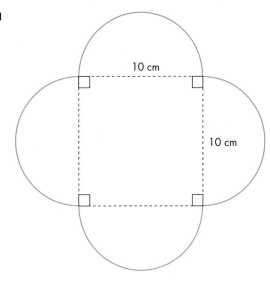

10 cm

10 cm

9 **a** Work out the area of a semicircle with radius 8 cm.

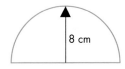

8 cm

b Work out the area of two semicircles with radii 4 cm.

4 cm

c Work out the area of four semicircles with radii 2 cm.

2 cm

d By looking at the pattern of areas of the answers to **a**, **b** and **c**, write down the area of eight semicircles with radii 1 cm.

FM **10** The diagram shows a plan of Mr Green's garden.

The flowerbed is in the shape of a semicircle and has a diameter of 8 m.

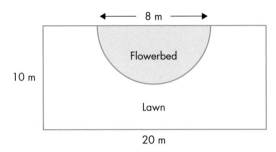

a Work out the area of the lawn, giving your answer in terms of π.

b His local garden centre sells grass seed in 500 g packets.

If 1 kg of grass seed covers up to 20 m², how many packets does he need to buy?

PS **11** A circle fits exactly inside a semicircle of diameter 12 cm.

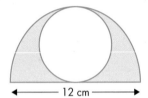

Find the area of the shaded region, giving your answer in terms of π.

AU **12** A shape is made from a rectangle and a quadrant of a circle.

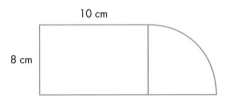

Which of these is the correct value of the area of the shape?

a $(80 + 4\pi)$ cm²

b $(80 + 8\pi)$ cm²

c $(80 + 16\pi)$ cm²

d $(80 + 32\pi)$ cm²

GRADE BOOSTER

G You know all the words associated with circles

G You can draw a circle if you know the radius

F You can draw shapes made from circles

D You can calculate the circumference of a circle

D You can calculate the area of a circle

C You can calculate perimeters and areas of compound shapes made with circles

---

What you should know now

- How to draw circles
- All the words associated with circles
- How to calculate the circumference of a circle
- How to calculate the area of a circle

1 In each diagram, O is the centre of the circle.

a Draw a diameter on a copy of this circle. (1)

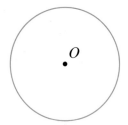

b Draw a tangent on a copy of this circle, at T. (1)

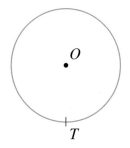

c A chord PQ has been drawn on the circle below.

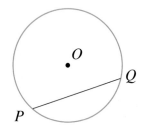

i Mark the midpoint of PQ and label it M. (1)

ii Join OM.

What do you notice about the angle between OM and PQ? (1)

(Total 4 marks)

AQA, June 2008, Module 5 Foundation, Question 4

 2 O is the centre of the circle.

A and B are two points on the circumference.

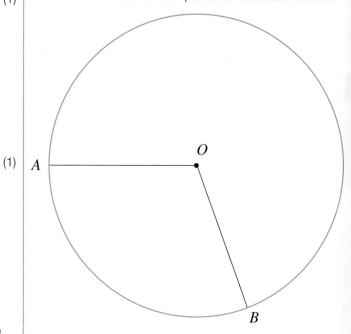

a Measure and write down the radius of the circle. (1)

b Measure and write down the size of the angle AOB. (1)

c Draw the line of symmetry of the sector AOB. (1)

d Draw the tangent to the circle at A. (1)

e Draw the chord AB. (1)

(Total 5 marks)

AQA, November 2008, Paper 2 Foundation, Question

 3 A wheel of a bicycle is shown.

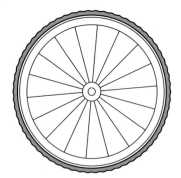

a The circumference is 70 cm.

Calculate the diameter. (2)

C   D   F   G

b The bicycle travels 50 metres.

How many complete revolutions does the wheel make? (3)

(Total 5 marks)

AQA, November 2008, Module 5 Higher, Question 1

 4 The diagram shows a circle of radius 5.4 metres.

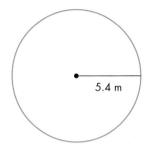

Not drawn accurately

Calculate the area of the circle.

State the units of your answer. (3)

(Total 3 marks)

AQA, June 2008, Module 5 Foundation, Question 16

5

Not drawn accurately

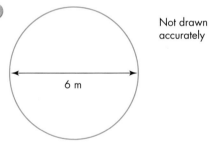

6 m

Calculate the area of a circle of diameter 6 metres.

Give your answer in terms of π. (2)

(Total 2 marks)

 6 The diagram shows a running track, made up of a rectangle plus two semicircles.

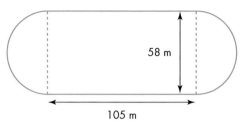

58 m

105 m

Not drawn accurately

Joel runs once round the perimeter of the track.

How far does he run? (4)

(Total 4 marks)

AQA, June 2007, Module 5 Foundation, Question 16

7 The diagram shows a square and two quarter circles.

The square has sides of 6 cm.

The radius of each circle is 3 cm.

Not drawn accurately

a Find the area of the shaded region. (3)

b Jane says that because a square has four lines of symmetry, the shaded area also has four lines of symmetry.

Is Jane correct?

Give a reason for your answer. (1)

(Total 4 marks)

AQA, June 2009, Paper 2 Foundation, Question 24

Worked Examination Questions

1 A circular pond has a radius of 2.2 m.

    a Calculate the circumference of the pond.

    b Calculate the area of the pond.

a $r = 2.2$, so $d = 4.4$ m

    $c = \pi d$

      $= \pi \times 4.4$ —————— 1 method mark is available for setting up an equation.

      $= 13.8$ m (1 decimal place) ——— You get 1 mark for accuracy for the correct answer.

2 marks

b $A = \pi r^2$

      $= \pi \times 2.2^2$ ——— 1 method mark is available for the first step of solving the equation. $\pi \times 4.84$ would also be acceptable.

      $= \pi \times 4.84$

      $= 15.2$ m^2 (1 decimal place) ——— You get 1 mark for accuracy for the correct answer.

2 marks

Total: 4 marks

In an examination, the answers could be given to any number of decimal places as the question does not state the accuracy required.

2 A semicircular flowerbed has a diameter of 2.6 m.

Calculate the area of the flowerbed.

Give your answer to one decimal place.

State the units of your answer.

← 2.6 m →

Area of whole circle is $\pi \times 1.3^2$ ——— You get 1 method mark for setting up an equation, remembering to halve the diameter.

So area of semicircle is $\pi \times \dfrac{1.3^2}{2}$

Area = 2.6546 … ——— You get 1 method mark for realising that the area of a semicircle must therefore divide $\pi \times 1.3^2$ by 2. You must write this equation down to get the mark. Now work this out on your calculator.

Area = 2.7

Area = 2.7 m^2

Total: 5 marks

You get 1 mark for accuracy for the answer to this equation. Now round the answer from your calculator display.

You get 1 mark for accuracy for the correct, rounded answer.

You can get 1 mark is given for the correct units, even if the answer is wrong.

Worked Examination Questions

PS **3** Four identical circles fit exactly in a square with side length x.

Work out the area of the shaded region.

Give your answer in terms of π.

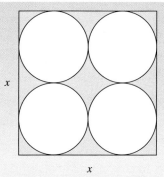

Area of square = x^2

Radius of each circle = $\dfrac{x}{4}$

You get 2 method marks for writing down the correct formulas.

Area of each circle = $\dfrac{\pi x^2}{16}$

You get 1 method mark for setting up the calculation to find the area of the circles.

Note $\dfrac{x}{4} \times \dfrac{x}{4} = \dfrac{x^2}{16}$

Area of four squares = $\dfrac{\pi x^2}{4}$

So area of shaded region = $x^2 - \dfrac{\pi x^2}{4}$

You get 1 method mark for setting up the calculation to find the area of the squares.

Note $4 \times \dfrac{\pi x^2}{16} = \dfrac{\pi x^2}{4}$

Total: 5 marks

You get 1 mark for accuracy for the final calculation. Note that if you make one error in the solution, then you would only lose 1 mark.

Problem Solving
Proving properties of circles

Circular shapes form the basis of many of the objects that we see and use every day. For example, we see circular shapes in DVDs, wheels, coins and jewellery. Where else do you see circles?

Given how frequently circles appear in our lives, it is important that we understand them mathematically.

In this task you will investigate angles in circles, using mathematical investigation to help you understand this shape more fully.

Getting started

- List the mathematical vocabulary that you know, that is related to circles. Explain each of the words you think of to a classmate.

- Select one fact that you know, that is related to circles. Explain your fact to your partner.

- Select a real-life object that is in the shape of a circle.

  What mathematical questions could you ask about this object?

Your task

Here are four statements about circles.

1 The sum of the opposite angles of a cyclic quadrilateral is 180°.

So, angles $a + c = 180°$ and angles $b + d = 180°$.

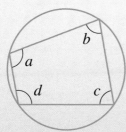

2 A radius bisects a chord at 90°.

Your task (continued)

3 An angle subtended by a chord at the centre is twice the angle subtended by the chord at the circumference. So, angle b is twice the size of angle a.

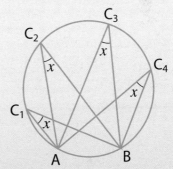

4 Angles subtended at the circumference in the same segment of a circle are equal.

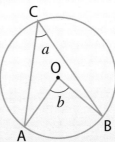

(You might like to think about what's special about angles subtended by the diameter)

Working in pairs, choose one or two of these statements and show whether you think it is true or false.

Write a presentation to explain your findings to the class. In your presentation you must:

- describe the approach that you took to the investigation
- explain your findings
- represent your findings mathematically, using expressions and diagrams
- support your findings with suitable measurements, calculations and diagrams
- explain what is meant by mathematical 'truth'.

Why this chapter matters

Where should a new by-pass be built? Where should new power lines be placed? Finding the best path is a challenge because there are a lot of different factors which need to be taken into account.

Putting power lines underground might be a good choice environmentally, but it will be a great deal more expensive than normal overhead cables. The cheapest route for a road might not be the best choice for other reasons, such as limiting access to local facilities, or creating unpleasant light or noise pollution.

Often, people don't want to have a major new road going past their window. Likewise, few want to see power lines constructed through attractive countryside. So the path traced out by the road or pylons between specified conditions such as a housing estate or a forest would need to satisfy these restrictions.

When a point moves subject to certain specified conditions, the path traced out by it is called a **locus**. This is just a Latin word which means 'place' (the plural of locus is loci). The question to answer, then, is where is the best place for the road or the power line? The loci can tell you this.

You are already familiar with the idea of a locus even if you do not call it by that name. For example, if you are holding a dog on a lead you are restricting where it can go by the length of its lead. If you pass another dog you may try to ensure that there is no overlap between the locus of your dog and the locus of the other dog – you will try to keep them apart.

In this chapter you will look at some other straightforward examples of the idea of a locus.

Chapter

Geometry: Constructions 2

1 Bisectors

2 Loci

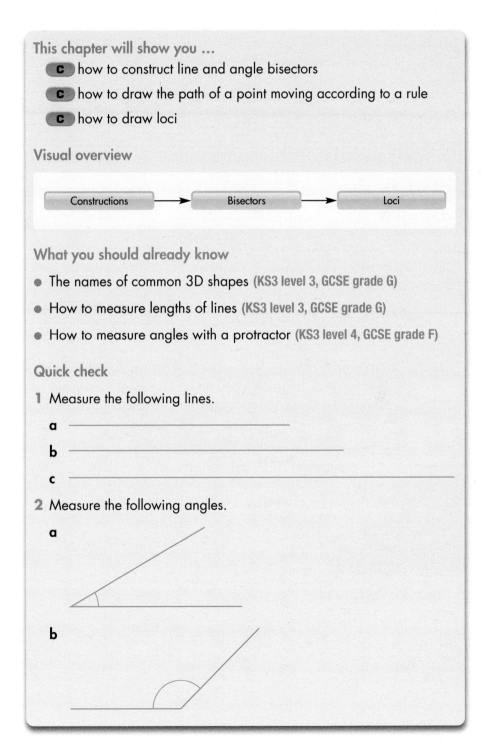

This chapter will show you …

- **c** how to construct line and angle bisectors
- **c** how to draw the path of a point moving according to a rule
- **c** how to draw loci

Visual overview

Constructions → Bisectors → Loci

What you should already know

- The names of common 3D shapes (KS3 level 3, GCSE grade G)
- How to measure lengths of lines (KS3 level 3, GCSE grade G)
- How to measure angles with a protractor (KS3 level 4, GCSE grade F)

Quick check

1 Measure the following lines.

a _____

b _____

c _____

2 Measure the following angles.

a

b

Bisectors

This section will show you how to:
- construct the bisectors of lines and angles
- construct angles of 60° and 90°

Key words

angle bisector
bisect
line bisector
perpendicular
   bisector

To **bisect** means to divide in half. So a bisector divides something into two equal parts.

- A **line bisector** divides a straight line into two equal lengths.

- An **angle bisector** is the straight line that divides an angle into two equal angles.

To construct a line bisector

It is usually more accurate to construct a line bisector than to measure its position (the midpoint of the line).

- **Step 1:** Here is a line to bisect.

- **Step 2:** Open your compasses to a radius of about three-quarters of the length of the line. Using each end of the line as a centre, and without changing the radius of your compasses, draw two intersecting arcs.

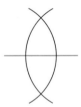

- **Step 3:** Join the two points at which the arcs intersect.

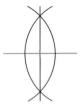

This line is the **perpendicular bisector** of the original line.

To construct an angle bisector

It is much more accurate to construct an angle bisector than to measure its position.

- **Step 1:** Here is an angle to bisect.

- **Step 2:** Open your compasses to any reasonable radius that is less than the length of the shorter line. If in doubt, go for about 3 cm. With the vertex of the angle as centre, draw an arc through both lines.

- **Step 3:** With centres at the two points at which this arc intersects the lines, draw two more arcs so that they intersect. (The radius of the compasses may have to be increased to do this.)

- **Step 4:** Join the point at which these two arcs intersect to the vertex of the angle.

This line is the **angle bisector**.

To construct an angle of 60°

It is more accurate to construct an angle of 60° than to measure and draw it with a protractor.

- **Step 1:** Draw a line and mark a point on it.
- **Step 2:** Open the compasses to a radius of about 4 centimetres. Using the point you have marked as the centre, draw an arc that crosses the line and extends almost above the point.

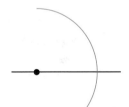

- **Step 3:** Keep the compasses set to the same radius. Using the point where the first arc crosses the line as a centre, draw another arc that intersects the first one.

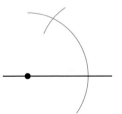

- **Step 4:** Join the original point to the point where the two arcs intersect.

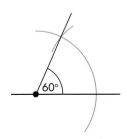

- **Step 5:** Use a protractor to check that the angle is 60°.

To construct a perpendicular from a point on a line

This construction will produce a perpendicular from a point A on a line.

- Open your compasses to about 2 or 3 cm.
  With point A as centre, draw two short arcs to intersect the line at each side of the point.

- Now extend the radius of your compasses to about 4 cm. With centres at the two points at which the arcs intersect the line, draw two arcs to intersect at X above the line.

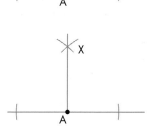

- Join AX.
  AX is perpendicular to the line.

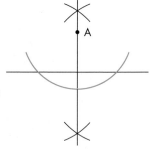

Note: If you needed to construct a 90° angle at the end of a line, you would first have to extend the line.

You could be even more accurate by also drawing two arcs *underneath* the line, which would give three points in line.

To construct a perpendicular from a point to a line

This construction will produce a perpendicular from a point A to a line.

- With point A as centre, draw an arc to intersect the line at two points.

- With centres at these two points of intersection, draw two arcs to intersect each other both above and below the line.

- Join the two points at which the arcs intersect. The resulting line passes through point A and is perpendicular to the line.

Examination note: When a question says *construct,* you must *only* use compasses – no protractor. When it says *draw,* you may use whatever you can to produce an accurate diagram. But also note that, when constructing, you may use your protractor to check your accuracy.

EXERCISE 15A

1 Draw a line 7 cm long and bisect it. Check your accuracy by seeing if each half is 3.5 cm.

2 Draw a circle of about 4 cm radius.

Draw a triangle inside the circle so that the corners of the triangle touch the circle.

Bisect each side of the triangle.

The bisectors should all meet at the same point, which should be the centre of the circle.

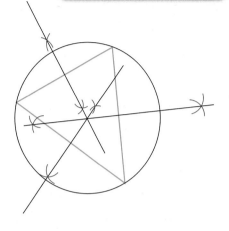

3 **a** Draw any triangle with sides of length between 5 cm and 10 cm.

b On each side construct the line bisector.

All your line bisectors should intersect at the same point.

c Using this point as the centre, draw a circle that goes through each vertex of the triangle.

4 Repeat question 3 with a different triangle and check that you get a similar result.

5 **a** Draw this quadrilateral.

b Construct the line bisector of each side. These all should intersect at the same point.

c Use this point as the centre of a circle that goes through the quadrilateral at each vertex. Draw this circle.

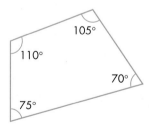

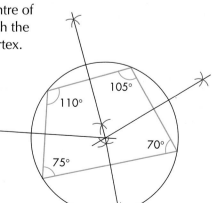

6 **a** Draw an angle of 50°.

b Construct the angle bisector.

c Check how accurate you have been by measuring each half. Both should be 25°.

7 Draw a circle with a radius of about 3 cm.

Draw a triangle so that the sides of the triangle
are tangents to the circle.

Bisect each angle of the triangle.

The bisectors should all meet at the same point,
which should be the centre of the circle.

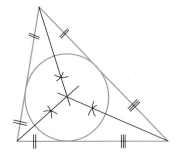

8 **a** Draw any triangle with sides of length between 5 cm and 10 cm.

b At each angle construct the angle bisector.
All three bisectors should intersect at the same point.

c Use this point as the centre of a circle that just touches the sides of the triangle.

9 Repeat question 8 with a different triangle.

FM 10 Pete and Babs have children living in Bristol and Norwich. Pete is about to start a
new job in Birmingham. They are looking on a map of Britain for places they might
move to.

Babs says, "I want to be the same distance from both children."
Pete says, "I want to be as close to Birmingham as possible."

Find a city that would suit both Pete and Babs.

Use a map of the UK to help you.

PS 11 Draw a circle with radius of about 4 cm.

Draw a quadrilateral, **not** a rectangle, inside the circle so that each vertex is on the
circumference.

Construct the bisector of each side of the quadrilateral.

Where is the point where these bisectors all meet?

AU 12 Write down instructions to explain how you would construct a triangle with angles 90°,
60° and 30°.

Loci

What is a locus?

A **locus** (plural **loci**) is the movement of a point according to a rule.

For example, a point that moves so that it is always at a distance of 5 cm from a fixed point, A, will have a locus that is a circle of radius 5 cm.

This is expressed mathematically as:

The locus of the point P is such that AP = 5 cm

Another point moves so that it is always **equidistant** or the same distance from two fixed points, A and B.

This is expressed mathematically as:

The locus of the point P is such that AP = BP

This is the same as the bisector of the line AB, which you have met in lesson 11.2.

Another point moves so that it is always 5 cm from a line AB.
The locus of the point P is given as a 'racetrack' shape.
This is difficult to express mathematically.

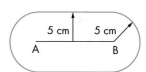

The three examples of loci just given occur frequently.

Imagine a grassy, flat field in which a horse is tied to a stake by a rope that is 10 m long. What is the shape of the area that the horse can graze?

In reality, the horse may not be able to reach the full 10 m if the rope is tied round its neck but you can ignore fine details like that.
The situation is 'modelled' by saying that the horse can move around in a 10 m circle and graze all the grass within that circle.

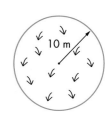

In this example, the locus is the whole of the area inside the circle.

This is expressed mathematically as:

The locus of the point P is such that AP ⩽ 10 m

EXERCISE 15B

1 A is a fixed point. Sketch the locus of the point P for these situations.

   a AP = 2 cm

   b AP = 4 cm

   c AP = 5 cm

HINTS AND TIPS

Sketch the situation before doing an accurate drawing.

2 A and B are two fixed points 5 cm apart. Sketch the locus of the point P for the following situations.

   a AP = BP

   b AP = 4 cm and BP = 4 cm

   c P is always within 2 cm of the line AB.

HINTS AND TIPS

If AP = BP this means the bisector of A and B.

FM **3** **a** A horse is tied in a field by a rope 4 m long. Describe or sketch the area that the horse can graze.

   b The same horse is still tied by the same rope but there is now a long, straight fence running 2 m from the stake. Sketch the area that the horse can now graze.

4 ABCD is a square of side 4 cm. In each of the following loci, the point P moves only inside the square. Sketch the locus in each case.

   a AP = BP   **b** AP < BP

   c AP = CP   **d** CP < 4 cm

   e CP > 2 cm   **f** CP > 5 cm

5 One of the following diagrams is the locus of a point on the rim of a bicycle wheel as it moves along a flat road. Which is it?

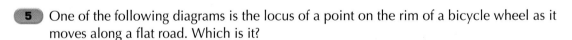

  a 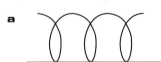   **b**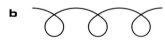

  c 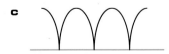   **d**

6 Draw the locus of the centre of the wheel for the bicycle in question 5.

PS 7 On a piece of plain paper, mark three points A, B and C, all about 5 cm to 7 cm away from each other. Find the locus of point P where:

a P is always closer to point A than point B

b P is always equidistant from points B and C.

AU 8 You do not have a pair of compasses or a protractor – only a ruler.

Explain how you could construct an equilateral triangle, with sides of length 3 cm.

Practical problems

Most of the loci problems in your GCSE examination will be of a practical nature, as shown in the next three examples.

EXAMPLE 1

Imagine that a radio company wants to find a site for a transmitter. The transmitter must be the same distance from both Doncaster and Leeds, and within 20 miles of Sheffield.

In mathematical terms, this means you are concerned with the perpendicular bisector between Leeds and Doncaster, and the area within a circle of radius 20 miles from Sheffield.

The map, drawn to a scale of 1 cm = 10 miles, illustrates the situation and shows that the transmitter can be built anywhere along the thick green line.

EXAMPLE 2

A radar station in Birmingham has a range of 150 km (that is, it can pick up any aircraft within a radius of 150 km). Another radar station in Norwich has a range of 100 km.

Can an aircraft be picked up by both radar stations at the same time?

The situation is represented by a circle of radius 150 km around Birmingham and another circle of radius 100 km around Norwich. The two circles overlap, so an aircraft could be picked up by both radar stations when it is in the overlap.

EXAMPLE 3

A dog is tied by a rope, 3 m long, to the corner of a shed, 4 m by 2 m. What is the area that the dog can guard effectively?

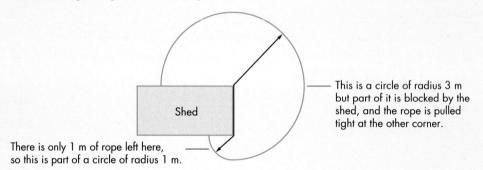

This is a circle of radius 3 m but part of it is blocked by the shed, and the rope is pulled tight at the other corner.

There is only 1 m of rope left here, so this is part of a circle of radius 1 m.

EXERCISE 15C

For questions 1 to 7, you should start by sketching the picture given in each question on a 6 × 6 grid, each square of which is 2 cm by 2 cm. The scale for each question is given.

1 A goat is tied to a stake by a rope, 7 m long, in a corner of a field with a fence at each side. What is the locus of the area that the goat can graze? Use a scale of 1 cm ≡ 1 m.

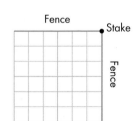

2 A horse in a field is tied to a stake by a rope 6 m long.
What is the locus of the area that the horse can graze?
Use a scale of 1 cm ≡ 1 m.

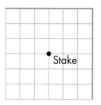

3 A cow is tied to a rail at the top of a fence 6 m long.
The rope is 3 m long. Sketch the area that the cow can graze.
Use a scale of 1 cm ≡ 1 m.

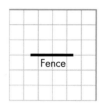

4 A horse is tied to a stake near a corner of a fenced field,
at a point 4 m from each fence. The rope is 6 m long.
Sketch the area that the horse can graze.
Use a scale of 1 cm ≡ 1 m.

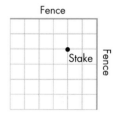

5 A horse is tied to a corner of a shed, 2 m by 1 m.
The rope is 2 m long. Sketch the area that the horse can graze.
Use a scale of 2 cm ≡ 1 m.

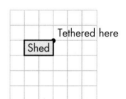

6 A goat is tied by a 4 m rope to a stake at one corner of a pen,
4 m by 3 m. Sketch the area of the pen on which the goat
cannot graze. Use a scale of 2 cm ≡ 1 m.

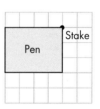

7 A puppy is tied to a stake by a rope, 1.5 m long, on a flat
lawn on which are two raised brick flower beds. The stake is
situated at one corner of a bed, as shown. Sketch the area in
which the puppy is free to roam. Use a scale of 1 cm ≡ 1 m.

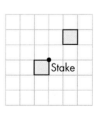

FM 8 The diagram, which is drawn to scale, shows two towns, A and B, which are 8 km apart.

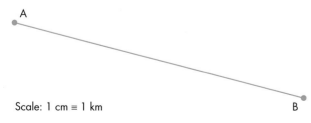

Scale: 1 cm ≡ 1 km

A phone company wants to erect a mobile phone mast.

It must be within 5 km of town A and within 4 km of town B.

Copy the diagram accurately.

Show the possible places where the mast could be.

FM **9** The map shows a field that is 100 m by 100 m.

There are two large trees in the field and a power line runs across it.

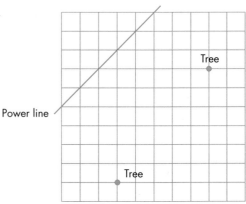

Power line

Tree

Tree

a Make an accurate scale drawing of the field, using a scale of 1 cm ≡ 10 m.

b Bernice wants to fly a kite.

She cannot fly the kite within 50 m of the power line.

She cannot fly the kite within 30 m of a tree.

Show the area where she can fly the kite.

FM **10** A radio station broadcasts from London on a frequency of 1000 kHz with a range of 300 km. Another radio station broadcasts from Glasgow on the same frequency with a range of 200 km.

a Sketch the area to which each station can broadcast. (See the map on the next page.)

b Will they interfere with each other?

c If the Glasgow station increases its range to 400 km, will they then interfere with each other?

FM **11** The radar at Leeds airport has a range of 200 km. The radar at Exeter airport has a range of 200 km. (See the map on the next page.)

a Will a plane flying over Glasgow be detected by the radar at Leeds?

b Sketch the area where a plane can be picked up by both radars at the same time.

FM **12** A radio transmitter is to be built according to the following rules.

HINTS AND TIPS

The same distance from York and Birmingham means on the bisector of the line joining York and Birmingham.

i It has to be the same distance from York and Birmingham. (See the map on the next page.)

ii It must be within 350 km of Glasgow.

iii It must be within 250 km of London.

a Sketch the line that is the same distance from York and Birmingham.

b Sketch the area that is within 350 km of Glasgow and 250 km of London.

c Show clearly the possible places at which the transmitter could be built.

Use a copy of this map, if you need to, to answer questions 10 to 17.

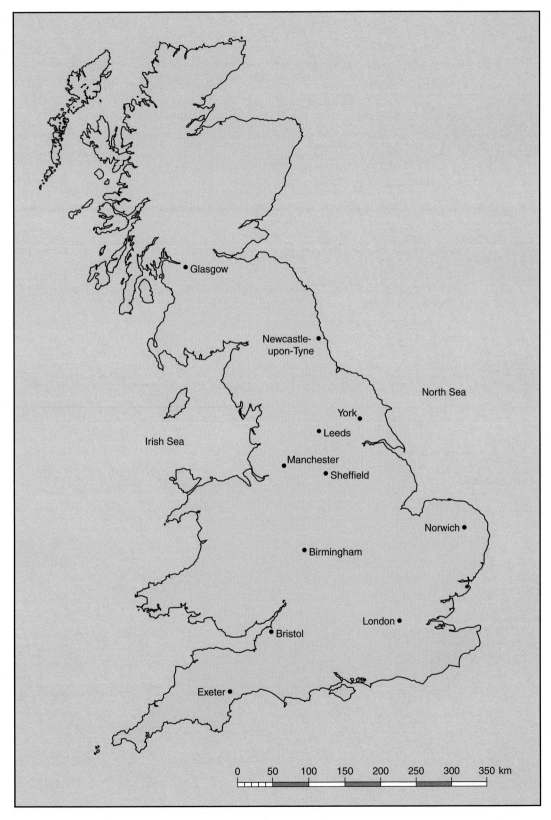

For each question, trace the map and mark on those points that are relevant to that question.

FM 13 A radio transmitter centred at Birmingham is designed to give good reception in an area greater than 150 km and less than 250 km from the transmitter. Sketch the area of good reception.

FM 14 Three radio stations pick up a distress call from a boat in the Irish Sea. The station at Glasgow can tell from the strength of the signal that the boat is within 300 km of the station. The station at York can tell that the boat is between 200 km and 300 km from York. The station at London can tell that it is less than 400 km from London. Sketch the area where the boat could be.

15 Sketch the area that is between 200 km and 300 km from Newcastle-upon-Tyne, and between 150 km and 250 km from Bristol.

16 The shape of Wathsea Harbour is as shown in the diagram. A boat sets off from point A and steers so that it keeps the same distance from the sea-wall and the West Pier. Another boat sets off from B and steers so that it keeps the same distance from the East Pier and the sea-wall. Copy the diagram and, on your diagram, show accurately the path of each boat.

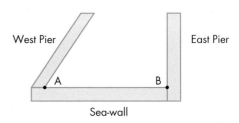

West Pier   East Pier

A   B

Sea-wall

PS 17 Trevor wanted to fly from the Isle of Wight north, towards Scotland. He wanted to stay the same distance from London as Bristol as far as he could.

Once he is past London and Bristol, which city should he aim towards, to keep him, as accurately as possible, the same distance from London and Bristol? Use the map to help you.

AU 18 A distress call is heard by coastguards in both Newcastle and Bristol. The signal strength suggests that the call comes from a ship the same distance from both places. Explain how the coastguards could find the area of sea to search.

GRADE BOOSTER

c You can construct line and angle bisectors, and draw the loci of points moving according to a rule

What you should know now

- How to draw loci of sets of points

1 In this question, you should use a ruler and compasses only.

  a Copy this diagram and construct the perpendicular bisector of the line joining the points A and B. (2)

  b Copy the diagram and bisect the angle ABC. (2)

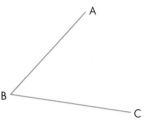

(Total 4 marks)

 2 The diagram shows a scale drawing of a straight road.

A walker is at point *P*.

P

Road ────────────────────────────────

Scale: 1 cm represents 0.5 km

  a Use a ruler and compasses to copy the diagram and construct the perpendicular from the point *P* to the road.

  You **must** show all your construction lines and arcs. (3)

  b Find the shortest real distance from the walker to the road. (2)

(Total 5 marks)

AQA, November 2007, Paper 2 Intermediate, Question 14

 3 **a** Using a ruler and compasses only, construct an angle of 60°.

  Show all your construction lines and arcs. (2)

  b Two lifeboat stations *A* and *B* receive a distress call from a boat.

  The boat is within 6 kilometres of station *A*.

  The boat is within 8 kilometres of station *B*.

  Shade the possible area in which the boat could be. (2)

Scale: 1 cm represents 1 km

(Total 4 marks)

AQA, November 2005, Paper 2 Intermediate, Question 22

 4 The positions of towns *A* and *B* are shown on the diagram.

The diagram is drawn to scale.

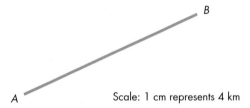

Scale: 1 cm represents 4 km

  a Work out the actual distance between towns *A* and *B*. (3)

  b A town *C* is 16 kilometres from *A* and 16 kilometres from *B*.

  Using compasses only, mark the **two** possible positions of town *C* on the diagram. (2)

(Total 5 marks)

AQA, November 2007, Module 5, Paper 1 Intermediate, Question 4

C D

5 The diagram shows an L shape.

Copy the diagram and draw the locus of all points 2 cm from the L shape. (3)

(Total 3 marks)

AQA, June 2005, Paper 1 Intermediate, Question 15

6 Use a ruler and compasses to construct a rhombus that has sides of 6 cm and whose shorter diagonal is 4 cm. (4)

(Total 4 marks)

AQA, June 2006, Paper 2 Foundation, Question 24

 7 The diagram, which is drawn to scale, shows a garden bordered by a house wall.

On the wall there is an electricity outlet (E) and a water outlet (W).

Ramesh is installing a pond.

The centre of the pond must be within 4 metres of the electricity outlet and within 6 metres of the water outlet.

Scale: 1 cm to 1 metre

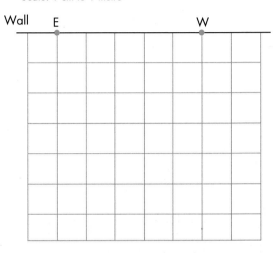

Find the area within which the centre of the pond may be located.

You already know that architects and engineers must construct accurate diagrams, to be certain that the buildings and constructions they have designed will be built correctly. However, did you know that the same principles used by architects and engineers to construct diagrams are also used when planning sports pitches, whether it is your local playing field or a Premier League football club?

In this task you will take on the role of the grounds staff of a football pitch. You will need to negotiate many variables to ensure that the pitch is drawn up correctly and can be maintained thoroughly.

Your task

As a member of the grounds staff you have been asked to prepare the pitch ready for the new football season.

The club needs the pitch designed according to FIFA's specifications.

1 Construct a scale drawing of the football pitch, to be used in laying out the pitch on the football field. Be sure to label all the dimensions of the pitch.

2 The pitch will need to be regularly watered in order to keep it in good condition. For this, you will need to design a comprehensive sprinkler system. On a copy of your drawing of the pitch, mark up where the sprinklers should go, ensuring that the maximum possible area of the pitch is watered at any one time.

Getting started
Discuss these questions with a partner.

- What shapes do you see on sports fields? Do these shapes vary, depending on the sports that take place on these pitches?
- What angles do you see on sports fields?
- How are shapes drawn on to sports fields?

FIFA specifications

The field

The field of play must be rectangular, divided into two halves by a halfway line. The centre mark is indicated at the midpoint of the halfway line and a circle with a radius of 9.15 m (10 yards) is marked around it.

The field dimensions should be as follows.

	Minimum	Maximum
Length	90 m (approx. 100 yards)	120 m (approx. 130 yards)
Width	45 m (approx. 50 yards)	90 m (approx. 100 yards)

The goal area

The goal mouth is 7.3 metres (8 yards) wide. The goal area is 5.5 m (6 yards) wide by 18.3 m (12 yards) long.

The penalty area

The penalty area is 16.5 m (18 yards) wide by 40.3 m (44 yards) long.

Within each penalty area there is a penalty mark 11 m (12 yards) from the midpoint between the goalposts and equidistant from them. An arc with a radius of 9.15 m (10 yards) from each penalty mark is drawn outside the penalty area.

The corner arc

A flagpost is placed in each corner. A quarter-circle with a radius of 1 m (approximately 1 yard) is drawn at each corner flag post, inside the field of play.

Hint: Use the internet to research football pitches and FIFA's specifications further.

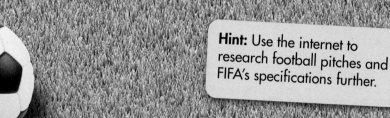

There are many curves that can be seen in everyday life. Did you know that all these curves can be represented mathematically?

Below are a few examples of simple curves that you may have come across. Can you think of any others?

Many road signs are circular.

A chain hanging freely between two supports forms a curve called a catenary.

The examples above both show circular-based curves. However, in mathematics curves can take many shapes. These can be demonstrated using a cone, as shown on the right and below. If you can make a cone out of plasticine or modelling clay, then you can prove this principle yourself. As you look at these curves, try to think of where you have seen them in your own life.

If you slice the cone parallel to the base, the shape you are left with is a **circle**.

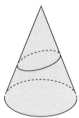

If you slice the cone at an angle to the base, the shape you are left with is an **ellipse**.

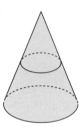

If you slice the cone vertically, the shape you are left with is a **hyperbola**.

The curve that will be particularly important in this chapter is the parabola. Car headlights are shaped like parabolas.

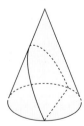

If you slice the cone parallel to its sloping side, the shape you are left with is a **parabola**.

All parabolas are quadratic graphs. During the course of this chapter you will be looking at how to use quadratic equations to draw graphs that have this kind of curve.

The suspension cables on the Humber Bridge are also parabolas.

16 Algebra: Quadratic graphs

1 Drawing quadratic graphs

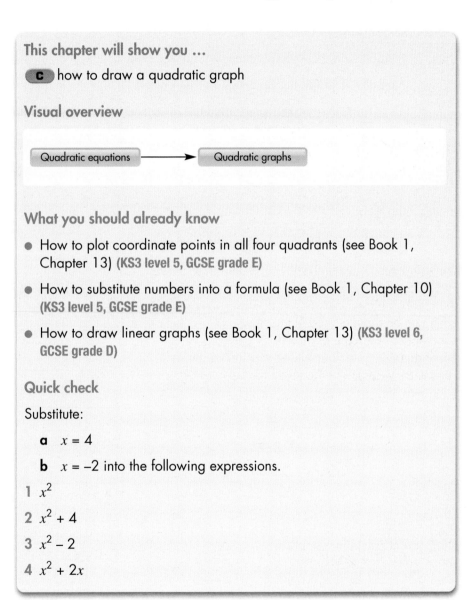

This chapter will show you ...

c how to draw a quadratic graph

Visual overview

Quadratic equations ⟶ Quadratic graphs

What you should already know

- How to plot coordinate points in all four quadrants (see Book 1, Chapter 13) **(KS3 level 5, GCSE grade E)**

- How to substitute numbers into a formula (see Book 1, Chapter 10) **(KS3 level 5, GCSE grade E)**

- How to draw linear graphs (see Book 1, Chapter 13) **(KS3 level 6, GCSE grade D)**

Quick check

Substitute:

a $x = 4$

b $x = -2$ into the following expressions.

1 x^2

2 $x^2 + 4$

3 $x^2 - 2$

4 $x^2 + 2x$

Drawing quadratic graphs

This section will show you how to:
- draw a quadratic graph, given its equation

Key words
parabola
quadratic equation
quadratic graph

A graph with a $\cup$ or $\cap$ shape is a quadratic graph.

A **quadratic graph** has an equation that involves x^2.

All of the following are **quadratic equations** and each would produce a quadratic graph.

$y = x^2$

$y = x^2 + 5$

$y = x^2 - 3x$

$y = x^2 + 5x + 6$

$y = x^2 + 2x - 5$

EXAMPLE 1

Draw the graph of $y = x^2$ for $-3 \leqslant x \leqslant 3$.

First make a table, as shown below.

x	-3	-2	-1	0	1	2	3
$y = x^2$	9	4	1	0	1	4	9

Now draw axes, with $-3 \leqslant x \leqslant 3$ and $0 \leqslant y \leqslant 9$, plot the points and join them to make a smooth curve.

This is the graph of $y = x^2$.
This type of graph is often referred to as a **parabola**.

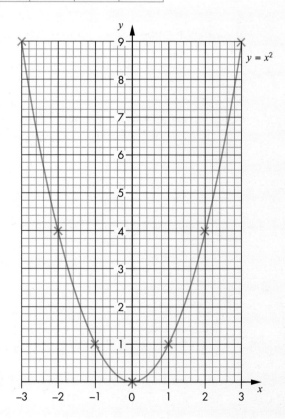

FM Functional Maths   **AU** (AO2) Assessing Understanding   **PS** (AO3) Problem Solving

Note that although it is difficult to draw accurate curves, examiners work to a *tolerance of only 2 mm*.

Here are some of the more common ways in which marks are lost in an examination.

- When the points are too far apart, a curve tends to 'wobble'.

Wobbly curve

- Drawing curves in small sections leads to 'feathering'.

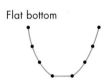

Feathering

- The place where a curve should turn smoothly is drawn 'flat'.

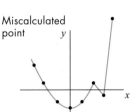

Flat bottom

- A curve is drawn through a point that, clearly, has been incorrectly plotted.

Miscalculated point

A quadratic curve drawn correctly will *always* be a smooth curve.

Here are some tips that will make it easier for you to draw smooth, curved graphs.

- If you are *right-handed*, you might like to turn your piece of paper or your exercise book round so that you draw from left to right. Your hand may be steadier this way than if you try to draw from right to left or away from your body. If you are *left-handed*, you may find drawing from right to left the more accurate way.

- Move your pencil over the points as a practice run without drawing the curve.

- Do one continuous curve and only stop at a plotted point.

- Use a *sharp* pencil and do not press too heavily, so that you may easily rub out mistakes.

Normally in an examination, grids are provided with the axes clearly marked. Remember that a tolerance of 2 mm is all that you are allowed. In the exercises, suitable ranges are suggested for the axes. Usually you will be expected to use 2 mm graph paper to draw the graphs.

EXAMPLE 2

a Draw the graph of $y = x^2 + 2x - 3$ for $-4 \leqslant x \leqslant 2$.

b Use your graph to find the value of y when $x = 1.6$.

c Use your graph to find the values of x that give a y-value of 1.

a Draw a table as follows to help work each step of the calculation.

x	−4	−3	−2	−1	0	1	2
x^2	16	9	4	1	0	1	4
$+2x$	−8	−6	−4	−2	0	2	4
−3	−3	−3	−3	−3	−3	−3	−3
$y = x^2 + 2x - 3$	5	0	−3	−4	−3	0	5

Generally, you do not need to work out all values in a table. If you use a calculator, you need only to work out the y-value. The other rows in the table are just working lines to break down the calculation.

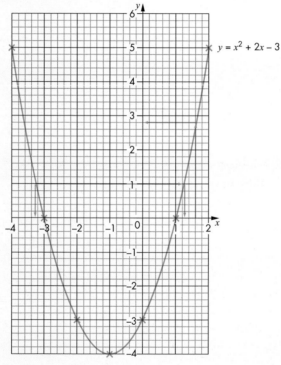

b To find the corresponding y-value for any value of x, you start on the x-axis at that x-value, go up to the curve, across to the y-axis and read off the y-value. This procedure is marked on the graph with arrows.

Always show these arrows because even if you make a mistake and misread the scales, you may still get a mark.

So when $x = 1.6$, $y = 2.8$.

c This time start at 1 on the y-axis and read off the two x-values that correspond to a y-value of 1.

Again, this procedure is marked on the graph with arrows.

So when $y = 1$, $x = -3.2$ or $x = 1.2$.

EXERCISE 16A

1 Copy and complete the table for the graph of $y = x^2$ for $-5 \leqslant x \leqslant 5$.

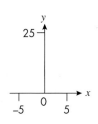

x	-5	-4.5	-4	-3.5	-3	-2.5	-2	-1.5	-1	-0.5	0
$y = x^2$	25	20.25	16						1	0.25	0

As the graph is symmetrical, you do not need to work out the values for 0 to 5.

2 Copy and complete the table for the graph of $y = 3x^2$ for $-3 \leqslant x \leqslant 3$.

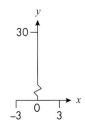

x	-3	-2	-1	0	1	2	3
$y = 3x^2$	27		3			12	

3 Copy and complete the table for the graph of $y = x^2 + 2$ for $-5 \leqslant x \leqslant 5$.

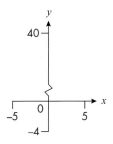

x	-5	-4	-3	-2	-1	0	1	2	3	4	5
$y = x^2 + 2$	27		11					6			

4 **a** Copy and complete the table for the graph of $y = x^2 - 3x$ for $-5 \leqslant x \leqslant 5$. Use your table to plot the graph.

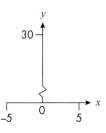

x	-5	-4	-3	-2	-1	0	1	2	3	4	5
x^2	25		9					4			
$-3x$	15							-6			
y	40							-2			

b Use your graph to find the value of y when $x = 3.5$.

c Use your graph to find the values of x that give a y-value of 5.

5 **a** Copy and complete the table for the graph of $y = x^2 - 2x - 8$ for $-5 \leqslant x \leqslant 5$. Use your table to plot the graph.

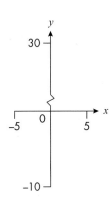

x	-5	-4	-3	-2	-1	0	1	2	3	4	5
x^2	25		9					4			
$-2x$	10							-4			
-8	-8							-8			
y	27							-8			

b Use your graph to find the value of y when $x = 0.5$.

c Use your graph to find the values of x that give a y-value of -3.

6 **a** Copy and complete the table for the graph of $y = x^2 - 5x + 4$ for $-2 \leqslant x \leqslant 5$. Use your table to plot the graph.

x	−2	−1	0	1	2	3	4	5
y	18		4			−2		

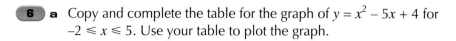

b Use your graph to find the value of y when $x = -0.5$.

c Use your graph to find the values of x that give a y-value of 3.

7 The diagram shows a side elevation of a cone and with a cut parallel to one side.

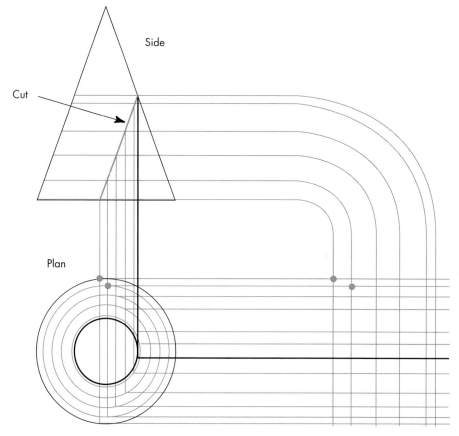

The cone is divided into horizontal sections.

A plan view of the cone is shown.

Construction lines have been drawn to link the elevation and the plan.

Two of the intersecting points have been drawn on the plan.

Two points have also been drawn where the construction lines from the side elevation intersect with the construction lines from the plan.

a Plot the rest of the points on the plan and join them with a smooth curve to see the plan view of the parabola.

b Plot the rest of the points on the intersecting lines and join them with a smooth curve to see the parabola.

FM **8** A car travelling at 30 metres per second, brakes at a set of traffic lights.

After t seconds the speed, v, in metres per second, is given by $v = 30 - 3t$.

The distance travelled, s, in metres, in time, t, after applying the brakes is given by the formula

$$s = 30t - 1.5t^2$$

a It takes the car 10 s to stop. How far has it travelled in this time?

b Complete the table for the speed and distance travelled for $0 \leqslant t \leqslant 10$.

t (s)	1	2	3	4	5	6	7	8	9	10
v (m/s)	27									
s (m)	28.5									

c Using a horizontal axis for v from 0 to 30 and a vertical axis for s from 0 to 150, plot the graph of v against s.

d Approximately how far had the car travelled when it had reduced its speed by half?

9 **a** Copy and complete the table for the graph of $y = x^2 + 2x - 1$ for $-3 \leqslant x \leqslant 3$. Use your table to plot the graph.

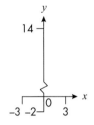

x	−3	−2	−1	0	1	2	3
x^2	9				1	4	
$+2x$	−6		−2			4	
-1	−1	−1				−1	
y	2					7	

b Use your graph to find the y-value when $x = -2.5$.

c Use your graph to find the values of x that give a y-value of 1.

10 **a** Copy and complete the table to draw the graph of $y = x^2 - 4$ for $-4 \leqslant x \leqslant 4$.

x	−4	−3	−2	−1	0	1	2	3	4
y	12			−3				5	

b Where does the graph cross the x-axis?

c Use your graph to find the y-value when $x = 1.5$.

d Use your graph to find the values of x that give a y value of 8.

11 **a** Copy and complete the table to draw the graph of $y = x^2 + 4x$ for $-5 \leqslant x \leqslant 2$.

x	-5	-4	-3	-2	-1	0	1	2
x^2	25			4			1	
$+4x$	-20			-8			4	
y	5			-4			5	

b Where does the graph cross the x-axis?

c Use your graph to find the y-value when $x = -2.5$.

d Use your graph to find the values of x that give a y-value of 3.

12 **a** Copy and complete the table to draw the graph of $y = x^2 - 6x + 3$ for $-1 \leqslant x \leqslant 7$.

x	-1	0	1	2	3	4	5	6	7
y	10			-5			-2		

b Where does the graph cross the x-axis?

c Use your graph to find the y-value when $x = 3.5$.

d Use your graph to find the values of x that give a y-value of 5.

FM 13 Rae drops objects from different heights and times how long they take to reach the ground.

Here are her results.

Height (m)	10	30	50	80	125	200
Time (s)	1.4	2.5	3.2	4	5	6.3

a Draw a graph to show these results, with a horizontal axis for time from 0 to 7 and a vertical axis for height from 0 to 250.

b Rae throws an object from the top of a cliff.

It takes 5.5 seconds to reach the base of the cliff.

Use the graph to estimate the height of the cliff.

GRADE BOOSTER

c You can draw a simple quadratic graph

c You can draw a more complex quadratic graph

c You can find x (or y) values from a quadratic graph given the y (or x) values

What you should know now

- How to draw a quadratic graph
- How to read values from a quadratic graph

1 a Complete the table of values for
$y = x^2 - x - 5$ (2)

x	−2	−1	0	1	2	3	4
y	1		−5	−5	−3	1	

b On a grid like the one below, draw the graph of $y = x^2 - x - 5$ for values of x from −2 to 4 (2)

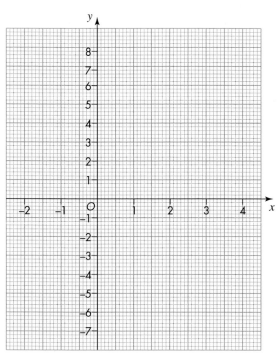

c Find the values of x when $y = 0$ (1)

(Total 5 marks)

AQA, June 2008, Paper 1 Foundation, Question 27

2 a Complete the table of values for
$y = x^2 - 3$ (1)

x	−3	−2	−1	0	1	2	3
y		1	−2	−3	−2		6

b On a grid like the one below, draw the graph of $y = x^2 - 3$ for values of x from −3 to +3 (2)

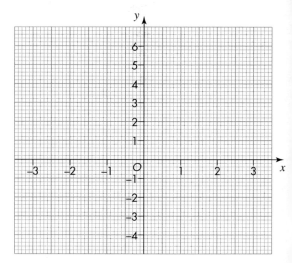

c Use your graph to find the values of x when $y = 0$ (2)

(Total 5 marks)

AQA, November 2005, Paper 1 Intermediate, Question 7

3 a Complete the table of values for
$y = x(x + 3)$ (2)

x	−4	−3	−2	−1	0	1	2
y	4	0		−2	0	4	

b On a grid like the one below, draw the graph of $y = x(x + 3)$ for values of x from −4 to +2 (2)

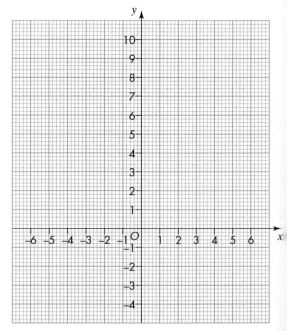

(Total 4 marks)

AQA, June 2009, Paper 2 Foundation, Question 13

c

Worked Examination Questions

1 a Complete the table of values for $y = x^2 - 4x + 1$.

x	−1	0	1	2	3	4	5
y	6		−2		−2		6

b Draw the graph of $y = x^2 - 4x + 1$ for $-1 \leqslant x \leqslant 5$.

c Use the graph to find the x-value when $y = 2$.

a When $x = 0$, $y = (0)^2 - 4(0) + 1 = 0 - 0 + 1 = 1$ — Substitute numbers into the equation to get 1 mark.

When $x = 2$, $y = (2)^2 - 4(2) + 1 = 4 - 8 + 1 = -3$

When $x = 4$, $y = (4)^2 - 4(4) + 1 = 16 - 16 + 1 = 1$

(1 mark)

b

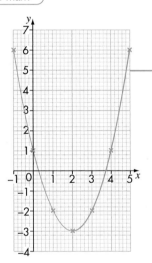

Plot the points. You get 1 mark for plotting the points and 1 mark for drawing a smooth curve.

(2 marks)

c

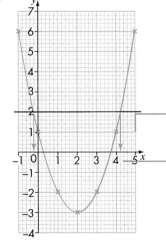

Draw a line $y = 2$ horizontally to meet the graph at two points.

Draw lines down to the x-axis from these points and read off the values to get 1 mark.

−0.2 and 4.2

(1 mark)   **Total: 4 marks**

The forms of many suspension bridges are based on quadratic functions. Their shape is a classic quadratic curve.

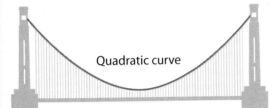

Quadratic curve

In this task you will investigate the quadratic functions that can be used to describe stable suspension bridges.

Getting started

Use these questions to familiarise yourself with how quadratic expressions can be used to represent bridges.

- What would you need to know, to be able to describe the shape of a bridge, in terms of a quadratic formula?
- Think about bridges and other landmarks in your local area. How many of these are based on quadratic equations and form parabolas. Use images to illustrate your findings.

Your task

Below you can see the dimensions of the Clifton Suspension Bridge. Use these dimensions to construct a diagram of the bridge.

Then, using your diagram, estimate the quadratic equation for the curve of the bridge. Represent this equation appropriately.

Write a report explaining the mathematical process that you used to solve this problem. State any assumptions that you made and explain whether you could have found different answers if you had changed your assumptions.

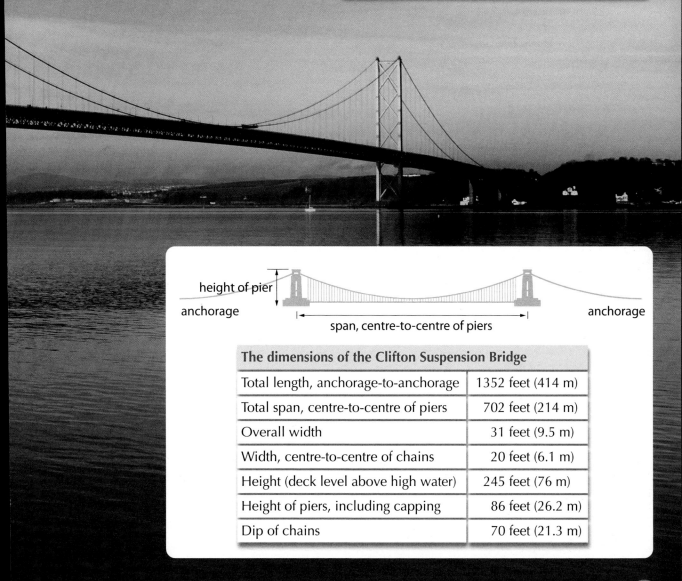

The dimensions of the Clifton Suspension Bridge	
Total length, anchorage-to-anchorage	1352 feet (414 m)
Total span, centre-to-centre of piers	702 feet (214 m)
Overall width	31 feet (9.5 m)
Width, centre-to-centre of chains	20 feet (6.1 m)
Height (deck level above high water)	245 feet (76 m)
Height of piers, including capping	86 feet (26.2 m)
Dip of chains	70 feet (21.3 m)

Why this chapter matters

Pythagoras was born on the island of Samos in 568 BC. He has been described as the first pure mathematician and is best known for developing the theorem that bears his name – Pythagoras' theorem.

The ancient world was filled with war and travel, so this gave Pythagoras the opportunity to visit different countries (sometimes as a prisoner!) and learn about different cultures and systems of knowledge, including different ways of thinking about mathematics.

Pythagoras

He learnt from the Egyptians and the Babylonians about science, religion, mathematics and astronomy, and ultimately settled in Italy where he established the Pythagorean Brotherhood (a secret society devoted to politics, mathematics and astronomy). All this helped him to develop the theorem we now know as Pythagoras' theorem.

His theorem centres on the fact that squares drawn on the sides of any right-angled triangle have a special relationship with each other – that is, that the smaller squares can be fitted into the larger square. This idea was known to the Babylonians a thousand years earlier, but it was Pythagoras who was the first actually to prove it.

It is said that when he discovered his famous theorem, he was so full of joy that he showed his gratitude to the gods by sacrificing a hundred oxen!

It is interesting to note that Pythagoras was only interested in the idea of the squares being drawn and being able to combine both smaller squares into the larger square. He was not interested in the numbers attached to these squares – this development came later. This may be because Pythagoras was not just a mathematician; he was a philosopher who thought of each new theorem as a platform for developing further and not an end in itself.

Pythagoras' theorem is now used in all sorts of fields, from geography to engineering. It is a cornerstone of much other mathematics and interestingly works in 3D blocks, too (although at GCSE you will only look at the theorem in 2D).

It is, therefore, a vital part of mathematics to get to grips with, both in life and for your GCSE examination.

Pythagoras' theorem can be used to find the height of a mountain.

Pythagoras' theorem can be used to make sure that structures, such as masts, are correctly positioned for maximum strength.

Chapter

17

Geometry: Pythagoras' theorem

1 Pythagoras' theorem

2 Finding a shorter side

3 Solving problems using Pythagoras' theorem

This chapter will show you ...

c how to use Pythagoras' theorem in right-angled triangles

c how to solve problems using Pythagoras' theorem

Visual overview

Right-angled triangles → Pythagoras' theorem → Solving problems

What you should already know

● How to find the square and square root of a number (KS3 level 4, GCSE grade F)

● How to round numbers to a suitable degree of accuracy (KS3 level 6, GCSE grade D)

Quick check

Use your calculator to evaluate the following, giving your answers to 1 decimal place.

1 2.3^2

2 15.7^2

3 0.78^2

4 $\sqrt{8}$

5 $\sqrt{260}$

6 $\sqrt{0.5}$

This section will show you how to:
- calculate the length of the hypotenuse in a right-angled triangle

Key words
hypotenuse
Pythagoras' theorem

Pythagoras, the mathematician, enjoyed geometry and playing with shapes. He was fascinated with right-angled triangles. He played around with squares on the sides of triangles until he discovered the now-famous Pythagoras' theorem about these squares.

The activity below will lead you to discover this rule for yourself and to show that it actually works.

ACTIVITY

Squares on triangles

1 Draw a right-angled triangle with sides of 3 cm and 4 cm, as shown.

2 Measure accurately the long side of the triangle (the **hypotenuse**).

3 Draw four more right-angled triangles, choosing your own lengths for the short sides.

3 cm

4 cm

4 When you have done this, measure the hypotenuse for each triangle.

5 Copy and complete the table below for your triangles.

Short side a	Short side b	Hypotenuse c	a^2	b^2	c^2
3	4	5	9	16	25

Is there a pattern in your results? Can you see that a^2, b^2 and c^2 are related in some way?

You should spot that the value of a^2 added to that of b^2 is very close to the value of c^2. (Why don't the values add up exactly?)

You have 'rediscovered' **Pythagoras' theorem**. His theorem can be expressed in several ways, two of which are given on the next page.

Consider squares being drawn on each side of a right-angled triangle, with sides 3 cm, 4 cm and 5 cm.

The longest side is called the hypotenuse and is always opposite the right angle.

Pythagoras' theorem can then be stated as follows:

> *For any right-angled triangle, the area of the square drawn on the hypotenuse is equal to the sum of the areas of the squares drawn on the other two sides.*

The form in which most of your parents would have learnt the theorem when they were at school – and which is still in use today – is as follows:

> *In any right-angled triangle, the square of the hypotenuse is equal to the sum of the squares of the other two sides.*

Pythagoras' theorem is more usually written as a formula:

$$c^2 = a^2 + b^2$$

Remember that Pythagoras' theorem can only be used in right-angled triangles.

Finding the hypotenuse

EXAMPLE 1

Find the length of the hypotenuse, marked x on the diagram.

Using Pythagoras' theorem gives:

$$x^2 = 8^2 + 5.2^2$$
$$= 64 + 27.04$$
$$= 91.04$$

So, $x = \sqrt{91.04} = 9.5$ cm (1 decimal place)

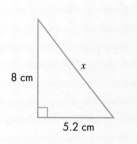

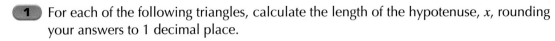

EXERCISE 17A

1 For each of the following triangles, calculate the length of the hypotenuse, x, rounding your answers to 1 decimal place.

a

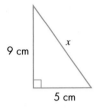

9 cm

x

5 cm

b

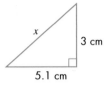

x

3 cm

5.1 cm

c

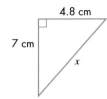

4.8 cm

7 cm

x

HINTS AND TIPS

In these questions you are finding the hypotenuse. Add the squares of the two shorter sides in every case.

---

d

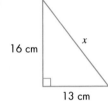

16 cm

x

13 cm

e

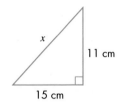

x

11 cm

15 cm

f

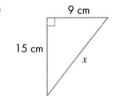

9 cm

15 cm

x

g

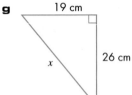

19 cm

26 cm

x

h

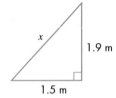

x

1.9 m

1.5 m

i

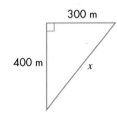

300 m

400 m

x

j

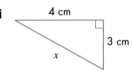

4 cm

3 cm

x

k

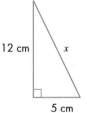

12 cm

x

5 cm

l

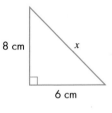

8 cm

x

6 cm

FM 2 Harold, the school groundsman, knew Pythagoras' theorem. He used lengths of rope to help him create right angles for the line markings on pitches.

He had ropes of lengths:

50 cm, 1 m, 1.2 m, 1.25 m, 1.3 m, 1.5 m, 2 m and 2.5 m.

Which of these rope lengths could be used to create a right angle?

PS 3 Find, by trial and improvement or otherwise, the length of x in this triangle.
Give your answer correct to 1 decimal place.

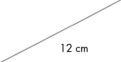

$x + 2$

x

12 cm

AU 4 Explain how you can tell that a triangle with sides of length 6 cm, 7 cm and 10 cm is not a right-angled triangle.

The last three examples in question 1 give whole-number answers. Sets of whole numbers that obey Pythagoras' theorem are called *Pythagorean triples*. Examples of these are:

3, 4, 5   5, 12, 13   and   6, 8, 10

Note that 6, 8, 10 are respectively multiples of 3, 4, 5.

17.2 Finding a shorter side

This section will show you how to:
- calculate the length of a shorter side in a right-angled triangle

Key word
Pythagoras' theorem

By rearranging the formula for **Pythagoras' theorem**, you can easily calculate the length of one of the shorter sides.

$$c^2 = a^2 + b^2$$

So: $a^2 = c^2 - b^2$ or $b^2 = c^2 - a^2$

EXAMPLE 2

Find the length x.

In the triangle, x is one of the shorter sides.

So, using Pythagoras' theorem gives:

$x^2 = 15^2 - 11^2$

$= 225 - 121$

$= 104$

So: $x = \sqrt{104} = 10.2$ cm (1 decimal place)

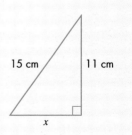

15 cm   11 cm

x

EXERCISE 17B

1 For each of the following triangles, calculate the length x to 1 decimal place.

a

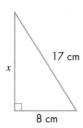

17 cm

x

8 cm

b

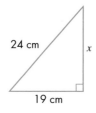

24 cm

x

19 cm

c

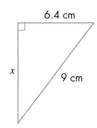

6.4 cm

x

9 cm

d

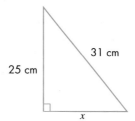

31 cm

25 cm

x

e

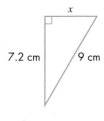

x

7.2 cm   9 cm

f

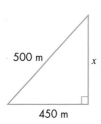

500 m

x

450 m

g

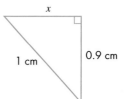

x

1 cm   0.9 cm

h

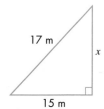

17 m

x

15 m

2 For each of the following triangles, calculate the length x to 1 decimal place.

a

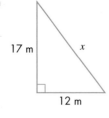

17 m x

12 m

b

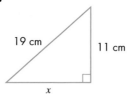

19 cm   11 cm

x

c

17 m

x

23 m

d

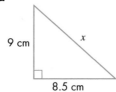

9 cm x

8.5 cm

e

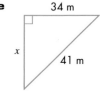

34 m

x 41 m

f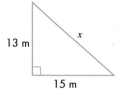

g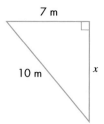

h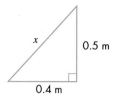

3 For each of the following triangles, find the length marked x.

a

b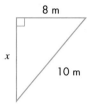

c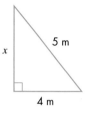

d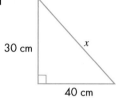

4 What is the length of the diagonal of the base of the back of the lorry?

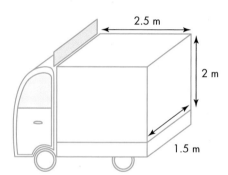

PS 5 Find three possible pairs of lengths for the sides marked a and b in this triangle.

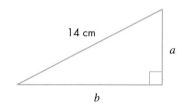

AU 6 Explain how you can tell that a triangle with sides of length 8 inches, 6 inches and 10 inches is a right-angled triangle.

Solving problems using Pythagoras' theorem

This section will show you how to:

● use Pythagoras' theorem to solve problems

Key word

Pythagoras' theorem

Pythagoras' theorem can be used to solve certain practical problems. When a problem involves two lengths only, follow these steps.

● Draw a diagram for the problem, making sure that it includes a right-angled triangle.

● Look at the diagram and decide which side has to be found: the hypotenuse or one of the shorter sides. Label the unknown side x.

● If it is the hypotenuse, then square both numbers, add the squares and take the square root of the sum.

● If it is one of the shorter sides, then square both numbers, subtract the smaller square from the larger square and take the square root of the difference.

EXAMPLE 3

A plane leaves Manchester airport and heads due east. It flies 160 km before turning due north. It then flies a further 280 km and lands. What is the distance of the return flight if the plane flies straight back to Manchester airport?

First, sketch the situation.

Using Pythagoras' theorem gives:

$$x^2 = 160^2 + 280^2$$

$$= 25\,600 + 78\,400$$

$$= 104\,000$$

So: $x = \sqrt{104\,000} = 322$ km

(nearest whole number)

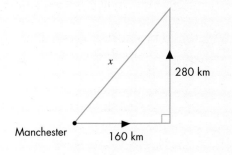

Remember the following tips when solving problems.

● Always sketch the right-angled triangle you need. Sometimes, the triangle is already drawn for you but some problems involve other lines and triangles that may confuse you. So identify which right-angled triangle you need and sketch it separately.

● Label the triangle with necessary information, such as the length of its sides, taken from the question. Label the unknown side x.

● Set out your solution as in the last example. Avoid shortcuts, since they often cause errors. You gain marks in your examination for showing clearly how you are applying Pythagoras' theorem to the problem.

● Round your answer to a suitable degree of accuracy.

EXERCISE 17C

1 A ladder, 12 m long, leans against a wall. The ladder reaches 10 m up the wall. How far away from the foot of the wall is the foot of the ladder?

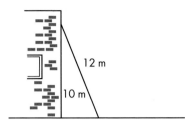

2 A model football pitch is 2 m long and 0.5 m wide. How long is the diagonal?

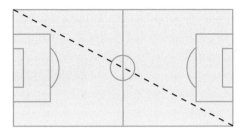

3 How long is the diagonal of a rectangle 6 m long and 9 m wide?

4 How long is the diagonal of a square with a side of 8 m?

5 In a hockey game, after a pass was made, the ball travelled 7 m up the field and 6 m across the field. How long was the actual pass?

6 A ship going from a port to a lighthouse steams 15 km east and 12 km north. How far is the lighthouse from the port?

7 A plane flies from London due north for 120 km before turning due west and flying for a further 85 km and landing at a secret location. How far from London is the secret location?

8 Some pedestrians want to get from point X on one road to point Y on another. The two roads meet at right angles.

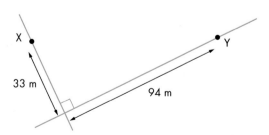

  a If they follow the roads, how far will they walk?

  b Instead of walking along the road, they take the shortcut, XY. Find the length of the shortcut.

  c How much distance do they save?

9 At the moment, three towns, A, B and C, are joined by two roads, as in the diagram. The council want to make a road that runs directly from A to C. How much distance will the new road save?

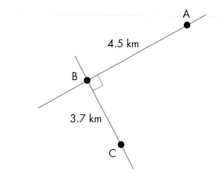

10 A mast on a sailboat is strengthened by a wire (called a stay), as shown on the diagram. The mast is 35 ft tall and the stay is 37 ft long. How far from the base of the mast does the stay reach?

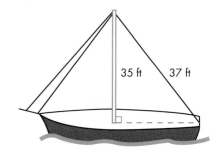

11 A four-metre ladder is put up against a wall.

a How far up the wall will it reach when the foot of the ladder is 1 m away from the wall?

b When it reaches 3.6 m up the wall, how far is the foot of the ladder away from the wall?

12 A pole, 8 m high, is supported by metal wires, each 8.6 m long, attached to the top of the pole. How far from the foot of the pole are the wires fixed to the ground?

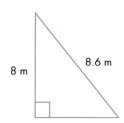

PS **13** How long is the line that joins two coordinates A(13, 6) and B(1, 1)?

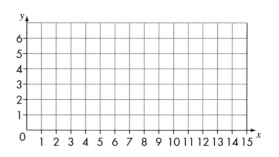

FM **14** The regulation for safe use of ladders states that, for a five-metre ladder: *The foot of the ladder must be placed between 1.6 m and 2.1 m from the foot of the wall.*

a What is the maximum height the ladder can safely reach up the wall?

b What is the minimum height the ladder can safely reach up the wall?

PS **15** A rectangle is 4.5 cm long. The length of its diagonal is 5.8 cm. What is the area of the rectangle?

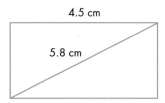

HINTS AND TIPS

First find the width, then the area.

16 Two large trees, 5.5 m and 6.8 m tall, stand 12 m apart. A bird flies directly from the top of one tree to the top of the other. How far has the bird flown?

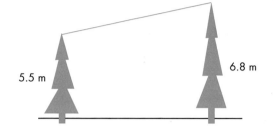

PS **17** Is the triangle with sides 7 cm, 24 cm and 25 cm, a right-angled triangle?

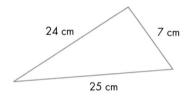

PS **18** The formula for Pythagoras' theorem in 3D is

$$a^2 + b^2 + c^2 = d^2$$

where d is the long diagonal through the inside of a cuboid, from one corner to the other.

Find some cuboids to test this theorem.

AU **19** If you are told the three sides of a triangle, how can you tell whether it is a right-angled triangle?

PS **20**

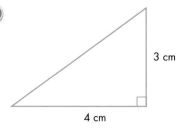

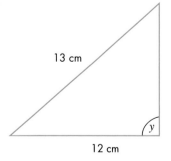

Will these triangles fit together exactly along the unknown edge?

Show all your workings.

GRADE BOOSTER

c You can use Pythagoras' theorem in right-angled triangles

c You can solve problems in 2D, using Pythagoras' theorem

What you should know now

- How to use Pythagoras' theorem to find the hypotenuse or one of the shorter sides of a right-angled triangle, given the other two sides
- How to solve problems, using Pythagoras' theorem

1 Calculate the length, x cm, in the triangle below. (3)

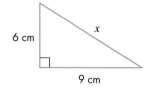

Not drawn accurately

6 cm

x

9 cm

(Total 3 marks)

AQA, June 2008, Paper 2 Foundation, Question 25

 2 *DEF* is a right-angled triangle.

DE = 15 cm, *DF* = 17 cm

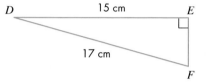

D   15 cm   E

Not drawn accurately

17 cm

F

Calculate the length of the side *EF*. (3)

(Total 3 marks)

 3 In the diagram, ABC is a right-angled triangle. AC = 19 cm and AB = 9 cm.

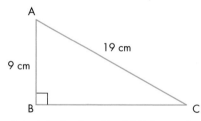

A

19 cm

9 cm

B   C

Calculate the length of BC (in centimetres).

(Total 3 marks)

AQA, June 2005, Paper 2 Higher, Question 10a

4 In the diagram, PQRS is a quadrilateral.

Angles RQS and QSP are right angles.

PS = 4 cm, QR = 12 cm and RS = 13 cm.

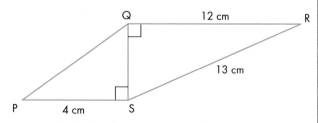

Q   12 cm   R

13 cm

P   4 cm   S

Calculate the length of PQ (in centimetres)

 5 A ladder of length 5 m rests against a wall.

The foot of the ladder is 1.7 m from the base of the wall.

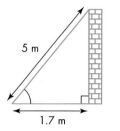

Not drawn accurately

5 m

1.7 m

How far up the wall does the ladder reach? (3)

(Total 3 marks)

AQA November 2008, Paper 2 Foundation, Question 25

 6 In triangle *XYZ*, angle *Y* = 90°, *XY* = 12.7 cm and *YZ* = 3.5 cm

12.7 cm

X   Y

Not drawn accurately

3.5 cm

Z

Calculate *XZ*. (3)

(Total 3 marks)

AQA, June 2008, Module 5, Paper 2 Foundation, Question 17

 7 The diagram shows a ship, S, out at sea.

It is 30 kilometres east and 25 kilometres north of a port, P.

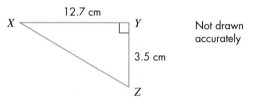

S

x

25 km

P

30 km

Calculate the direct distance from the port to the ship. This distance is marked *x* on the diagram.

Give your answer to 1 decimal place.

Worked Examination Questions

1 The sketch shows triangle ABC.
AB = 40 cm, AC = 41 cm and CB = 9 cm.

By calculation, show that triangle ABC is a
right-angled triangle.

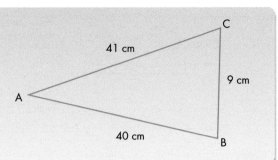

For triangle ABC to be right-angled,
the right angle must be at B and
AC must be the hypotenuse.

$$40^2 + 9^2 = 1681$$

$$41^2 = 1681$$

so $40^2 + 9^2 = 41^2$ making this
a right-angled triangle

Total: 3 marks

> Selecting the two smallest squares and adding
> them together earns 1 method mark.

> Squaring all three numbers will earn 1 method
> mark.

> Showing that the two smaller squares add up to
> the largest square, then stating it is a right-
> angled triangle, earns 1 mark.

2 By looking at the side lengths of a triangle, how can you tell whether the triangle is:

    a a right-angle triangle

    b an obtuse-angled triangle?

Look at the lengths of the two smaller sides,
a and b.

Square them both and add them together,
$a^2 + b^2$.

Square the length of the longest side, c^2.

If $a^2 + b^2 = c^2$ then the triangle
is right-angled.

If $a^2 + b^2 < c^2$ then the triangle
is obtuse-angled.

Total: 3 marks

> Recognising that you need to use the
> squares of all three sides earns 1 method
> mark.

> 1 mark is available for recognising the
> conditions for a right-angled triangle.

> 1 mark is available for recognising the
> conditions for an obtuse-angled triangle.

Worked Examination Questions

AU **3** A 3, 4, 5 triangle is a right-angled triangle.

Explain how you know that this is not the only triangle that has a hypotenuse of 5.

Given the hypotenuse is 5.

Suppose one of the smaller sides was, say, 2 cm.

> You earn 1 mark for using another possible side length with the hypotenuse of 5 cm.

I could use Pythagoras' theorem to find the length of the other side as:

$\sqrt{(5^2 - 2^2)} = 4.58$ cm

> 1 mark is available for showing you can have sides that are not 3 and 4.

Total: 2 marks

When you are out walking on the hills it can be very useful to be able to estimate various distances that you have to cover. Of course, you can just use the scale on the map, but another way is by using Pythagoras' theorem with small right-angled triangles.

Getting started

- You know that the **square** of 2 is $2^2 = 4$. Write down the **square** of each of these numbers.

  5   0.1   0.4   0.03

  Think of a number that has a square between 0.1 and 0.001.

- You know that the **square root** of 4 is $\sqrt{4} = 2$. Now write down the **square root** of each of these numbers.

  16   81   0.01   0.025

  Think of a number that has a square root between 50 and 60.

- On a set of coordinate axes, draw the points A(1, 2) and B(4, 6).

  What is the distance from point A to point B?

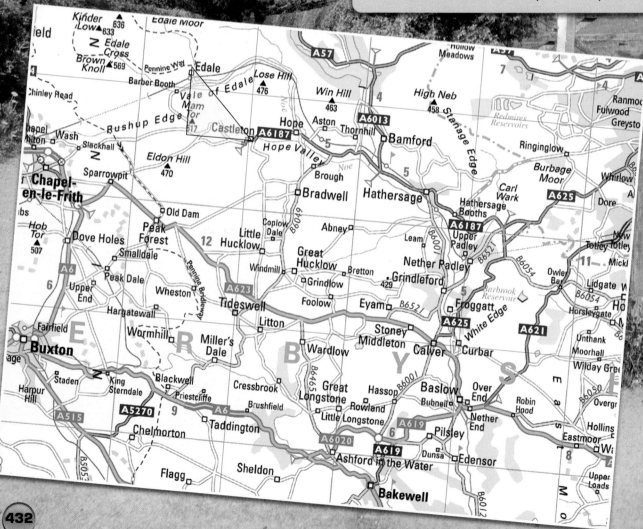

Your task

Freya and Chris often go out walking in the Peak District. On the left is a copy of the map they use. Use this map extract to complete these tasks.

1 Write five questions similar to the examples given on the right. Swap them with the person next to you.

 Answer each other's questions, making sure you show your working clearly.

 Now swap again and mark each other's answers. Give constructive feedback.

2 Plan a walk with a circular route that is between 20 and 35 km long.

 If the average person walks at approximately 4.5 miles per hour, estimate the time it would take to complete your route.

Example

Freya and Chris were at Edale. They wanted to know the rough distance to Castleton. Freya decided to set herself a maths problem, using Pythagoras' theorem.

She looked at the map and imagined the yellow right-angled triangle.

Using the fact that each square on the map represents an area 5 km by 5 km, she estimated each small side of the triangle to be 3 km.

Then she applied Pythagoras' theorem.

$3^2 + 3^2 = 9 + 9 = 18$

On the hillside, without a calculator, she estimated the square root of 18 to be just over 4, giving a distance of 4 km.

Another day, Freya and Chris were at Hucklow and wanted to know the distance to Hathersage.

Use Freya's method to estimate the distance from Hucklow to Hathersage.

Additional information

When working in distances, you need to work in either miles and other imperial units or kilometres and other metric units.

To change between these units there are some key conversion facts. Either use a textbook or the internet to find these.

Answers to Chapter 1

1.1 Basic calculations and using brackets

Exercise 1A

1 **a** 45 **b** 52 **c** 66

2 **a** 200 **b** 115 **c** 236

3 **a** 56 **b** 157 **c** 76 **d** 193

4 **a** 144 **b** 108

5 **a** 12.54 **b** 27.45

6 **a** 26.7 **b** 24.5 **c** 145.3 **d** 1.5

7 Sovereign is 102.47p per litre, Bridge is 102.73p per litre so Sovereign is cheaper.

8 Abby 1.247, Bobby 2.942, Col 5.333, Donna 6.538 Col is correct.

9 $31 \times 3600 \div 1610 = 69.31677 \approx 70$

10 **a** 167.552 **b** 196.48

11 **a** 2.77 **b** 6

12 **a** 497.952 **b** 110.978 625

1.2 Using a calculator to add and subtract fractions

Exercise 1B

1 **a** $1\frac{11}{20}$ **b** $1\frac{8}{15}$ **c** $1\frac{1}{4}$ **d** $\frac{147}{200}$ **e** $\frac{43}{80}$ **f** $1\frac{63}{80}$

   g $\frac{11}{30}$ **h** $\frac{29}{48}$ **i** $\frac{17}{96}$ **j** $\frac{167}{240}$ **k** $\frac{61}{80}$ **l** $\frac{277}{396}$

2 **a** $6\frac{11}{20}$ **b** $8\frac{8}{15}$ **c** $16\frac{1}{4}$ **d** $11\frac{147}{200}$ **e** $7\frac{43}{80}$ **f** $11\frac{63}{80}$

   g $3\frac{11}{30}$ **h** $2\frac{29}{48}$ **i** $3\frac{17}{96}$ **j** $7\frac{167}{240}$ **k** $7\frac{61}{80}$ **l** $1\frac{277}{396}$

3 $\frac{1}{12}$

4 **a** $12\frac{1}{4}$ **b** $3\frac{1}{4}$

5 Use the fraction key to input $\frac{3}{25}$, then key in + and then use the

   fraction key again to input $\frac{7}{10}$

6 $\frac{47}{120}$

7 **a** $-\frac{77}{1591}$ **b** Answer is negative

8 **a** $\frac{223}{224}$ **b** $\frac{97}{1248}$ **c** $-\frac{97}{273}$

   d One negative and one positive so $\frac{5}{7}$ is less than $\frac{14}{39}$ and is

   less than $\frac{5}{7}$.

9 **a** Answers will vary
   b Yes, always true, unless fractions are equivalent

10 $18\frac{11}{12}$ cm

11 $\frac{5}{12}$ (anticlockwise) or $\frac{7}{12}$ (clockwise)

1.3 Using a calculator to multiply and divide fractions

Exercise 1C

1 **a** $\frac{3}{5}$ **b** $\frac{7}{12}$ **c** $\frac{9}{25}$ **d** $\frac{27}{200}$

   e $\frac{21}{320}$ **f** $\frac{27}{128}$ **g** $5\frac{2}{5}$ **h** $5\frac{1}{7}$

   i $2\frac{1}{16}$ **j** $\frac{27}{40}$ **k** $3\frac{9}{32}$ **l** $\frac{11}{18}$

2 $\frac{1}{6}$ m²

3 15

4 **a** $\frac{27}{64}$ **b** $\frac{27}{64}$

5 **a** $\frac{4}{5}$ **b** $\frac{4}{5}$ **c** $\frac{16}{21}$ **d** $\frac{16}{21}$

6 **a** $8\frac{11}{20}$ **b** $18\frac{1}{60}$ **c** $65\frac{91}{100}$ **d** $22\frac{1}{8}$

   e $7\frac{173}{320}$ **f** $52\frac{59}{160}$ **g** $2\frac{17}{185}$ **h** $2\frac{22}{103}$

   i $1\frac{305}{496}$ **j** $5\frac{17}{65}$ **k** $7\frac{881}{4512}$ **l** $5\frac{547}{1215}$

7 $18\frac{5}{12}$ m²

8 $3\frac{11}{32}$ cm³

9 $90\frac{5}{8}$ miles

10 3

11 3

Examination questions

1 $\frac{14}{3} = 4\frac{2}{3}$ litres milk
   So 3 bottles needed.

2 $73\frac{13}{19}$ or 73.7 mph

3 16

4 $20\frac{5}{12}$ cm²

5 **a** $\frac{3}{4}$ metre **b** 100 strides
   c 2250 metres = 2.25 km

6 **a** $56\frac{1}{4}$ cm² **b** $10\frac{1}{4}$ cm

7 **a** 19.85454545… **b** 19.9

8 a 23.76153642... **b** 23.8 or 24

9 a 30.94694426... **b** 30.95

10 a 3.586440678... **b** 3.59

11 a 77 cm^3
   b Two numbers with a product of 20, for example 5 cm and 4 cm

Answers to Chapter 2

2.1 Multiplying and dividing by powers of 10

Exercise 2A

1 a 31 **b** 310 **c** 3100 **d** 31 000

2 a 65 **b** 650 **c** 6500 **d** 65 000

3 Factors of 10 are the same, e.g. 100 = 10^2

4 a 7.3×10 **b** 7.3×10^2 **c** 7.3×10^3
   d 7.3×10^5

5 a 0.31 **b** 0.031 **c** 0.0031
   d 0.00031

6 a 0.65 **b** 0.065 **c** 0.0065
   d 0.00065

7 Factors of 10 are the same, e.g. 1000 = 10^3

8 a 7.3 ÷ 10 **b** 7.3 ÷ 10^2 **c** 7.3 ÷ 10^3
   d 7.3 ÷ 10^5

9 a 250 **b** 34.5 **c** 4670
   d 346 **e** 207.89 **f** 56 780
   g 246 **h** 0.76 **i** 76
   j 89 700 **k** 865 **l** 10 050
   m 999 000 **n** 23 456 **o** 98 765.4
   p 43 230 000 **q** 7867.9 **r** 2036.7
   s 764.3 **t** 3 457 800 **u** 345.78

10 a 0.025 **b** 0.345 **c** 0.004 67
   d 3.46 **e** 0.207 89 **f** 0.056 78
   g 0.0246 **h** 0.0076 **i** 0.000 076
   j 0.000 008 97 **k** 0.000 865 **l** 1.005
   m 0.000 000 999 **n** 2.3456 **o** 0.098 765 4
   p 0.000 043 23 **q** 0.786 79 **r** 20.367
   s 7.643

11 a 60 000 **b** 120 000 **c** 10 000
   d 42 000 **e** 21 000 **f** 300
   g 150 **h** 1400 **i** 100 000
   j 200 000 **k** 28 000 **l** 900
   m 400 **n** 8000 **o** 160 000
   p 4500 **q** 8000 **r** 250 000

12 a 5 **b** 50 **c** 25 **d** 30 **e** 7 **f** 300
   g 6 **h** 30 **i** 4 **j** 5 **k** 2 **l** 100
   m 40 **n** 200 **o** 20 **p** 20 **q** 2 **r** 1
   s 16 **t** 150 **u** 12 **v** 15 **w** 40 **x** 5
   y 40 **z** 320

13 a 54 400 **b** 16 000

14 30 × 90 000 = 2 700 000
   600 × 8000 = 4 800 000
   5000 × 4000 = 20 000 000
   200 000 × 700 = 140 000 000

15 1400 million

2.2 Prime factors, LCM and HCF

Exercise 2B

1 a 84 = 2 × 2 × 3 × 7 **b** 100 = 2 × 2 × 5 × 5
   c 180 = 2 × 2 × 3 × 3 × 5 **d** 220 = 2 × 2 × 5 × 11
   e 280 = 2 × 2 × 2 × 5 × 7
   f 128 = 2 × 2 × 2 × 2 × 2 × 2 × 2
   g 50 = 2 × 5 × 5 **h** 1000 = 2 × 2 × 2 × 5 × 5 × 5
   i 576 = 2 × 2 × 2 × 2 × 2 × 2 × 3 × 3
   j 650 = 2 × 5 × 5 × 13

2 a $2^2 × 3 × 7$ **b** $2^2 × 5^2$ **c** $2^2 × 3^2 × 5$
   d $2^2 × 5 × 11$ **e** $2^3 × 5 × 7$ **f** 2^7
   g $2 × 5^2$ **h** $2^3 × 5^3$ **i** $2^6 × 3^2$
   j $2 × 5^2 × 13$

3 1, 2, 3, 2^2, 5, 2 × 3, 7, 2^3, 3^2, 2 × 5, 11, 2^2 × 3, 13, 2 × 7, 3 × 5, 2^4, 17, 2 × 3^2, 19, 2^2 × 5, 3 × 7, 2 × 11, 23, 2^3 × 3, 5^2, 2 × 13, 3^3, 2^2 × 7, 29, 2 × 3 × 5, 31, 2^5, 3 × 11, 2 × 17, $5 × 7$, $2^2 × 3^2$, 37, 2 × 19, 3 × 13, 2^3 × 5, 41, 2 × 3 × 7, 43, 2^2 × 11, 3^2 × 5, 2 × 23, 47, 2^4 × 3, 7^2, 2 × 5^2

4 a Each is double the previous number
   b 64, 128
   c 81, 243, 729
   d 256, 1024, 4096
   e 3, 3^2, 3^3, 3^4, 3^5, 3^6, ...; 4, 4^2, 4^3, 4^4, 4^5, ...

5 a 2 × 2 × 3 × 5
   b 2^2 × 3 × 5
   c 120 = 2^3 × 3 × 5, 240 = 2^4 × 3 × 5, 480 = 2^5 × 3 × 5

6 a $7^2 × 11^2 × 13^2$
   b $7^3 × 11^3 × 13^3$
   c $7^{10} × 11^{10} × 13^{10}$

7 Because 3 is not a factor of 40 so it does not divide exactly

Exercise 2C

1 a 20 **b** 56 **c** 6 **d** 28 **e** 10 **f** 15
   g 24 **h** 30

2 It is their product

3 a 8 **b** 18 **c** 12 **d** 30

4 No. Because the numbers in each part have common factors

5 a 168 **b** 105 **c** 84 **d** 84 **e** 96 **f** 54
   g 75 **h** 144

6 3 packs of cheese slices and 4 packs of bread rolls

7 a 8 **b** 7 **c** 4 **d** 14 **e** 4 **f** 9
   g 5 **h** 4 **i** 3 **j** 16 **k** 5 **l** 9

8 a i no   **ii** yes   **iii** yes   **iv** no
  b i no   **ii** no   **iii** yes   **iv** no

9 18 and 24

2.3   Rules for multiplying and dividing powers

Exercise 2D

1 a 5^4 **b** 5^{10} **c** 5^5 **d** 5^3 **e** 5^{15} **f** 5^9
  g 5^6 **h** 5^9 **i** 5^8

2 a x^8 **b** x^9 **c** x^8 **d** x^5 **e** x^{12} **f** x^{13}
  g x^{11} **h** x^{10} **i** x^{16}

3 a 6^3 **b** 6^5 **c** 6^1 **d** 6^0 **e** 6^1 **f** 6^3
  g 6^2 **h** 6^1 **i** 6^2

4 a x^4 **b** x^5 **c** x^3 **d** x^3 **e** x^6 **f** x^5
  g x^2 **h** x^6 **i** x^9

5 a 1   **b** 6^0 **c** 1

6 a 1   **b** 5^0 **c** 1

7 Answers for power 0 are always 1.

8 Two values with a sum of 7, for example $a = 5$ and $b = 2$

9 a $a = 1$ and $x = 2$
  b $x = 1$

Examination questions

1 a i 8   **ii** 27   **b** 3, 4, 100, 3^3, 4^3

2 a c^4 **b** d^5 **c** $1/e^7$

3 a $2^2 \times 7$ **b** 84

4 40 beats

5 a 5^3 **b** $a = 2, b = 3$ **c** 27

6 a 14, 21, 42
  b 5
  c 23, 43, 13, 31

7 a 2^6
  b 640 000 = 64 × 10 000; 64 has one prime factor (2) and 10 000 has one other prime factor (5).

8 2 packs of bread sticks, 5 packs of small pies.

Answers to Chapter 3

3.1   Equivalent percentages, fractions and decimals

Exercise 3A

1 a $\frac{2}{25}$ **b** $\frac{1}{2}$ **c** $\frac{1}{4}$ **d** $\frac{7}{20}$ **e** $\frac{9}{10}$ **f** $\frac{3}{4}$

2 a 0.27   **b** 0.85   **c** 0.13   **d** 0.06   **e** 0.8   **f** 0.32

3 a $\frac{3}{25}$ **b** $\frac{2}{5}$ **c** $\frac{9}{20}$ **d** $\frac{17}{25}$ **e** $\frac{1}{4}$ **f** $\frac{5}{8}$

4 a 29%   **b** 55%   **c** 3%   **d** 16%   **e** 60%   **f** 125%

5 a 28%   **b** 30%   **c** 95%   **d** 34%   **e** 27.5% **f** 87.5%

6 a 0.6   **b** 0.075   **c** 0.76   **d** 0.3125 **e** 0.05   **f** 0.125

7 150

8 none

9 20

10 a 77%   **b** 39%   **c** 63%

11 27%

12 61.5%

13 a 50%   **b** 20%   **c** 80%

14 12.5%, 25%, 37.5%, 50%, 75%

15 a 20%   **b** 25%   **c** 75%   **d** 45%   **e** 14%   **f** 50%
  g 60%   **h** 17.5% **i** 55%   **j** 130%

16 a 33.3% **b** 16.7% **c** 66.7% **d** 83.3% **e** 28.6% **f** 78.3%
  g 68.9% **h** 88.9% **i** 81.1% **j** 20.9%

17 a 7%   **b** 80%   **c** 66%   **d** 25%   **e** 54.5% **f** 82%
  g 30%   **h** 89.1% **i** 120%   **j** 278%

18 a $\frac{3}{5}$ **b** 0.6   **c** 60%

19 a 63%, 83%, 39%, 62%, 77%   **b** English

20 6.7%

21 25.5%

22 34%, 0.34, $\frac{17}{50}$; 85%, 0.85, $\frac{17}{20}$; 7.5%, 0.075, $\frac{3}{40}$

23 $\frac{9}{10}$ or 90%

3.2   Calculating a percentage of a quantity

Exercise 3B

1 a 0.88   **b** 0.3   **c** 0.25   **d** 0.08   **e** 1.15

2 a 78%   **b** 40%   **c** 75%   **d** 5%   **e** 110%

3 **a** £45   **b** £6.30   **c** 128.8 kg
   d 1.125 kg   **e** 1.08 h   **f** 37.8 cm
   g £0.12   **h** 2.94 m   **i** £7.60
   j 33.88 min   **k** 136 kg   **l** £162

4 48

5 £2410

6 **a** 86%   **b** 215

7 8520

8 287

9 Each team: 54 000, referees: 900, other teams: 9000 (100 to each, FA: 18 000, celebrities: 8100

10 990

11 Mon: 816, Tue: 833, Wed: 850, Thu: 799, Fri: 748

12 **a** £3.25   **b** 2.21 kg   **c** £562.80
    d £6.51   **e** 42.93 m   **f** £24

13 480 cm³ nitrogen, 120 cm³ oxygen

480 cm^3 nitrogen, 120 cm^3 oxygen

14 13

15 £270

16 More this year as it was 3% of a higher amount than last year.

3.3   Increasing or decreasing quantities by a percentage

Exercise 3C

1 **a** 1.1   **b** 1.03   **c** 1.2   **d** 1.07   **e** 1.12

2 **a** £62.40   **b** 12.96 kg   **c** 472.5 g
    d 599.5 m   **e** £38.08   **f** £90
    g 391 kg   **h** 824.1 cm   **i** 253.5 g
    j £143.50   **k** 736 m   **l** £30.24

3 £29 425

4 1 690 200

5 **a** Bob: £17 325, Anne: £18 165, Jean: £20 475, Brian: £26 565
    b 5% of different amounts is not a fixed amount. The more pay to start with, the more the increase (5%) will be.

6 £83.05

7 193 800

8 575 g
9 918

10 60

11 TV: £287.88, microwave: £84.60, CD: £135.13, stereo: £34.66

12 £20

Exercise 3D

1 **a** 0.92   **b** 0.85   **c** 0.75   **d** 0.91   **e** 0.88

2 **a** £9.40   **b** 23 kg   **c** 212.4 g
    d 339.5 m   **e** £4.90   **f** 39.6 m
    g 731 m   **h** 83.52 g   **i** 360 cm
    j 117 min   **k** 81.7 kg   **l** £37.70

3 £5525

4 **a** 52.8 kg   **b** 66 kg   **c** 45.76 kg

5 Mr Speed: £176, Mrs Speed: £297.50, James: £341, John: £562.50

6 448

7 705

8 £18 975

9 **a** 66.5 mph   **b** 73.5 mph

10 £16.72, £22.88

11 **a** 524.8 units
    b Less gas since 18% of the smaller amount of 524.8 units (94.464 units) is less than 18% of 640 units (115.2 units). I use 619.264 units now.

12 TV £222.31, DVD player £169.20

13 10% off £50 is £45, 10% off £45 is £40.50; 20% off £50 is £40

3.4   Expressing one quantity as a percentage of another

Exercise 3E

1 **a** 25%   **b** 60.6%   **c** 46.3%   **d** 12.5%   **e** 41.7%   **f** 60%
    g 20.8%   **h** 10%   **i** 1.9%   **j** 8.3%   **k** 45.5%   **l** 10.5%

2 32%

3 6.5%

4 33.7%

5 **a** 49.2%   **b** 64.5%   **c** 10.6%

6 17.9%

7 4.9%

8 90.5%

9 **a** Brit Com: 20.9%, USA: 26.5%, France: 10.3%, Other 42.3%
    b Total 100%, all imports
10 Calum 41.7%, Stacey 42.9%, Stacey has greater percentage increase.

11 Takings £144.94, cost wholesale £108. Percentage profit = 34.2%, so not achieved target.

Exercise 3F

1 **a** 0.6, 60%   **b** $\frac{7}{10}$, 70%   **c** $\frac{11}{20}$, 0.55

2 **a** £10.20   **b** 48 kg   **c** £1.26

3 **a** 56%   **b** 68%   **c** 37.5%

4 **a** 276   **b** 3204

5 **a** 20%   **b** 30%   **c** £13.20

6 **a** £6400   **b** £5440

7 **a** 70.4 kg   **b** iii

3.5 Ratio

Exercise 3G

1 **a** 1 : 3 **b** 3 : 4 **c** 2 : 3 **d** 2 : 3 **e** 2 : 5 **f** 2 : 5
   g 5 : 8 **h** 25 : 6 **i** 3 : 2 **j** 8 : 3 **k** 7 : 3 **l** 5 : 2
   m 1 : 6 **n** 3 : 8 **o** 5 : 3 **p** 4 : 5

2 **a** 1 : 3 **b** 3 : 2 **c** 5 : 12 **d** 8 : 1 **e** 17 : 15 **f** 25 : 7
   g 4 : 1 **h** 5 : 6 **i** 1 : 24 **j** 48 : 1 **k** 5 : 2 **l** 3 : 14
   m 2 : 1 **n** 3 : 10 **o** 31 : 200 **p** 5 : 8

3 $\frac{7}{10}$

4 $\frac{10}{25} = \frac{2}{5}$

5 **a** $\frac{2}{5}$ **b** $\frac{3}{5}$

6 **a** $\frac{7}{10}$ **b** $\frac{3}{10}$

7 Amy $\frac{3}{5}$, Katie $\frac{2}{5}$

8 Fruit crush $\frac{5}{32}$, lemonade $\frac{27}{32}$

9 $13\frac{1}{2}$ litres

10 **a** $\frac{1}{2}$ **b** $\frac{7}{20}$ **c** $\frac{3}{20}$

11 James $\frac{1}{2}$, John $\frac{3}{10}$, Joseph $\frac{1}{5}$

12 sugar $\frac{5}{22}$, flour $\frac{3}{11}$, margarine $\frac{2}{11}$, fruit $\frac{7}{22}$

13 3 : 1

14 1 : 4

Exercise 3H

1 **a** 160 g, 240 g   **b** 80 kg, 200 kg   **c** 150, 350
   d 950 m, 50 m   **e** 175 min, 125 min
   f £20, £30, £50   **g** £36, £60, £144
   h 50 g, 250 g, 300 g   **i** £1.40, £2, £1.60
   j 120 kg, 72 kg, 8 kg

2 **a** 175   **b** 30%

3 **a** 40%   **b** 300 kg

4 21

---

5 **a** Mott: no, Wright: yes, Brennan: no, Smith: no, Kaye: yes
   b For example: W26, H30; W31, H38; W33, H37

6 **a** 1 : 400 000   **b** 1 : 125 000   **c** 1 : 250 000
   d 1 : 25 000   **e** 1 : 20 000   **f** 1 : 40 000
   g 1 : 62 500   **h** 1 : 10 000   **i** 1 : 60 000

7 **a** 1 : 1 000 000   **b** 47 km   **c** 8 mm

8 **a** 1 : 250 000   **b** 2 km   **c** 4.8 cm

9 **a** 1 : 20 000   **b** 0.54 km   **c** 40 cm

10 **a** 1 : 1.6   **b** 1 : 3.25   **c** 1 : 1.125
   d 1 : 1.44   **e** 1 : 5.4   **f** 1 : 1.5
   g 1 : 4.8   **h** 1 : 42   **i** 1 : 1.25

11 100°

12 141°

Exercise 3I

1 **a** 3 : 2   **b** 32   **c** 80

2 1000 g

3 10 125

4 **a** 14 min   **b** 75 min

5 **a** 11 pages   **b** 32%

6 Kevin £2040, John £2720

7 **a** lemonade 20 litres, ginger 0.5 litres
   b This one, one-thirteenth is greater than one-fiftieth.

8 100

9 40 cc

Examination questions

1 50%, $\frac{7}{10}$, 0.03

2 **a i** 0.4375   **ii** 0.27   **b** 0.095, $\frac{6}{10}$, 65%, 0.7

3 **a** £30   **b** £220

4 $\frac{3}{5}$ of £25 = £15 so is larger than 40% of £30 which is £12

5 £20

6 £141

7 50%

8 **a** £495   **b** £594

9 65 824

10 **a** £120   **b i** 220   **ii** 54p   **c** 1%

11 20% off £300 = £240, 10% off 240 = £216, 30% off 300 = £210

12 **a** 0.6   **b** £55.48   **c** 72%

Answers to Chapter 4

4.1 The language of algebra

Exercise 4A

1 **a** $x + 2$ **b** $x - 6$ **c** $k + x$ **d** $x - t$ **e** $x + 3$
f $d + m$ **g** $b - y$ **h** $p + t + w$ **i** $8x$

j hj **k** $x \div 4$ or $\frac{x}{4}$ **l** $2 \div x$ or $\frac{2}{x}$

m $y \div t$ or $\frac{y}{t}$ **n** wt **o** a^2 **p** g^2

2 **a** **i** $P = 4, A = 1$ **b** **i** $P = 4s$ cm
    ii $P = 4x, A = x^2$ **ii** $A = s^2$ cm^2
    iii $P = 12, A = 9$
    iv $P = 4t, A = t^2$

3 **a** $x + 3$ yr **b** $x - 4$ yr

4 $F = 2C + 30$

5 Rule **c**

6 **a** $C = 100M$ **b** $N = 12F$ **c** $W = 4C$ **d** $H = P$

7 **a** $3n$ **b** $3n + 3$ **c** $n + 1$ **d** $n - 1$

8 Rob: $2n$, Tom: $n + 2$, Vic: $n - 3$, Will: $2n + 3$

9 **a** $P = 8n, A = 9n^2$ **b** $P = 24n, A = 36n^2$

10 **a** £4 **b** £$(10 - x)$ **c** £$(y - x)$ **d** £$2x$

11 **a** 75p **b** $15x$ p **c** $4A$ p **d** Ay p

12 £$(A - B)$

13 £$A \div 5$ or $\frac{£A}{5}$

14 **a** Dad: $(72 + x)$ yr, me: $(T + x)$ yr **b** 31

15 **a** $T \div 2$ or $\frac{T}{2}$ **b** $T \div 2 + 4$ or $\frac{T}{2} + 4$ **c** $T - x$

16 **a** $8x$ **b** $12m$ **c** $18t$

17 Andrea: $3n - 3$, Bert: $3n - 1$, Colin: $3n - 6$ or $3(n - 2)$, Davina: 0, Emma: $3n - n = 2n$. Florinda: $3n - 3m$

18 For example, $2 \times 6m$, $1 \times 12m$, $6m + 6m$, etc

19 Any values picked for l and w and substituted into the formulae to give the same answers

20 13

21 15p

22 **a** expression **b** formula **c** equation

4.2 Simplifying expressions

Exercise 4B

1 **a** $6t$ **b** $15y$ **c** $8w$ **d** $5b^2$ **e** $2w^2$ **f** $8p^2$
g $6t^2$ **h** $15t^2$ **i** $2mt$ **j** $5qt$ **k** $6mn$ **l** $6qt$
m $10hk$ **n** $21pr$

2 **a** All except $2m \times 6m$
  b 2 and 0

3 $4x$ cm

4 **a** y^3 **b** $3m^3$ **c** $4t^3$ **d** $6n^3$ **e** t^4 **f** h^5
g $12n^5$ **h** $6a^7$ **i** $4k^7$ **j** t^3 **k** $12d^3$ **l** $15p^6$
m $3mp^2$ **n** $6m^2n$ **o** $8m^2p^2$

5 The number of people who get told the rumour doubles each day ie 2, 4, 8, 16, 32, 64, 128, 256, 512, 1024, 2048, but the number who know the rumour is 3, 7, 15, 31, 63, 127, 255, 511, 1023, 2047 so by the 10th day everyone in the school would know, plus 47 other people!

Exercise 4C

1 **a** £t **b** £$(4t + 3)$

2 **a** $10x + 2y$ **b** $7x + y$ **c** $6x + y$

3 **a** $5a$ **b** $6c$ **c** $9e$ **d** $6f$ **e** $4j$ **f** $3q$
g 0 **h** $-w$ **i** $6x^2$ **j** $5y^2$ **k** 0

4 **a** $7x$ **b** $3t$ **c** $-5x$ **d** $-5k$ **e** $2m^2$ **f** 0

5 **a** $7x + 5$ **b** $5x + 6$ **c** $5p$
d $5x + 6$ **e** $5p + t + 5$ **f** $8w - 5k$
g c **h** $8k - 6y + 10$

6 **a** $2c - 3d$ **b** $5d - 2e$ **c** $f + 3g + 4h$
d $6u - 3v$ **e** $7m - 7n$ **f** $3k + 2m + 5p$
g $2v$ **h** $2w - 3y$ **i** $11x^2 - 5y$
j $-y^2 - 2z$ **k** $x^2 - z^2$

7 **a** $8x + 6$ **b** $3x + 16$ **c** $2x + 2y + 8$

8 Any acceptable answers, e.g. $x + 4x + 2y + 2y$ or $6x - x + 6y - 2y$

9 **a** $2x$ and $2y$ **b** a and $7b$

10 **a** $3x - 1 - x$ **b** $10x$ **c** 25 cm

11 **a** $12p + 6s$ **b** 13 m and 50 cm

12 Maria is correct, as the two short horizontal lengths are equal to the bottom length and the two short vertical lengths are equal to the side length.

4.3   Expanding brackets

Exercise 4D

1 a $6 + 2m$ **b** $10 + 5l$ **c** $12 - 3y$

   d $20 + 8k$ **e** $12d - 8n$ **f** $t^2 + 3t$

   g $m^2 + 5m$ **h** $k^2 - 3k$ **i** $3g^2 + 2g$

   j $5y^2 - y$ **k** $5p - 3p^2$ **l** $3m^2 + 12m$

   m $15t - 12t^2$ **n** $6d^2 + 12de$ **o** $6y^2 + 8ky$

   p $15m^2 - 10mp$

2 a $y^3 + 5y$ **b** $h^4 + 7h$ **c** $k^3 - 5k$
   d $3t^3 + 12t$ **e** $15d^3 - 3d^4$ **f** $6w^3 + 3tw$
   g $15a^3 - 10ab$ **h** $12p^4 - 15mp$ **i** $5m^2 + 4m^3$
   j $t^4 + 2t^5$ **k** $5g^2t - 4g^4$ **l** $15t^3 + 3mt^2$

3 a $5(t - 1)$ and
    $5t - 5$
   b Yes as $5(t - 1)$ when $t = 4.50$ is $5 \times 3.50 = £17.50$

4 He has worked out
3×5 as 8 instead of 15 and he has not multiplied the second
term by 3. Answer should be $15x - 12$.

5 a $3(2y + 3)$
   b $2(6z + 4)$ or
    $4(3z + 2)$

Exercise 4E

1 a $7t$ **b** $3y$ **c** $9d$ **d** $3e$ **e** $3p$ **f** $2t$
   g $5t^2$ **h** $5ab$ **i** $3a^2d$

2 a $22 + 5t$ **b** $21 + 19k$ **c** $10 + 16m$
   d $16 + 17y$ **e** $22 + 2f$ **f** $14 + 3g$
   g $10 + 11t$ **h** $22 + 4w$

3 a $2 + 2h$ **b** $9g + 5$ **c** $6y + 11$
   d $7t - 4$ **e** $17k + 16$ **f** $6e + 20$
   g $7m + 4$ **h** $3t + 10$

4 a $4m + 3p + 2mp$ **b** $3k + 4h + 5hk$ **c** $3n + 2t + 7nt$
   d $3p + 7q + 6pq$ **e** $6h + 6j + 13hj$ **f** $6t + 8y + 21ty$
   g $24p + 12r + 13pr$ **h** $20k - 6m + 19km$

5 a $5(f + 2s) + 2(2f + 3s) = 9f + 16s$
   b $£(270f + 480s)$ **c** $42\,450 - 30\,000 = £12\,450$

6 for x-coefficients 3 and 1 or 1 and 4; For y-coefficients 5 and 1
or 3 and 4 or 1 and 7

7 $5(3x + 2) - 3(2x - 1) = 9x + 13$

4.4   Factorisation

Exercise 4F

1 a $6(m + 2t)$ **b** $3(3t + p)$ **c** $4(2m + 3k)$
   d $4(r + 2t)$ **e** $m(n + 3)$ **f** $g(5g + 3)$
   g $2(2w - 3t)$ **h** $2(4p - 3k)$ **i** $2(8h - 5k)$
   j $2m(p + k)$ **k** $2b(2c + k)$ **l** $2a(3b + 2c)$
   m $y(3y + 2)$ **n** $t(4t - 3)$ **o** $2d(2d - 1)$
   p $3m(m - p)$

2 a $3p(2p + 3t)$ **b** $2p(4t + 3m)$ **c** $4b(2a - c)$
   d $4a(3a - 2b)$ **e** $3t(3m - 2p)$ **f** $4at(4t + 3)$
   g $5bc(b - 2)$ **h** $2b(4ac + 3ed)$ **i** $2(2a^2 + 3a + 4)$
   j $3b(2a + 3c + d)$ **k** $t(5t + 4 + a)$
   l $3mt(2t - 1 + 3m)$ **m** $2ab(4b + 1 - 2a)$
   n $5pt(2t + 3 + p)$

3 a Not possible   **b** $m(5 + 2p)$ **c** $t(t - 7)$
   d Not possible   **e** $2m(2m - 3p)$ **f** Not possible
   g $a(4a - 5b)$ **h** Not possible   **i** $b(5a - 3bc)$

4 a Mary has taken out a common factor
   b Because the bracket adds up to £10
   c £30

5 Bella. Aidan has not taken out the largest possible common
factor. Craig has taken m out of both terms but there isn't an m
in the second term.

6 There are no common factors.

4.5   Substitution

Exercise 4G

1 a 8   **b** 17   **c** 32

2 a 3   **b** 11   **c** 43

3 a 9   **b** 15   **c** 29

4 a 9   **b** 5   **c** -1
5 a 13   **b** 33   **c** 78

6 a 10   **b** 13   **c** 58

7 a £4   **b** 13 km
   c No, 5 miles is 8 km so fare would be £6.50

8 a $2 \times 8 + 6 \times 11 - 3 \times 2 = 76$
   b $5 \times 2 - 2 \times 11 + 3 \times 8 = 12$

9 Any values such that $lw = \frac{1}{2}bh$ or $bh = 2lw$

10 a 32   **b** 64   **c** 160

11 a 6.5   **b** 0.5   **c** -2.5

12 a 2   **b** 8   **c** -10

13 a 3   **b** 2.5   **c** -5

14 a 6   **b** 3   **c** 2

15 a 12 **b** 8 **c** $1\frac{1}{2}$

16 a $\dfrac{1050}{n}$ **b** £925

17 a i odd
ii odd
iii even
iv odd
b Any valid expression such as $xy + z$

18 a £20
b i −£40
ii Delivery cost will be zero.
c 40 miles

Examination questions

1 a $4p$ **b** $12qr$ **c** $-3t$

2 a $4a + 7b$
b $5p + 10q - 15r$

3 a i £2.40 **ii** £7.60

4 a $y + 5$ **b** $1-2(y + 5)$

5 a $13x$ **b** 4

6 a $s = (3)^2 - 6(3) + 9 = 9 - 18 + 9 = 0$

7 a 64 **b** 2 **c** $5p + 2q$

8 19.1

9 a $7x + 16$
b $2x + 2y$ or $2(x + y)$

10 a $4x - 12$
b $x(x + 5)$

11 a $4Q + 3R$
b $2R - 4Q$

12 a $6x - 42$
b $-2x^2 + 3x + 4$

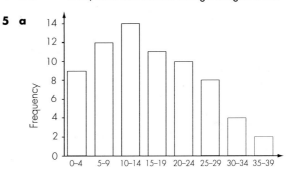

Answers to Chapter 5

5.1 The mode

1 a 4 **b** 48 **c** −1 **d** $\frac{1}{4}$ **e** no mode **f** 3.21

2 a red **b** sun **c** β **d** ★

3 a 32 **b** 6 **c** no
d no; boys generally take larger shoe sizes

4 a 5 **b** no; more than half the form got a higher mark

5 a

(bar chart: Frequency vs Number of e-mails)

Number of e-mails	Frequency
0–4	9
5–9	12
10–14	14
15–19	11
20–24	10
25–29	8
30–34	4
35–39	2

b 70 **c** 24
d cannot tell; know only that 9 households had between 0 and 4 e-mails
e 10–14

6 The mode will be the most popular item or brand sold in a shop.

7 a 28 **b i** brown **ii** blue **iii** brown
c Both students had blue eyes.

8 a May lose count.
b Put in a table, or arrange in order **c** 4

5.2 The median

1 a 5 **b** 33 **c** $7\frac{1}{2}$ **d** 24 **e** $8\frac{1}{2}$ **f** 0
g 5.25

2 a £2.20 **b** £2.25 **c** median, because it is the central value

3 a 5 **b i** 15 **ii** 215 **iii** 10 **iv** 10

4 a 13, Ella **b** 162 cm, Pat **c** 40 kg, Elisa
d Ella, because she is closest to the 3 medians

5 a 12 **b** 13

6 a 21 **b** 16
c

Mark	12	13	14	15	16	17	18
Frequency	1	3	4	3	6	3	1

d 15

7 Answers will vary

8 a 22 s   **b** 25 s

9 a 56   **b** 48   **c** 49   **d** 45.5
   e Boys have higher average and highest score.

10 12, 14, 14, 16, 20, 22, 24

11

2	4						
3	8	8	9				
4	1	2	5	7	7	7	8
5	0	3	5	8			
6	0	3	4	4	8	9	
7	1	3	5				
8	9						

key: 2 | 4 = 24   median = 53

12 a Possible answer: 11, 15, 21, 21 (one below or equal to 12 and three above or equal)
   b Any four numbers higher than or equal to 12, and any two lower or equal
   c Eight, all 4 or under

13 A median of £8 does not take into account the huge value of the £3000 so is in no way representative.

5.3   The mean

Exercise 5C

1 a 6   **b** 24   **c** 45   **d** 1.57   **e** 2

2 a 55.1   **b** 324.7   **c** 58.5   **d** 44.9   **e** 2.3

3 a 61   **b** 60   **c** 59   **d** Brian   **e** 2

4 42 min

5 a £200   **b** £260   **c** £278
   d Median, because the extreme value of £480 is not taken into account

6 a 35   **b** 36

7 a 6
   b 16; all the numbers and the mean are 10 more than those in part a
   c i 56   **ii** 106   **iii** 7

8 Possible answers: Speed – Kath, James, John, Joseph; Roberts – Frank, James, Helen, Evie. Other answers are possible.

9 36

10 24

5.4   The range

Exercise 5D

1 a 7   **b** 26   **c** 5   **d** 2.4   **e** 7

2 a 5°, 3°, 2°, 7°, 3°
   b Variable weather over England

3 a £31, £28, £33   **b** £8, £14, £4
   c Not particularly consistent

4 a 82 and 83   **b** 20 and 12
   c Fay, because her scores are more consistent

5 a 5 min and 4 min   **b** 9 min and 13 min
   c Number 50, because times are more consistent

6 a Issac, Oliver, Andrew, Chloe, Lilla, Billy and Isambard
   b 70 cm to 92 cm

7 a Teachers because they have a high mean and students could not have a range of 20.
   b Year 11 students as the mean is 15–16 and the range is 1.

5.5   Which average to use

Exercise 5E

1 a i 29   **ii** 28   **iii** 27.1   **b** 14

2 a i Mode 3, median 4, mean 5   **ii** 6, 7, $7\frac{1}{2}$ **iii** 4, 6, 8
   b i Mean: balanced data   **ii** Mode: 6 appears five times
   iii Median: 28 is an extreme value

3 a Mode 73, median 76, mean 80
   b The mean, because it is the highest average

4 a 150   **b** 20

5 a i 6   **ii** 16   **iii** 26   **iv** 56   **v** 96
   b Units are the same
   c i 136   **ii** 576   **iii** 435   **iv** 856
   d i 5.6   **ii** 15.6   **iii** 25.6   **iv** 55.6   **v** 95.6

6 a Mean   **b** Median   **c** Mode
   d Median   **e** Mode   **f** Mean

7 No. Mode is 31, median is 31, and mean is $31\frac{1}{2}$

8 a 9   **b** 7

9 a i £20 000   **ii** £28 000   **iii** £34 000
   b i The 6% rise, because it gives a greater increase in salary for the higher paid employees.
   ii 6% increase: £21 200, £29 680, £36 040;
   +£1500: £21 500, £29 500, £35 500

10 a Median   **b** Mode   **c** Mean

11 Tom mean, David median, Mohaned mode

12 a $6x$ **b** 8

13 Possible answers: **a** 1, 6, 6, 6, 6   **b** 2, 5, 5, 6, 7

14 $2x + 3$

15 Boss chose the mean while worker chose the mode.

16 11.6

17 42.7 kg

5.6   Frequency tables

Exercise 5F

1 a i 7   **ii** 6   **iii** 6.4   **b i** 4   **ii** 4   **iii** 3.7
   c i 8   **ii** 8.5   **iii** 8.2   **d i** 0   **ii** 0   **iii** 0.3

2 a 668   **b** 1.9   **c** 0   **d** 328

3 a 2.2, 1.7, 1.3   **b** Better dental care

4 a 0   **b** 0.96

5 a 7   **b** 6.5   **c.** 6.5

6 a 1   **b** 1   **c** 0.98

7 a Roger 5, Brian 4   **b** Roger 3, Brian 8
  c Roger 5, Brian 4   **d** Roger 5.4, Brian 4.5
  e Roger, because he has the smaller range
  f Brian, because he has the better mean

8 Possible answers: 3, 4, 15, 3 or 3, 4, 7, 9 …

9 Add up the weeks to see she travelled in 52 weeks of the year, the median is in the 26th and 27th week. Looking at the weeks in order, the 23rd entry is the end of 2 days in a week so the median must be in the 3 days in a week.

5.7   Grouped data

Exercise 5G

1 a i $30 < x \leq 40$ **ii** 29.5   **b i** $0 < y \leq 100$ **ii** 158.3
  c i $5 < z \leq 10$ **ii** 9.43   **d i** 7–9   **ii** 8.4 weeks

2 a $100 < w \leq 120$ g   **b** 10 860 g   **c** 108.6 g

3 a 207   **b** 19–22 cm   **c** 20.3 cm

4 a 160   **b** 52.6 min   **c** modal group   **d** 65%

5 a $175 < h \leq 200$ **b** 31%   **c** 193.25   **d** No

6 Average price increases: Soundbuy 17.7p, Springfields 18.7p, Setco 18.2p

7 a Yes average distance is 11.7 miles per day.
  b Because shorter runs will be completed faster, which will affect the average.
  c Yes because the shortest could be 1 mile and the longest 25 miles.

8 The first 5 and the 10 are the wrong way round.

9 Find the midpoint of each group, multiply that by the frequency and add those products. Divide that total by the total frequency.

10 a As we do not know what numbers are in each group, we cannot say what the median is.
  b Yes. The lowest number could be, for example, 28, and the highest could be 52, giving a range of 34.

5.8   Frequency polygons

Exercise 5H

1 a
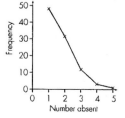
  b 1.72

2 a
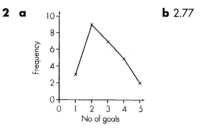
  b 2.77

3 a
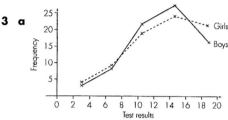
  b boys 12.9, girls 13.1

4 a
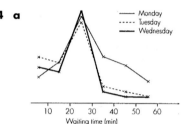
  b Mon 28.4, Tue 20.9, Wed 21.3
  c There are more people on a Monday as they became ill over the weekend.

5 a i 17, 13, 6, 3, 1   **ii** £1.45
  b i
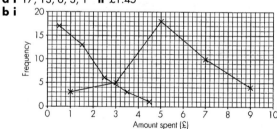
  ii £5.35
  c There is a much higher mean, first group of people just want a paper or a few sweets. Later people are buying food for the day.

6 a

Age, x	$20 < x \leq 30$	$30 < x \leq 40$	$40 < x \leq 50$	$50 < x \leq 60$	$60 < x \leq 70$
Frequency	14	16	14	12	4

  b 41

7 2.17 hours

8 That is the middle value of the time group 0 to 1 minute, it would be very unusual for most of them to be exactly in the middle at 30 seconds.

Examination questions

1 a 25   **b** 2   **c** 2   **d** 2.4

2 a 6   **b** 9

3 16   **4** 21 °C   **5 a** 16   **b** 35.4   **c** 60

6 31.3 or answer that rounds to 31.3

7 a 8   **b** $40 < x \leq 60$
   c Points plotted at (10, 2), (30, 12), (50, 16), (70, 6), (90, 2) and a line connecting these points.

8 a i 74   **ii** 25
   b i They are quick growing   **ii** They are of even height

9 a 37   **b** 52   **c** 48

10 a 58   **b** 13   **c** 15
   d $288 \div 13 = 22.2$

11 a 44   **b** 33
   c Not affected by particularly low (or high) mark

12 1   **13** 1.9   **14** 3.55 km

Answers to Chapter 6

6.1 Perimeter

Exercise 6A

1 a 10 cm **b** 8 cm   **c** 14 cm **d** 12 cm **e** 16 cm **f** 6 cm

2 a 10 cm **b** 12 cm **c** 12 cm **d** 14 cm **e** 12 cm **f** 12 cm

3 18 m

4 No, the perimeter is 30 cm as two sides will touch.

5 False, it is 28 cm

6.2 Area of an irregular shape

Exercise 6B

1 a 10 cm^2 **b** 11 cm^2 **c** 13 cm^2 **d** 12 cm^2 (estimates only)

2 Student's answer

3 Student's answer

4 18–24 cm^2

5 26–30 m^2

6 The area of the rectangle around the shape is 24 cm^2

6.3 Area of a rectangle

Exercise 6C

1 a 35 cm^2, 24 cm   **b** 33 cm^2, 28 cm   **c** 45 cm^2, 36 cm
   d 70 cm^2, 34 cm   **e** 56 cm^2, 30 cm   **f** 10 cm^2, 14 cm

2 a 53.3 cm^2, 29.4 cm   **b** 84.96 cm^2, 38 cm

3 a 20 cm, 21 cm^2 **b** 18 cm, 20 cm^2 **c** 2 cm, 8 cm^2
   d 3 cm, 15 cm^2 **e** 3 mm, 18 mm   **f** 4 mm, 22 mm
   g 5 m, 10 m^2 **h** 7 m, 24 m

4 39

5 a 4   **b** 1 h 52 min

6 £839.40

7 40 cm

8 96

9 B, 44 cm^2

10 Never (the area becomes four times greater).

11 a 100 mm^2
   b i 300 mm^2 **ii** 500 mm^2 **iii** 630 mm^2

12 a 10 000 cm^2
   b i 20 000 cm^2 **ii** 40 000 cm^2 **iii** 56 000 cm^2

6.4 Area of a compound shape

Exercise 6D

1 a 30 cm^2 **b** 40 cm^2 **c** 51 cm^2
   d 35 cm^2 **e** 43 cm^2 **f** 51 cm^2
   g 48 cm^2 **h** 33 cm^2

2 24

3 The correct answer is 44 cm^2, the length of the bottom rectangle is 6 cm (10 − 4).

4 72 cm^2

5 48 cm

6 Yes, the area to paint is 9.1 m^2

6.5 Area of a triangle

Exercise 6E

1 a 6 cm^2, 12 cm   **b** 120 cm^2, 60 cm   **c** 30 cm^2, 30 cm

2 40 cm^2

3 84 m^2

4 a 21 cm^2 **b** 55 cm^2 **c** 165 cm^2

5 c 75 cm^2

6 32 cm, 36 cm^2

7 108 cm^2

8 a 5.5 m^2 **b** 4

Exercise 6F

1 a 21 cm^2 **b** 12 cm^2 **c** 14 cm^2
   d 55 cm^2 **e** 90 cm^2 **f** 140 cm^2

2 a 28 cm^2 **b** 8 cm   **c** 4 cm
   d 3 cm   **e** 7 cm   **f** 44 cm^2

3 64 cm^2

4 a 40 cm^2 **b** 65 m^2 **c** 80 cm^2

5 a 65 cm^2 **b** 50 m^2

6 For example: height 10 cm, base 10 cm; height 5 cm, base 20 cm; height 25 cm, base 4 cm; height 50 cm, base 2 cm

7 a 1500 cm^2 **b** 1800 cm^2

8 Triangle c; a and b each have an area of 15 cm^2 but c has an area of 16 cm^2.

6.6   Area of a parallelogram

Exercise 6G

1 a 96 cm^2 **b** 70 cm^2 **c** 20 cm^2
   d 125 cm^2 **e** 10 cm^2 **f** 112 m^2

2 No, it is 24 cm^2, she used the slanting side instead of the perpendicular height.

3 16 cm

4 a 500 cm^2 **b** 15

6.7   Area of a trapezium

Exercise 6H

1 a 30 cm^2 **b** 77 cm^2 **c** 24 cm^2
   d 42 cm^2 **e** 40 cm^2 **f** 6 cm
   g 3 cm

2 a 27.5 cm, 36.25 cm^2 **b** 33.4 cm, 61.2 cm^2
   c 38.6 m, 88.2 m^2

3 Any pair of lengths that add up to 10 cm
For example: 1 cm, 9 cm; 2 cm, 8 cm; 3 cm, 7 cm; 4 cm, 6 cm; 4.5 cm, 5.5 cm

4 Shape c. Its area is 25.5 cm^2

5 Shape a. Its area is 28 cm^2

6 a

7 2 cm

8 1.4 m^2

Examination questions

1 a 7 cm^2
   b Check students' diagrams. Rectangles should be either 1 cm × 8 cm or 2 cm × 4 cm.
   c 12–14 cm^2

2 a 24 cm   **b** 32 cm^2

3 a 900 cm^2 **b** 50 cm   **c** 60

4 a 4.5 squares   **b** 8 squares   **c** 12 squares

5 a 10 cm^2 **b** 31.5 cm^2

6 39 cm^2

7 59.9 cm^2

Answers to Chapter 7

7.1   Speed, time and distance

Exercise 7A

1 18 mph

2 280 miles

3 52.5 mph

4 11.50 am

5 500 s

6 a 75 mph   **b** 6.5 h   **c** 175 miles
   d 240 km   **e** 64 km/h   **f** 325 km
   g 4.3 h (4 h 18 min)

7 a 7.75 h   **b** 52.9 mph

8 a 2.25 h   **b** 99 miles

9 a 1.25 h   **b** 1 h 15 min

10 a 48 mph   **b** 6 h 40 min

11 a 120 km   **b** 48 km/h

12 a 30 min   **b** 6 mph

13 a 10 m/s   **b** 3.3 m/s   **c** 16.7 m/s
   d 41.7 m/s   **e** 20.8 m/s

14 a 90 km/h   **b** 43.2 km/h   **c** 14.4 km/h
   d 108 km/h   **e** 1.8 km/h

15 a 64.8 km/h   **b** 28 s   **c** 8.07 (37 min journey)

16 a 6.7 m/s   **b** 66 km   **c** 5 minutes
   d 133.3 metres

17 6.6 minutes

7.2 Direct proportion problems

Exercise 7B

1 60 g

2 £5.22

3 45

4 £6.72

5 a £312.50   **b** 8

6 a 56 litres   **b** 350 miles

7 a 300 kg   **b** 9 weeks

8 40 s

9 a i 100 g margarine, 200 g sugar, 250 g flour, 150 g ground rice
  ii 150 g margarine, 300 g sugar, 375 g flour, 225 g ground rice
  iii 250 g margarine, 500 g sugar, 625 g flour, 375 g ground rice
b 24

10 Peter's shop as I can buy 24. At Paul's shop I can only buy 20.

7.3 Best buys

Exercise 7C

1 a £4.50 for a 10-pack   **b** £1.08 for 6
  c £2.45 for 1 litre   **d** Same value
  e 29p for 250 grams   **f** £1.39 for a pack of 6
  g £4 for 3 cartons

2 a Large jar   **b** 600 g tin   **c** 5 kg bag
  d 75 ml tube   **e** Large box   **f** Large box
  g 400 ml bottle

3 a £5.11   **b** Large tin

4 a 95p   **b** Family size

5 Bashir's

6 Mary

7 Kelly

Examination questions

1 £75

2 a 60%   **b** 210 ml

3 Small pack, as £9.60 for 12
4 a 14 days   **b** 153 miles

5 a 3 hours 35 minutes   **b** 64 mph

6 Holiday shop 3.2p per ml
Southern Pharmacy 3p per ml
Southern Pharmacy is better value

7 a $2\frac{1}{2}$ hours   **b** 90 mph   **c** 4 hours

8 a 12 mph   **b** 14.4 mph

Answers to Chapter 8

8.1 Solving simple linear equations

Exercise 8A

1 a $x = 4$ **b** $w = 14$ **c** $y = 5$
  d $p = 10$ **e** $x = 5$ **f** $x = 6$
  g $z = 24$ **h** $x = 2.5$ **i** $q = 4$
  j $x = 1$ **k** $r = 28$ **l** $s = 12$

2 Any valid equation such as $4x = 16$ or $x + 3 = 7$

3 a Because it only has even numbers.
  b Because it has a minus sign.
  c Because the answer is not 6.

4 $x + 3 = 17, x = 14$

5 $6y = 180, y = 30, 30p$

Exercise 8B

1 a $\leftarrow \div 3 \leftarrow -5 \leftarrow, x = 2$ **b** $\leftarrow \div 3 \leftarrow + 13 \leftarrow, x = 13$
  c $\leftarrow \div 3 \leftarrow + 7 \leftarrow, x = 13$ **d** $\leftarrow \div 4 \leftarrow + 19 \leftarrow, y = 6$
  e $\leftarrow \div 3 \leftarrow - 8, a = 1$ **f** $\leftarrow \div 2 \leftarrow - 8 \leftarrow, x = 3$
  g $\leftarrow \div 2 \leftarrow - 6 \leftarrow, y = 6$ **h** $\leftarrow \div 8 \leftarrow - 4 \leftarrow, x = 1$
  i $\leftarrow \div 2 \leftarrow + 10 \leftarrow, x = 9$ **j** $\leftarrow \times 5 \leftarrow - 2 \leftarrow, x = 5$
  k $\leftarrow \times 3 \leftarrow + 4 \leftarrow, t = 18$ **l** $\leftarrow \times 4 \leftarrow - 1 \leftarrow, y = 24$
  m $\leftarrow \times 2 \leftarrow + 6 \leftarrow, k = 18$ **n** $\leftarrow \times 8 \leftarrow + 4 \leftarrow, h = 40$
  o $\leftarrow \times 6 \leftarrow - 1 \leftarrow, w = 18$ **p** $\leftarrow \times 4 \leftarrow - 5 \leftarrow, x = 8$
  q $\leftarrow \times 2 \leftarrow + 3 \leftarrow, y = 16$ **r** $\leftarrow \times 5 \leftarrow - 2 \leftarrow, f = 30$

2 2

3 27p

Exercise 8C

1 a 56   **b** 2   **c** $6\frac{1}{2}$
  d 3   **e** 4   **f** $2\frac{1}{2}$
  g $3\frac{1}{2}$ **h** $2\frac{1}{2}$ **i** 4
  j 21   **k** 72   **l** 56
  m 0   **n** −7   **o** −18
  p 36   **q** 36   **r** 60

2 a −4
  b 15

3 Any valid equation such as
$3x + 8 = 2, 6x + 20 = 8$

4 a Betsy
  b Second line: Amanda subtracts 1 instead of adding 1; fourth line: Amanda subtracts 2 instead of dividing by 2.

Exercise 8D

1
  a 1   **b** 3   **c** 2
  d 2   **e** 9   **f** 5
  g 6   **h** 4   **i** 2
  j −2   **k** 24   **l** 10
  m 21   **n** 72   **o** 56

2   **a** 33   **b** 48

3   **a** 5   **b** 28   **c** 5
    d 35   **e** 33   **f** 23

4   25

8.2   Solving equations with brackets

Exercise 8E

1
  a 3   **b** 7   **c** 5
  d 3   **e** 4   **f** 6
  g 8   **h** 1   **i** 1.5
  j 2.5   **k** 0.5   **l** 1.2
  m −4   **n** −2

2   Any values that work, e.g. $a = 2$, $b = 3$ and $c = 30$

3   55

4   3.25

8.3   Equations with the variable on both sides

Exercise 8F

1
  a 2   **b** 1   **c** 7
  d 4   **e** 2   **f** −1
  g −2   **h** 2

2 $3x − 2 = 2x + 5$, $x = 7$

3   **a** 6   **b** 11   **c** 1
    d 4   **e** 9   **f** 6

4 $8x + 7 + x + 5 = 11x$
    $+ 5 − x − 5$, $x = 12$

5 $6x + 3 = 6x + 10$,
    $6x − 6x = 10 − 3$, $0 = 7$ which is obviously false. Both sides
    have $6x$, which cancels out.

6   When both sides are expanded you get $12x + 18 = 12x + 18$
    so no matter what value is put in for x it will work.

8.4   Rearranging formulae

Exercise 8G

1 $k = \dfrac{T}{3}$

2 $m = P − 7$

3 $y = X + 1$

4 $p = 3Q$

5   **a** $m = p − t$ **b** $t = p − m$

6 $k = \dfrac{t − 7}{2}$

7 $m = gv$

8 $m = \sqrt{t}$

9 $r = \dfrac{C}{2\pi}$

10 $b = \dfrac{A}{h}$

11 $I = \dfrac{P − 2w}{2}$

12 $p = \sqrt{m − 2}$

13   **a** $5x = 9y + 75$, $y = \dfrac{5x − 75}{9}$
                            b 25p

14   Average speed on first journey
     = 72 k/h. Return 63 k/h,
     taking 2 hours. Held up for 15 min.

15   **a** $−40 − 32 = −72$, $−72 \div 9 = −8$, $5 \times −8 = −40$
     b $68 − 32 = 36$, $36 \div 9 = 4$, $4 \times 5 = 20$
     c $F = \dfrac{9}{5} C + 32$

8.5   Solving linear inequalities

Exercise 8H

1
  a $x < 3$ **b** $t > 8$ **c** $p \geq 10$ **d** $x < 5$ **e** $y \leq 3$ **f** $t > 5$
  g $x < 6$ **h** $y \leq 15$ **i** $t \geq 18$ **j** $x < 7$ **k** $x \leq 3$ **l** $t \geq 5$

2   **a** 8   **b** 6   **c** 16   **d** 3   **e** 7

3   **a** 11   **b** 16   **c** 16   **d** 3   **e** 7

4 $2x + 3 < 20$, $x < 8.50$ so the most was £8.49

5   **a** Because $3 + 4 = 7$ which is less than the third side 8
     b $x + x + 2 > 10$, $2x + 2 > 10$, $2x > 8$, $x > 4$, so smallest
       value of x is 5.

6   **a** $x = 6$ and $x < 3$ scores $−1$ (nothing in common), $x < 3$ and
       $x > 0$ scores $+1$ (1 in common, for example), $x > 0$ and $x =$
       2 scores $+ 1$ (2 in common), $x = 2$ and $x \geq 4$ scores $−1$
       (nothing in common), $−1 + 1 + 1 − 1 = 0$
     b $x > 0$ and $x = 6$ scores $+1$ (6 in common), $x = 6$ and $x \geq 4$
       scores $+1$ (6 in common), $x \geq 4$ and $x = 2$ scores $−1$
       (nothing in common), $x = 2$ and $x < 3$ scores $+1$ (2 in
       common). $+1 + 1 − 1 + 1 = 2$
     c Any acceptable combination for example:
       $x = 2$, $x < 3$, $x > 0$, $x \geq 4$, $x = 6$

Exercise 8I

1
  a $x > 1$ **b** $x \leq 3$ **c** $x < 2$
  d $x \geq −1$ **e** $x \leq −1$ **f** $1 < x \leq 4$

2 **a** $x \leqslant 3$

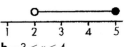

b $x > -2$

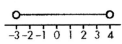

c $x \geqslant 0$

d $x < 5$

e $x \geqslant -1$

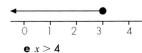

f $2 < x \leqslant 5$

g $-1 \leqslant x \leqslant 3$

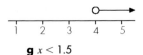

h $-3 < x < 4$

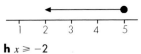

3 **a** $x \geqslant 4$

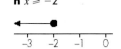

b $x < -2$

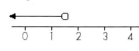

c $x \leqslant 3$

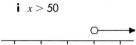

d $x > 3$

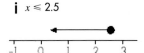

e $x > 4$

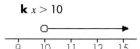

f $x \leqslant 5$

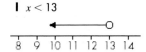

g $x < 1.5$

h $x \geqslant -2$

i $x > 50$

j $x \leqslant 2.5$

k $x > 10$

l $x < 13$

4 Any two inequalities that overlap only on the integers -1, 0, 1 and 2; for example $x \geqslant -1$ and $x < 3$

5 **a** Because 3 apples plus the chocolate bar cost more than £1.20, $x > 22$
   b Because 2 apples plus the chocolate bar left Max with at least 16p change, $x \leqslant 25$

c
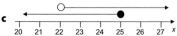

   d Apples could cost 23p, 24p or 25p.

6 Set 1: $3x + 5 > 2$, $\{0, 1, 2\}$, $4x - 1 < 7$
   Set 2: $3x + 1 < 7$, $\{-1, 0, 1\}$, $2x + 7 \geqslant 5$

Examination questions

1 **a** 5   **b** 2   **c** 6   **d** 5

2 **a** 11   **b** 5

3 **a** $z + 4$ (pence)
   b $4z + 4$ (pence)
   c $4z + 4 = 60$ 14p

4 **a i** 4.5   **ii** 7   **iii** 3
   b i $13q$ **ii** $7n + 3p$

5 **a** $2(x + 10.5) = 28$
   b 3.5 cm

6 **a** 7   **b** 5   **c** 6   **d** -2 **e** 12

7 **a** 5   **b** 4   **c** 2.5

8 **a** $2(3x - 5)$
   b 1.25
   c -1
   d $-2, -1, 0, 1$

9 **a** 9
   b False, False, True
   c 10 or any integer greater than 10

10 **a** 3.5   **b** 17

11 **a** $x(x + 4)$
   b $y < 1.5$ **c** $r = \dfrac{p - 3}{2}$

12 $t = 3u$

13 4

14 **a** $r = \dfrac{C}{6}$
   b $r = \sqrt{(A/3)}$

15 **a** $x \leqslant 2$
   b $-2, 1, 0, 1$

16 Left Boundary:
   $-3 \leqslant$ open circle < -2 or $-3 \leqslant$ closed circle < -2
   Right Boundary:
   Closed circle on 3 or line beyond 3 with any termination (e.g. arrow, circle, nothing)

Answers to Chapter 9

9.1 Basic algebra

Exercise 9A

1 a Stu $3n$; Tamara $n + 5$; Ursula $n - 4$; Vic $3n + 6$ **b** $9n + 7$

2 a $35 + z$, $Y + z$

3 a $20x$ **b** $3w^2$ **c** $18h^2$ **d** $12x$ **e** $5z$ **f** $7y^2$
g $11a + 5b$ **h** $3x - 2$

4 50

5 Any expressions such as $2 \times 15w$ or $10w + 20w$

6 45p

7 a $300L + 6S$ **b** 271.2 m

8 a 21 **b** 13 **c** 41 **d** 15 **e** −60 **f** 96

9 a expression (E) **b** formula (F) **c** equation (Q)

10 Any equivalent expressions such as $6x + 9y + 2x - 12y$

11 a £7.10 **b** 4 km **c** No, he needs £13.10

12 a All of them **b** $\frac{1}{2}$

13 a $15 - 5m$ **b** $6x + 21$ **c** $x^2 + 2x$
d $10m - 2m^2$ **e** $5s^2 + 15s$ **f** $3nm - 3np$

14 a $3(6 - m)$ **b** $6(x + 2)$ **c** $x(x + 5)$
d $m(10 - m)$ **e** $3(5s^2 + 1)$ **f** $n(3 - p)$

15 a $-3x - 8y$ **b** $-2a + 4b$

16 a Side AF − side DE $= 4x - 1 - x = 3x - 1$
b $14x$ **c** 84 cm

17 4 cm × 12 cm

18 Any values that work, e.g. $x = 8$, $b = 4$, $h = 32$

19 a $6 + 3 \times 9 - 15 = 18$ **b** $2 \times 6 - 9 + 3 \times 3 = 12$

20 a $\frac{450}{n}$ **b** £390

21 a $12p^3 - 4p^2q$ **b** $10t^4 + 35t^2$ **c** $10x^2 + 35xy$
d $10m^2 - 2m^5$ **e** $8s^4 + 24s^3t$ **f** $6nm^3 - 6n^2m^2$

22 a $4(t - 2)$ and $4t - 8$ **b** £26

23 a $23x + 11$ **b** $9y + 7$ **c** $2x - 8$
d $22x + 9$ **e** $14x^2 - 10x$ **f** $2x^3 + 17x^2 - 9$

24 a $3p(3p + 2t)$ **b** $4m(3p - 2m)$ **c** $4ab(4a + 1)$
d $2(2a^2 - 3a + 1)$ **e** $5xy(4y + 2x + 1)$ **f** $4mt(2t - m)$

25 Darren has added 2 and 3 instead of multiplying and added 2 and −5 instead of multiplying. The correct answer is $6x - 10$

26 a $4(2y + 4)$ **b** $3(2z + 1)$

27 Number off each day continues 81, 243, 729, 2187
Total number off continues 121, 364, 1093, 3280
So by the 8th day there are no students left in school.

28 a $21f + 21s$ **b** $315f + 504s$ **c** £240 profit

29 $6(3x + 5) - 2(x - 2) = 18x + 30 - 2x + 4 = 16x + 34$

30 a Both calculations give the cost of 5 main courses and 5 deserts.
b Easier to work out as bracket evaluates to 10
c £50

31 a B **b** They do not take out the highest common factor.

32 No common factors

9.2 Substitution using a calculator

Exercise 9B

1 a 14.3 **b** 38.8 **c** 5.4

2 a 7.2 **b** 11.4 **c** 9.7

3 a 9.36 **b** 6.69 **c** 3

4 a 28 **b** 10.8 **c** 18

5 a 7.44 **b** 0.61 **c** 1.16

6 a 684 **b** 342 **c** 792

7 a 8.6 **b** 4.8 **c** 8.4

8 a 4.68 **b** 5.02 **c** −3.1

9 a £477.90 **b** £117.90 still owed (debit)

10 a One odd value and one even value, different from each other.
b Any valid combination, e.g. $x = 1$, $y = 2$

11 a 4.1 **b** 8 **c** 4.525

12 a £767.50 **b** £107.50 in debit

13 a x must be 2, y can be any other prime number.
b x must be an odd prime, y can be any other prime number.

14 a First term is cost of petrol, each mile is a tenth of £0.98. Second term is the hire cost divided by the miles.
b 29.8p per mile

9.3 Solving linear equations

Exercise 9C

1 a 5 **b** 17 **c** $7\frac{1}{2}$ **d** 50

2 a $2\frac{1}{2}$ **b** 10 **c** −2 **d** $2\frac{1}{2}$ **e** 20 **f** 60

3 a 13 **b** 19 **c** −1 **d** 41

4 a −3 **b** $2\frac{1}{2}$ **c** 0 **d** $5\frac{1}{2}$

5 a 2   **b** 15   **c** 7   **d** 1

6 a −12   **b** 5   **c** −1   **d** 1

9.4   Setting up equations

Exercise 9D

1 a 9   **b** 8

2 a $\frac{M}{4} - 5 = 7$ **b** £48

3 8 m^2

4 a $8c - 10 = 56$ **b** £8.25

5 a B: 450 cars, C: 450 cars, D: 300 cars   **b** 800   **c** 750

6 Length: 5.5 m, width: 2.5 m, area: 13.75 m^2. Carpet costs £123.75

7 90p

8 a 1.5   **b** 2

9 a 1.5 cm   **b** 6.75 cm^2

10 17

11 3 years old

12 9 years old

13 3 cm

14 5

15 a $4x + 40 = 180$ **b** $x = 35°$

16 a 15   **b** −1   **c** $2(n + 3)$, $2(n + 3) - 5$
   d $2(n + 3) - 5 = n$, $2n + 6 - 5 = n$, $2n + 1 = n$, $n = -1$

17 No as $x + x + 2 + x + 4 + x + 6 = 360$ gives $x = 87°$ so the consecutive numbers (87, 89, 91, 93) are not even but odd.

18 $4x + 18 = 3x + 1 + 50$, $x = 33$. Large bottle 1.5 litres, small bottle 1 litre

9.5   Trial and improvement

Exercise 9E

1 a 4 and 5   **b** 4 and 5   **c** 2 and 3

2 $x = 3.5$

3 $x = 3.7$
4 $x = 2.5$

5 $x = 1.5$

6 a $x = 2.4$ **b** $x = 2.8$ **c** $x = 3.2$

7 a $x = 7.8$ cm   **b** width is 7.8 cm and length is 12.8 cm

8 $x = 5.8$

9 Volume = $x \times 2x(x + 8) = 500$, $x^3 + 8x^2 = 250$, $4 \Rightarrow 192$, $5 \Rightarrow 325$, $4.4 \Rightarrow 240.064$, $4.5 \Rightarrow 253.125$, $4.45 \Rightarrow 246.541125$, so dimensions are 4.5 cm, 9 cm and 12.5 cm

10 a Volume of cube is x^3, volume of hole is $\frac{x}{2} \times \frac{x}{2} \times 8 = 2x^3$. Cube minus hole is 1500

   b $12 \Rightarrow 1440$, $13 \Rightarrow 1859$, $12.1 \Rightarrow 1478.741$, $12.2 \Rightarrow 1518.168$, $12.15 \Rightarrow 1498.368375$ so the value of $x =$ 12.2 (to 1 dp)

11 2.76 and 7.24

Examination questions

1 a T   **b** T   **c** F

2 a 6   **b** $5(x + 2)$

3 a i 2 kg   **ii** 1 kg   **b** 2.5 kg

4 a $5c$
   b i 12   **ii** 20   **iii** 32   **iv** 2.5

5 a $6x + 5$
   b 15
   c i $1 \times 3 + 4 = 7$, $1 \times (3 + 4) = 7$
     ii $a(b + c) = a \times b + a \times c = ab + ac \neq ab + c$

6 a $x + 5$ cm
   b $x - 2$ cm
   c $2x$ cm
   d 90 cm

7 a 20.1 cm^2
   b i $3R + S$
     ii $2R = S$

8 −3.34

9 a $y + 5$
   b $2y$
   c $4y + 5$
   d $4y + 5 = 77$, $4y = 72$, $y = 18$

10 $y = 5$, $5z = 20$

11 a £220
   b 250

12 $\frac{2}{3}$

13 2.4

14 3.7

Answers to Chapter 10

10.1 Addition rule for events

1 a $\frac{1}{6}$ **b** $\frac{1}{6}$ **c** $\frac{1}{3}$

2 a $\frac{1}{4}$ **b** $\frac{1}{4}$ **c** $\frac{1}{2}$

3 a $\frac{2}{11}$ **b** $\frac{4}{11}$ **c** $\frac{6}{11}$

4 a $\frac{1}{3}$ **b** $\frac{2}{5}$ **c** $\frac{11}{15}$ **d** $\frac{11}{15}$ **e** $\frac{1}{3}$

5 a 0.6 **b** 120

6 a 0.8 **b** 0.2

7 a $\frac{17}{20}$ **b** $\frac{2}{5}$ **c** $\frac{3}{4}$

8 Because these are three separate events. Also, probability cannot exceed 1.

9 $\frac{3}{4}$

10 $\frac{8}{45}$

11 The probability for each day stays the same, at $\frac{1}{4}$.

12 a The choices of drink and snack are not connected.
b (C, D), (T, G), (T, R), (T, D), (H, G), (H, R), (H, D)
c A possibility is not the same as a probability.
Nine possibilities do not mean a probability of $\frac{1}{9}$.
d i (C, D) = 0.12, (T, G) = 0.15, (T, R) = 0.05,
(T, D) = 0.3, (H, G) = 0.09, (H, R) = 0.03, (H, D) = 0.18
ii The total is 1 as this covers all the possibilities.

10.2 Experimental probability

1 a B **b** B **c** C **d** A **e** B **f** A
g B **h** B

2 a 0.2, 0.08, 0.1, 0.105, 0.148, 0.163, 0.1645
b 6 **c** 1 **d** $\frac{1}{6}$ **e** 1000

3 a 0.095, 0.135, 0.16, 0.265, 0.345
b 40 **c** No; all numbers should be close to 40.

4 a 0.2, 0.25, 0.38, 0.42, 0.385, 0.3974 **b** 8

5 a 6 **b, c** Student to provide own answers.

6 a Caryl, threw the greatest number of times.
b 0.39, 0.31, 0.17, 0.14
c Yes; all answers should be close to 0.25.

7 a not likely **b** impossible **c** not likely
d certain **e** impossible **f** 50–50 chance
g 50–50 chance **h** certain **i** quite likely

8 The missing top numbers are 4 and 5, the bottom two numbers are both likely to be close to 20.

9 Thursday

10 Although he might expect the probability to be close to $\frac{1}{2}$, giving 500 heads, the actual number of heads is unlikely to be exactly 500, but should be close to it.

10.3 Combined events

1 a 7 **b** 2 and 12
c $\frac{1}{36}, \frac{1}{18}, \frac{1}{12}, \frac{1}{9}, \frac{5}{36}, \frac{1}{6}, \frac{5}{36}, \frac{1}{9}, \frac{1}{12}, \frac{1}{18}, \frac{1}{36}$
d i $\frac{1}{12}$ **ii** $\frac{1}{3}$ **iii** $\frac{1}{2}$ **iv** $\frac{7}{36}$ **v** $\frac{5}{12}$ **vi** $\frac{5}{18}$

2 a $\frac{1}{12}$ **b** $\frac{11}{36}$ **c** $\frac{1}{6}$ **d** $\frac{5}{9}$

3 a $\frac{1}{36}$ **b** $\frac{11}{36}$ **c** $\frac{5}{18}$

4 **Score on second dice**

6	5	4	3	2	1	0	
5	4	3	2	1	0	1	
4	3	2	1	0	1	2	
3	2	1	0	1	2	3	
2	1	0	1	2	3	4	
1	0	1	2	3	4	5	
	1	2	3	4	5	6	

Score on first dice

a $\frac{5}{18}$ **b** $\frac{1}{6}$ **c** $\frac{1}{9}$ **d** 0 **e** $\frac{1}{2}$

5 a $\frac{1}{4}$ **b** $\frac{1}{2}$ **c** $\frac{3}{4}$ **d** $\frac{1}{4}$

6 **Score on second spinner**

5	6	7	8	9	10
4	5	6	7	8	9
3	4	5	6	7	8
2	3	4	5	6	7
1	2	3	4	5	6
	1	2	3	4	5

Score on first spinner

a 6 **b i** $\frac{4}{25}$ **ii** $\frac{13}{25}$ **iii** $\frac{1}{5}$ **iv** $\frac{3}{5}$

7 $\frac{8}{64} = \frac{1}{8}$

8 It will show all the possible products.

9 a

Type of rose

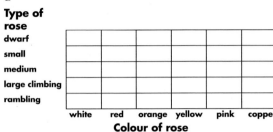

dwarf
small
medium
large climbing
rambling

white   red   orange   yellow   pink   copper
Colour of rose

b i $\frac{6}{30} = \frac{1}{5}$ **ii** $\frac{4}{5}$

10 impossible: no dice; very unlikely: 5 or more dice; unlikely: 4 dice; evens: not possible; likely: 3 dice; very likely: 2 dice; certain: 1 dice

10.4   Expectation

1 a $\frac{1}{6}$ **b** 25

2 a $\frac{1}{2}$ **b** 1000

3 a i $\frac{1}{2}$ **ii** $\frac{1}{13}$ **iii** $\frac{1}{4}$ **iv** $\frac{1}{52}$

  b i 260   **ii** 40   **iii** 130   **iv** 10

4 a $\frac{1}{37}$ **b** 5

5 a 150   **b** 100   **c** 250   **d** 0

6 a 167   **b** 833

7 1050

8 a 10, 10, 10, 10, 10, 10   **b** 3.5
  c Find the average of the scores (21 ÷ 6)

9 a 0.111   **b** 40

10 281 days

11 Multiply the number of plants by 0.003.

10.5   Two-way tables

1 a Everton **b** Man Utd, Everton, Liverpool **c** Leeds

2 a

		Shaded	Unshaded
Shape	Circle	3	3
	Triangle	2	2

b $\frac{1}{2}$

3 a 40   **b** 16   **c** 40%   **d** 40%   **e** 16

4 a

		No. on disc		
		4	**5**	**6**
Letter on card	**A**	3	4	5
	B	4	5	6
	C	5	6	7

b $\frac{4}{9}$ **c** $\frac{1}{3}$

5 a 23   **b** 20%   **c** $\frac{4}{25}$ **d** 480

6 a 10   **b** 7   **c** 14%   **d** 15%

7 a

		Spinner A			
		1	**2**	**3**	**4**
Spinner B	**5**	6	7	8	9
	6	7	8	9	10
	7	8	9	10	11
	8	9	10	11	12

b 4   **c i** $\frac{1}{4}$ **ii** $\frac{3}{16}$ **iii** $\frac{1}{4}$

8 a 6   **b** 16   **c** 34   **d** $\frac{13}{15}$

9 a

		Number on dice					
		1	**2**	**3**	**4**	**5**	**6**
Coin	**H**	1	2	3	4	5	6
	T	2	4	6	8	10	12

b 2 (1 and 4)   **c** $\frac{1}{4}$

10

	Men	Women	Children
Left footed	4	3	5
Right footed	21	12	15

11 a Those from the greenhouse have a larger mean diameter.
  b Those from the garden have a smaller range, so are more consistent.

12 Either Reyki, because she had bigger tomatoes, or Daniel, because he had more tomatoes.

13 $\frac{22}{36} = \frac{11}{18}$

14 a Score on
second
spinner

	10	10	11	13	15	17	19
	8	8	9	11	13	15	17
	6	6	7	9	11	13	15
	4	4	5	7	9	11	13
	2	2	3	5	7	9	11
	0	0	1	3	5	7	9
		0	1	3	5	7	9

Score on first spinner

b 9   **c** 0   **d** $\frac{15}{36} = \frac{5}{12}$ **e** $\frac{30}{36} = \frac{5}{6}$

Examination questions

1 a

	C	D
Boy	3	2
Girl	1	4

b 0.7

2 a 1H, 1T, 2H, 2T, 3H, 3T, 4H, 4T, 5H, 5T, 6H, 6T
  b $\frac{1}{4}$ **c** $\frac{3}{4}$

3 a $\frac{3}{10}$ **b** $\frac{13}{20}$ **c** 6

4 a W $\frac{1}{4}$, G $\frac{9}{20}$, B $\frac{3}{10}$ **b** Second, as there are more trials

Answers to Chapter 11

11.1 Patterns in number

Exercise 11A

1 $11111 \times 11111 = 123454321$,
  $111111 \times 111111 = 12345654321$

2 $99999 \times 99999 = 9999800001$,
  $999999 \times 999999 = 999998000001$

3 $7 \times 8 = 7^2 + 7, 8 \times 9 = 8^2 + 8$

4 $50 \times 51 = 2550, 60 \times 61 = 3660$

5 $1 + 2 + 3 + 4 + 5 + 4 + 3 + 2 + 1 = 25 = 5^2$,
  $1 + 2 + 3 + 4 + 5 + 6 + 5 + 4 + 3 + 2 + 1 = 36 = 6^2$

6 $21 + 23 + 25 + 27 + 29 = 125 = 5^3$,
  $31 + 33 + 35 + 37 + 39 + 41 = 216 = 6^3$

7 $1 + 6 + 15 + 20 + 15 + 6 + 1 = 64$,
  $1 + 7 + 21 + 35 + 35 + 21 + 7 + 1 = 128$

8 $12\,345\,679 \times 45 = 555\,555\,555$,
  $12\,345\,679 \times 54 = 666\,666\,666$

9 $1^3 + 2^3 + 3^3 + 4^3 = (1 + 2 + 3 + 4)^2 = 100$,
  $1^3 + 2^3 + 3^3 + 4^3 + 5^3 = (1 + 2 + 3 + 4 + 5)^2 = 225$

10 $36^2 + 37^2 + 38^2 + 39^2 + 40^2 = 41^2 + 42^2 + 43^2 + 44^2$,
  $55^2 + 56^2 + 57^2 + 58^2 + 59^2 + 60^2 = 61^2 + 62^2 + 63^2 + 64^2 + 65^2$

11 12345678987654321

12 999999998000000001

13 $12^2 + 12$

14 8190

15 $81 = 9^2$

16 $512 = 8^3$
17 512

18 999 999 999

19 $(1 + 2 + 3 + 4 + 5 + 6 + 7 + 8 + 9)^2 = 2025$

20 X. There are 351 (1 + 2 + ... + 25 + 26) letters from A to Z.
  $3 \times 351 = 1053$. $1053 - 26 = 1027$, $1027 - 25 = 1002$, so,
  as Z and Y are eliminated, the 1000th letter must be X.

11.2 Number sequences

Exercise 11B

1 a 9, 11, 13: add 2
  b 10, 12, 14: add 2
  c 80, 160, 320: double
  d 81, 243, 729: multiply by 3
  e 28, 34, 40: add 6
  f 23, 28, 33: add 5
  g 20 000, 200 000, 2 000 000: multiply by 10
  h 19, 22, 25: add 3
  i 46, 55, 64: add 9
  j 405, 1215, 3645: multiply by 3
  k 18, 22, 26: add 4
  l 625, 3125, 15 625: multiply by 5

2 a 16, 22   **b** 26, 37   **c** 31, 43
  d 46, 64   **e** 121, 169   **f** 782, 3907
  g 22 223, 222 223   **h** 11, 13   **i** 33, 65
  j 78, 108

3 a 48, 96, 192   **b** 33, 39, 45   **c** 4, 2, 1
  d 38, 35, 32   **e** 37, 50, 65   **f** 26, 33, 41
  g 14, 16, 17   **h** 19, 22, 25   **i** 28, 36, 45
  j 5, 6, 7   **k** 0.16, 0.032, 0.0064
  l 0.0625, 0.031 25, 0.015 625

4 a 21, 34: add previous 2 terms
  b 49, 64: next square number
  c 47, 76: add previous 2 terms
  d 216, 343: cube numbers

5 15, 21, 28, 36

6 61, 91, 127

7 364: Daily totals are 1, 3, 6, 10, 15, 21, 28, 36, 45, 55, 66,
  78 (these are the triangle numbers). Cumulative totals are: 1, 4,
  10, 20, 35, 56, 84, 120, 165, 220, 286, 364.

8 29, 41

9 No, they both increase by the same number (3).

10 a 89 km
  b 42 miles

11.3 The nth term of a sequence

Exercise 11C

1 **a** 3, 5, 7, 9, 11
 b 1, 4, 7, 10, 13
 c 7, 12, 17, 22, 27
 d 1, 4, 9, 16, 25
 e 4, 7, 12, 19, 28

2 **a** 4, 5, 6, 7, 8
 b 2, 5, 8, 11, 14
 c 3, 8, 13, 18, 23
 d 0, 3, 8, 15, 24
 e 9, 13, 17, 21, 25

3 $\frac{2}{4}, \frac{3}{5}, \frac{4}{6}, \frac{5}{7}, \frac{6}{8}$

4 **a** 6, 10, 15, 21, 28
 b Triangular numbers

5 **a** £305
 b £600
 c 3
 d 5

6 **a** $\frac{3}{4}, \frac{5}{7}, \frac{7}{10}$

 b i 0.6666667778
 ii $\frac{2}{3}$

 iii ignore the constant term (1) and just cancel the ns,
 i.e. $\frac{2n+1}{3n+1} \approx \frac{2n}{3n} = \frac{2}{3}$

7 Write $3n + 7 = 4n - 2$ and solve for n.

8 **a** 2, 6, 24, 720
 b 69! (but this does depend on the calculator)

Exercise 11D

1 **a** 13, 15, $2n + 1$
 b 25, 29, $4n + 1$
 c 33, 38, $5n + 3$
 d 32, 38, $6n - 4$
 e 20, 23, $3n + 2$
 f 37, 44, $7n - 5$
 g 21, 25, $4n - 3$
 h 23, 27, $4n - 1$
 i 17, 20, $3n - 1$
 j 42, 52, $10n - 8$
 k 24, 28, $4n + 4$
 l 29, 34, $5n - 1$

2 **a** $3n + 1$, 151
 b $2n + 5$, 105
 c $5n - 2$, 248
 d $4n - 3$, 197
 e $8n - 6$, 394
 f $n + 4$, 54
 g $5n + 1$, 251
 h $8n - 5$, 395
 i $3n - 2$, 148
 j $3n + 18$, 168
 k $7n + 5$, 355
 l $8n - 7$, 393

3 **a i** $4n + 1$ **ii** 401
 b i $2n + 1$ **ii** 201
 c i $3n + 1$ **ii** 301
 d i $2n + 6$ **ii** 206
 e i $4n + 5$ **ii** 405
 f i $5n + 1$ **ii** 501
 g i $3n - 3$ **ii** 297
 h i $6n - 4$ **ii** 596
 i i $8n - 1$ **ii** 799
 j i $2n + 23$ **ii** 223

4 **a** $8n + 2$
 b $8n + 1$
 c $8n$
 d £8

5 **a** 31, 33, 35, ... goes up in 2s, so the nth term is $2n \pm a$
 and $2 + 29 = 31$
 b $n + 108$
 c $\frac{2n + 29}{n + 108} \approx \frac{2n}{n} = 2$
 d The 79th term, i.e. when $2n + 29 = n + 108$, $n = 79$

6 **a** 36, 49, 64, 81, 100
 b i $n^2 + 1$ **ii** $2n^2$ **iii** $n^2 - 1$

11.4 General rules from given patterns

Exercise 11E

1 **a** appropriate diagram
 b $4n - 3$
 c 97
 d 50th diagram

2 **a** appropriate diagram
 b $2n + 1$
 c 121
 d 49th set

3 **a** 18
 b $4n + 2$
 c 12

4 **a i** 20 cm **ii** $(3n + 2)$ cm **iii** 152 cm
 b 332

5 **a i** 20 **ii** 162
 b 79.8 km

6 Formula is $\left(\frac{3}{4}\right)^n$.
 Picking a large value of n gives an answer of 0 on a calculator,
 so eventually the whole area would be covered.

7 Yes, as the number of matches is 12, 21, 30, 39, ..., which is $9n$
 $+ 3$, so he will need $9 \times 20 + 3 = 183$ matches for the 20th
 step and he has $5 \times 42 = 210$ matches.

8 **a** 2^n
 b i The quantity doubles **ii** 1600 ml

9 **a** They are the same

10 **a** even
 b odd
 c odd

Examination questions

1 **a** 15, 11
 b Subtract 4
 c Sequence continues 7, 3, -1, -5

2 **a i** 30, 24 **ii** 32, 64
 b Add 2, 4, 8, 16, etc.

3 **a** Correct drawing with 7 squares, 4 on top and 3 below
 b 11

4 **a** Always even
 b Could be either odd or even.

5 6, 9, 14

6 **a** 1, -1
 b i 7, 6.5 **ii** Yes, as this is the same as $(a + b) \div 2$

7 **a i** 5, 9, 13 **ii** All numbers are odd
 b $(1 + 3)^2 - 9 = 4^2 - 9 = 16 - 9 = 7$

8 **a** Even
 b Odd

9 **a** $\frac{5 \times 6}{2}$

 b $1 + 2 + 3 + 4 + 5 + 6 = \frac{6 \times 7}{2}$

 c $1 + ... + 24 = \frac{24 \times 25}{2} = 300$

10 a i 25
   b square numbers
   c 100

11 The square of any positive fraction less than 1, for example

12 a For any value of n, $2n$ is even, so $2n + 1$ is odd
   b The square of an odd number will be odd, the square of an even number will be even

13 S, A, N

14 a i 31, add 6   **ii** -1, subtract 3
   b $4n + 2$

Answers to Chapter 12

12.1   Units of volume

Exercise 12A

1 a 12 cm³   **b** 20 m³
   c 23 cm³   **d** 32 cm³

2 She reached this conclusion by multiplying 5 3 5 3 4.

3 65 cm²

4 6

12.2   Surface area and volume of a cuboid

Exercise 12B

1 a i 198 cm³   **ii** 234 cm²
  b i 90 cm³   **ii** 146 cm²
  c i 1440 cm³   **ii** 792 cm²
  d i 525 cm³   **ii** 470 cm²

2 24 litres

3 a 160 cm³   **b** 480 cm³   **c** 150 cm³

4 a i 64 cm³   **ii** 96 cm²
  b i 343 cm³   **ii** 294 cm²
  c i 1000 mm³   **ii** 600 mm²
  d i 125 m³   **ii** 150 m²
  e i 1728 m³   **ii** 864 m²

5 86

6 a 180 cm³   **b** 5 cm   **c** 6 cm
   d 10 cm   **e** 81 cm³

7 1.6 m

8 48 m²

9 a 3 cm   **b** 5 m   **c** 2 mm
   d 1.2 m

10 a 148 cm³   **b** 468 cm³

11 If this was a cube, the side length woud be 5 cm, so total surface area would be $5 \times 5 \times 6 = 150$ cm²; no this particular cuboid is not a cube.

12 600 cm²

12.3   Surface area and volume of a prism

Exercise 12C

1 a i a 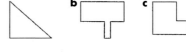   **b**   **c**

   d   **e**   **f**

  ii a 21 cm²   **b** 48 cm²   **c** 36 m²
     d 108 m²   **e** 25 m²   **f** 111 m²
  iii a 63 cm³   **b** 432 cm³   **c** 324 m³
     d 432 cm³   **e** 225 cm³   **f** 1332 m³

2 a i 21 cm² **ii** 210 cm²   **b i** 54 cm² **ii** 270 cm²

3 525 000 litres

4 solid **b** has greater volume (900 cm³) than solid **a** (594 cm³)

5 a 75 m³   **b** 75 000 litres   **c** to be supplied

6 384 cm²

7 Multiply the volume by the weight of 1 cm³ of the metal, then convert to kg.

12.4   Volume of a cylinder

Exercise 12D

1 a 251 cm³   **b** 445 cm³   **c** 2150 cm³   **d** 25 m³

2 a 226 cm³   **b** 15 cm³   **c** 346 cm³   **d** 1060 cm³

3 £80

4 332 litres

5 366 ml

6 2827 cm³

7 a 360π cm³   **b** 300π cm³

8 17.19 km

9 Volume of A $= 2^2 \times \pi \times 5 = 20\pi$, volume of B $= 1^2 \times \pi \times 21 = 21\pi$, B has the larger volume.

Examination questions

1 a 4 cm^3 **b** 18 cm^2

2 34 cm^2

3 3.75 cm

4 a $\frac{1}{2} \times \pi \times 1.4^2 = 3.1$ m^2 **b** $3.1 \times 0.5 = 1.5$ m^3

5 $\pi r^2 h = \pi \times 8^2 \times 5 = 320\pi$ cm^3

6 a 200 cm^2 **b** 132 cm^2

Answers to Chapter 13

13.1 Translations

Exercise 13A

1 Check students' translations.

2 Check students' translations.

3 a i $\begin{pmatrix} 1 \\ 3 \end{pmatrix}$ **ii** $\begin{pmatrix} 4 \\ 2 \end{pmatrix}$ **iii** $\begin{pmatrix} 2 \\ -1 \end{pmatrix}$

iv $\begin{pmatrix} 5 \\ 1 \end{pmatrix}$ **v** $\begin{pmatrix} -1 \\ 6 \end{pmatrix}$ **vi** $\begin{pmatrix} 4 \\ 6 \end{pmatrix}$

b i $\begin{pmatrix} -1 \\ -3 \end{pmatrix}$ **ii** $\begin{pmatrix} 3 \\ -1 \end{pmatrix}$ **iii** $\begin{pmatrix} 1 \\ -4 \end{pmatrix}$

iv $\begin{pmatrix} 4 \\ -2 \end{pmatrix}$ **v** $\begin{pmatrix} -2 \\ 3 \end{pmatrix}$ **vi** $\begin{pmatrix} 3 \\ 3 \end{pmatrix}$

c i $\begin{pmatrix} -4 \\ -2 \end{pmatrix}$ **ii** $\begin{pmatrix} -3 \\ 1 \end{pmatrix}$ **iii** $\begin{pmatrix} -2 \\ -3 \end{pmatrix}$

iv $\begin{pmatrix} 1 \\ -1 \end{pmatrix}$ **v** $\begin{pmatrix} -5 \\ 4 \end{pmatrix}$ **vi** $\begin{pmatrix} 0 \\ 4 \end{pmatrix}$

d i $\begin{pmatrix} 3 \\ 2 \end{pmatrix}$ **ii** $\begin{pmatrix} -4 \\ 2 \end{pmatrix}$ **iii** $\begin{pmatrix} 5 \\ -4 \end{pmatrix}$

iv $\begin{pmatrix} -2 \\ -7 \end{pmatrix}$ **v** $\begin{pmatrix} 5 \\ 0 \end{pmatrix}$ **vi** $\begin{pmatrix} 1 \\ -5 \end{pmatrix}$

4

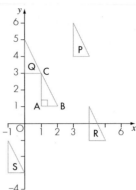

5 a $\begin{pmatrix} -3 \\ -1 \end{pmatrix}$ **b** $\begin{pmatrix} 4 \\ -4 \end{pmatrix}$ **c** $\begin{pmatrix} -5 \\ -2 \end{pmatrix}$ **d** $\begin{pmatrix} 4 \\ 7 \end{pmatrix}$

e $\begin{pmatrix} -1 \\ 5 \end{pmatrix}$ **f** $\begin{pmatrix} 1 \\ 6 \end{pmatrix}$ **g** $\begin{pmatrix} -4 \\ 4 \end{pmatrix}$ **h** $\begin{pmatrix} -4 \\ -7 \end{pmatrix}$

6 $10 \times 10 = 100$ (including $\begin{pmatrix} 0 \\ 0 \end{pmatrix}$)

7 Check students' designs for a Snakes and ladders board.

8 $\begin{pmatrix} -x \\ -y \end{pmatrix}$

13.2 Reflections

Exercise 13B

1 a

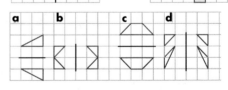

b

c

d

2

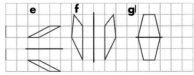

3 **a** **b** **c**

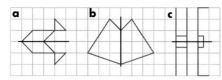

4

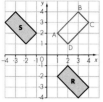

c congruent

5 a-e

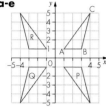

f reflection in y-axis

6

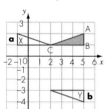

7 **a** **b** **c**

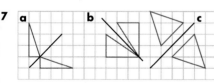

8 **a** **b** **c**

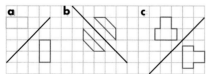

9 Possible answer:

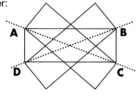

0 Possible answer: Take the centre square as ABCD then reflect this square each time in the line, AB, then BC, then CD and finally AD.

1 Possible answer: A reflection of a reflection in a line will always return to its starting position as each reflected point is the same perpendicular distance from the mirror line as the original.

12 a–i

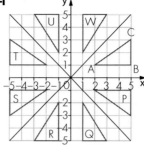

j A reflection in the line $y = x$.

13 **b** yes **c** yes

13.3 Rotations

Exercise 13C

1 **a** **b** **c** **d**

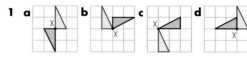

2

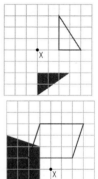

d 90° turn clockwise about O

3

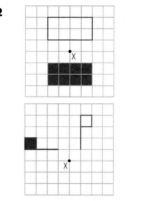

4 **a** A(1, 2), B(3, 1), C(4, 3) **b** (2, −1), (1, −3), (3, −4)
   c (−1, −2), (−3, −1), (−4, −3)
   d (−2, 1), (−1, 3), (−3, 4)
   e Corresponding vertices have same pairs of numbers switching round and changing signs.

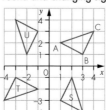

5 a

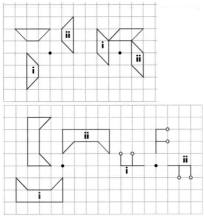

b Rotation 90° anticlockwise.
Rotation 270° clockwise.

6

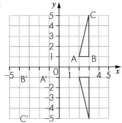

Centre point is C

7 Possible answer: If ABCD is the centre square with A being bottom left and B bottom right, rotate about A 90° anticlockwise, rotate about new B 180°, now rotate about new C 180° and finally rotate about new D 180°.

8 4 × 90° is 360°, a complete turn brings the shape back to the original place.

9 a, b

c a rotation of 90° about the point (0, −1)

10 a–c

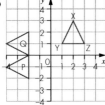

d rotation 180° about O   **e** yes   **f** yes

13.4 Enlargements

1

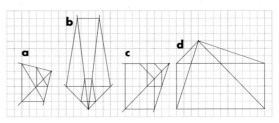

2 a

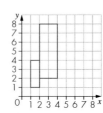

c

3 a

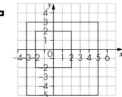

b

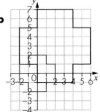

4 a-d They are all congruent.

5

6 No, it is false, congruent shapes must be identical in size.

7 9

8

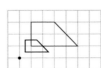

9 a

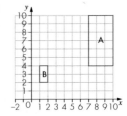

b 3 : 1
c 3 : 1
d 9 : 1

Examination questions

1 a A and C
b Check students' diagrams.

2 a Rotation, clockwise 90° (or anticlockwise 270°), centre (1, 1)
b Check students' diagrams. Ensure that centre is (0, 7).

3 a Reflection in the line $x = 0$ (the y-axis).
b Check students' diagrams.
c Check students' diagrams. Ensure that centre is (0, 1).

4 a Rotation through 90° anticlockwise (or 270° clockwise) about (0, 0).
b Check students' diagrams. Ensure shape B is reflected in the line $y = -1$

5 a Check students' diagrams.
b Ensure shape B is reflected in the line $y = 1$
c Check students' diagrams.

Answers to Chapter 14

14.1 Drawing circles

Exercise 14A

1 a $1\frac{1}{2}$ cm, 3 cm **b** 2 cm, 4 cm **c** 3 cm, 6 cm

2 Student's circles

3 a diameter **b** chord **c** radius
d sector

4 Student's drawings

5 Student's drawings

6 a–c Student's drawing **d** regular hexagon

7 a–c Student's drawing **d** 90°
e Student's drawings
f A radius is perpendicular to a tangent at a point
It may be worth some discussion to point out that it is always true that a radius meets a tangent at right angles.

8 a 6 cm **b** 12 cm

14.2 The circumference of a circle

Exercise 14B

1 a 25.1 cm **b** 15.7 cm **c** 44.0 cm
d 22.0 cm **e** 18.8 cm **f** 47.1 cm
g 28.9 cm **h** 14.8 cm

2 a 6.3 cm **b** 8.2 cm **c** 5.3 cm
d 7.5 cm

3 a 31.4 cm **b** 18.8 cm **c** 9.4 cm
d 25.1 cm **e** 5.7 cm **f** 15.7 cm
g 81.7 cm **h** 39.6 cm

4 a 198 cm **b** 505

5 a A 188.5 m, B 194.8 m, C 201.1 m, D 207.3 m
b 18.7 or 18.8 m

6 879.6 or 880 cm

7 a 37.7 cm **b** 3770 cm **c** 37.7 m
d 37.7 km

8 100 cm

9 24.2 cm

10 15.9 cm

11 b, 25.7 cm

12 a Sue 62.8 cm, Julie 69.1 cm, Dave 75.4 cm, Brian 81.7 cm
b The difference between the distances round the waists of two people is 2π times the difference between their radii.
c 6.28 m

13 a Perimeters of shapes A and B are both 25.1 cm
b 25.1 cm

14 $4a = 2\pi r$, so $2a = \pi r$, therefore $r = \dfrac{2a}{\pi}$

15 11

14.3 The area of a circle

Exercise 14C

1 a 78.5 cm² **b** 28.3 cm² **c** 7.1 cm²
d 50.3 cm² **e** 2.5 cm² **f** 19.6 cm²
g 530.9 cm² **h** 124.7 cm²

2 a 3.1 cm² **b** 5.3 cm² **c** 2.3 cm²
d 4.5 cm²

3 a 50.3 cm² **b** 19.6 cm² **c** 153.9 cm²
d 38.5 cm² **e** 28.3 cm² **f** 176.7 cm²
g 66.5 cm² **h** 17.3 cm²

4 a 9.1 cm² **b** 138 **c** 2000 cm²
d 1255.8 cm² or 1252.9 cm² using unrounded answer from **a**
e 744.2 cm² or 747.1 cm², using unrounded answer from **a**

5 3848.5 m²

6 a i 56.5 cm **ii** 254.5 cm²
b i 69.1 cm **ii** 380.1 cm²
c i 40.8 cm **ii** 132.7 cm²
d i 88.0 cm **ii** 615.8 cm²

7 a 19.1 cm **b** 9.5 cm
c 286.5 cm² (or 283.5 cm²)

8 962.9 cm² (or 962.1 cm²)

9 a 56.5 cm² **b** 19.6 cm²

10 a 50.3 m² **b** 44.0 cm² **c** 28.3 cm²

11 141.4 cm²; $A = \pi \times 9^2 - \pi \times 6^2 = 141.4$ cm²

12 $a^2 = \pi r^2$, so $r^2 = \dfrac{a^2}{\pi}$, therefore $r = \dfrac{a}{\sqrt{\pi}}$

13 21.5 cm²

14.4 Answers in terms of π

Exercise 14D

1 10π cm

2 a 4π cm   **b** 20π cm   **c** 15π cm
    d 4π cm

3 a 16π cm^2 **b** 25π cm^2 **c** 9π cm^2
    d 81π cm^2

4 25 cm

5 10 cm

6 $\frac{200}{\pi}$ cm

7 $\frac{5}{\sqrt{\pi}}$ cm

8 a 12.5π cm^2 **b** 16π cm^2 **c** (16π + 80) cm^2
    d (50π + 100) cm^2

9 a 32π cm^2 **b** 16π cm^2 **c** 8π cm^2
    d 4π cm^2

10 a (200 − 8π) m^2 **b** 18

11 9π cm^2; $\frac{1}{2}(\pi \times 6^2) - \pi \times 3^2 = 18\pi - 9\pi = 9\pi$

12 c; $80 + \frac{1}{4}(\pi \times 8^2) = 80 + 16\pi$

Examination questions

1 a Check students' diagrams. The diameter should be a straight
line drawn from one side of the circle to the other, passing
through the centre point, 0.
   b Check students' diagrams. The tangent should be a straight
line that just touches the circumference at one point only.
   c i Check students' diagrams. Ensure midpoint is marked
accurately.
     ii it is a right angle

2 a 4 cm
   b 110°
   c Check students' diagrams. The line of symmetry should be the
bisector of angle AOB (midway between A and B).
   d Check students' diagrams. The tangent should touch the
circumference at A and be at 90° to AO.
   e Check students' diagrams. The chord should be a straight line
joining points A and B.

3 a 22.3 cm   **b** 224

4 91.6 m^2

5 9π m^2

6 392 m

7 a 14.1 cm^2 **b** No, it has 1 line of symmetry

15.1 Bisectors

Exercise 15A

1–9 No answers required.

10 Coventry or Leicester.

11 The centre of the circle.

12 Start with a base line AB, then construct a perpendicular to the
line at point A. At the point B, construct an angle of 60°. Ensure
that the line for this 60° angle crosses the perpendicular line;
where they meet will be the final point C.

15.2 Loci

Exercise 15B

1 Circle with radius   **a** 2 cm   **b** 4 cm   **c** 5 cm

2 a   **b**

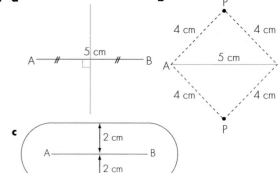

   c

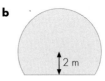

3 a Circle with radius 4m   **b**

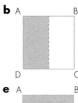

4 a   **b**   **c**
   d   **e**   **f**

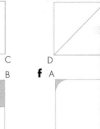

5 Diagram **c**

6 -----------------------------

7 a All points such that AP < PB.
  b The perpendicular bisector of the line BC.

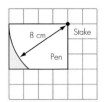

8 Start with a base line, AB, 3 cm long. At point A, draw a few points all 3 cm away from A towards the upper right side. Lightly join these dots with an arc. You can now find the point C that is 3 cm away from point B and draw your equilateral triangle.

Exercise 15C

1

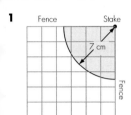

2

3

4

5

6

7

8

9

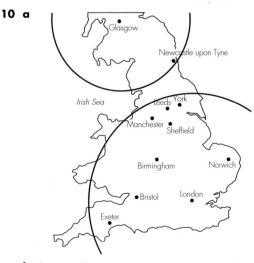

10 a

b No   **c** Yes

11 a No

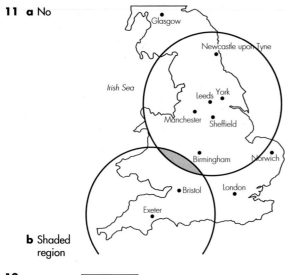

b Shaded region

12

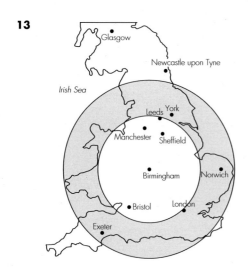

a The line   **b** The region   **c** This part of line

13

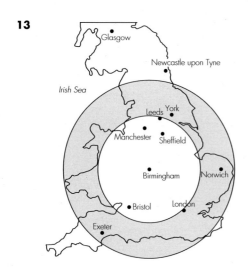

14

15

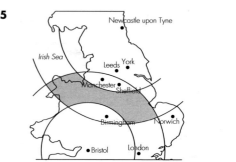

16

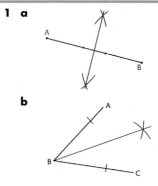

17 Leeds

18 On a map, draw a straight line from Newcastle to Bristol, construct the line bisector, then the search will be anywhere on the sea along than line.

Examination questions

1 a

b

2 a Check students' diagrams.
b 1.85 kg

3 a

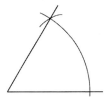

b Check students' diagrams. The diagram should comprise a circular arc with a radius of 6 km, with the centre at point A and another circular arc with a radius of 8 km, with the cetnre at point B. The intersection must be shaded.
Scale: 1 cm represents 1 km

4 a 20 km
b Check students' diagrams.

5 Check students' diagrams.

6 Check students' diagrams.

7

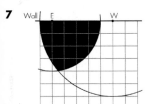

---

7 Points plotted and joined should give parabolas

8 a 150 m
b

t (s)	1	2	3	4	5	6	7	8	9	10
v (m/s)	27	24	21	18	15	12	9	6	3	0
s (m)	28.5	54	76.5	96	112.5	126	136.5	138	148.5	150

c Check students' graphs. **d** 5.5

9 a x: −3, −2, −1, 0, 1, 2, 3
  x^2: 9, 4, 1, 0, 1, 4, 9
  +2x: −6, −4, −2, 0, 2, 4, 6
  −1: −1, −1, −1, −1, −1, −1, −1
  y: 2, −1, −2, −1, 2, 7, 14
b 0.25 or 0.3 **c** −2.7, 0.7

10 a x: −4, −3, −2, −1, 0, 1, 2, 3, 4
  y: 12, 5, 0, −3, −4, −3, 0, 5, 12
b $x = \pm 2$ **c** −1.8 **d** ±3.5

11 a

x	−5	−4	−3	−2	−1	0	1	2
x^2	25	16	9	4	1	0	1	4
+4x	−20	−16	−12	−8	−4	0	4	8
y	5	0	−3	−4	−3	0	5	12

b $x = -4$ and 0
c −3.8 **d** 0.6, −4.6

12 a x: −1, 0, 1, 2, 3, 4, 5, 6, 7
  y: 10, 3, −2, −5, −6, −5, −2, 3, 10
b $x = 0.5$ or 5.5 **c** −5.8 **d** −0.3, 6.3

13 a Check students' graphs. **b** 150 m

Examination questions

1 a −3, 7
b Check students' graphs. Ensure following co-ordinates are used: (−2, 1), (−1, −3), (0, −5), (1, −5), (2, −3), (3, 1), (4, 7), x-axis from −2 to 4, y-axis from −3 to 4.
c 2.8 and −1.8

2 a 6, 1
b Check students' graphs. Ensure following co-ordinates are used: (−3, 6), (−2, 1), (−1, −2), (0, −3), (1, −2), (2, 1), (3, 5), x-axis from −3 to 3, y-axis from −4 to 7.
c −1.7 and 1.7

3 a −2, 10
b Check students' graphs. Ensure following co-ordinates are used: (−4, 4), (−3, 0), (−2, 2), (−1, −2), (0, 0), (1, 4), (2, 10), x-axis from −4 to 2, y-axis from −3 to 11.

Answers to Chapter 16

16.1 Drawing quadratic graphs

Exercise 16A

1 x: −5, −4.5, −4, −3.5, −3, −2.5, −2, −1.5, −1, −0.5, 0
  y: 25, 20.25, 16, 12.25, 9, 6.25, 4, 2.25, 1, 0.25, 0

2 x: −3, −2, −1, 0, 1, 2, 3
  y: 27, 12, 3, 0, 3, 12, 27

3 x: −5, −4, −3, −2, −1, 0, 1, 2, 3, 4, 5
  y: 27, 18, 11, 6, 3, 2, 3, 6, 11, 18, 27

4 a x: −5, −4, −3, −2, −1, 0, 1, 2, 3, 4, 5
  x^2: 25, 16, 9, 4, 1, 0, 1, 4, 9, 16, 25
  −3x: 15, 12, 9, 6, 3, 0, −3, −6, −9, −12, −15
  y: 40, 28, 18, 10, 4, 0, −2, −2, 0, 4, 10
b 1.8 **c** −1.2, 4.2

5 a x: −5, −4, −3, −2, −1, 0, 1, 2, 3, 4, 5
  x^2: 25, 16, 9, 4, 1, 0, 1, 4, 9, 16, 25
  −2x: 10, 8, 6, 4, 2, 0, −2, −4, −6, −8, −10
  −8: −8, −8, −8, −8, −8, −8, −8, −8, −8, −8, −8
  y: 27, 16, 7, 0, −5, −8, −9, −8, −5, 0, 7
b −7.9 **c** −2.5, 4.5

6 a x: −2, −1, 0, 1, 2, 3, 4, 5
  y: 18, 10, 4, 0, −2, −2, 0, 4
b 6.8 **c** 0.2, 4.8

Answers to Chapter 17

17.1   Pythagoras' theorem

Exercise 17A

1 **a** 10.3 cm   **b** 5.9 cm   **c** 8.5 cm
     d 20.6 cm   **e** 18.6 cm   **f** 17.5 cm
     g 32.2 cm   **h** 2.4 m   **i** 500 m
     j 5 cm   **k** 13 cm   **l** 10 cm

2 50 cm, 1.2 m and 1.3 m or 1.5 m, 2 m and 2.5 m

3 7.43 cm

4 Because $6^2 + 7^2$ does not equal 10^3.

17.2   Finding a shorter side

Exercise 17B

1 **a** 15 cm   **b** 14.7 cm   **c** 6.3 cm
     d 18.3 cm   **e** 5.4 cm   **f** 217.9 m
     g 0.4 cm   **h** 8 m

2 **a** 20.8 m   **b** 15.5 cm   **c** 15.5 m
     d 12.4 cm   **e** 22.9 m   **f** 19.8 m
     g 7.1 m   **h** 0.64 m

3 **a** 5 m   **b** 6 m   **c** 3 m
     d 50 cm

4 **a** 3.53 m

5 Many different combinations are possible, such as: 8 cm and
     11.5 cm, 10 cm and 9.8 cm, 12 cm and 7.2 cm.

6 Because $8^2 + 6^2 = 10^2$ or because the lengths are double those
     of a 3, 4, 5 triangle.

17.3   Solving problems using Pythagoras' theorem

Exercise 17C

1 6.6 m

2 2.1 m

3 10.8 m

4 11.3 m

5 9.2 m

6 19.2 km

7 147 km

8 **a** 127 m   **b** 99.6 m   **c** 27.4 m

9 2.4 km

10 12 ft

11 **a** 3.9 m   **b** 1.7 m

12 3.2 m

13 13 units

14 **a** 4.7 m   **b** 4.5 m

15 16.5 cm^2

16 12.07 m

17 Yes, $25^2 = 24^2 + 7^2$

18 Yes, Pythagoras' theorem works in 3D, diagonal$^2 = a^2 + b^2 + c^2$.

19 Check if the sum of the squares of the two smallest sides is equal to the square of the longest side

20 In the triangle on the left, the hypotenuse is $\sqrt{(3^2 = 4^2)} = 5$. In the triangel on the right, the hypotenuse is $\sqrt{(12^2 + 13^2)} = 17.7$, so no, they will not match.

Examination questions

1 10.8 cm

2 8 cm

3 16.7 cm (1 dp)

4 6.4 cm (1 dp)

5 4.7 m (1 dp)

6 13.17 cm or 13.2 cm (1 dp)

7 39.1 km (1 dp)

INDEX